# The Nuclear Fuel Cycle

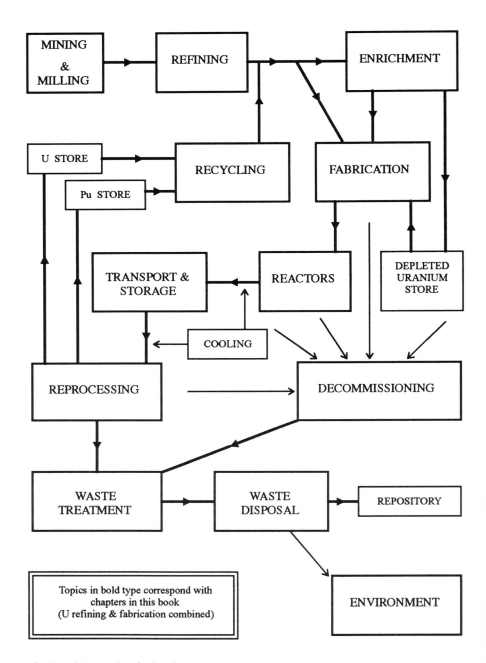

Outline of the nuclear fuel cycle

# The Nuclear Fuel Cycle
## from Ore to Wastes

edited by
**P. D. Wilson**
*British Nuclear Fuels plc, Sellafield*

OXFORD • NEW YORK • TOKYO
OXFORD UNIVERSITY PRESS
1996

Oxford University Press, Walton Street, Oxford OX2 6DP

Oxford  New York
Athens Auckland Bangkok Bombay
Calcutta Cape Town Dar es Salaam Delhi
Florence Hong Kong Istanbul Karachi
Kuala Lumpur Madras Madrid Melbourne
Mexico City Nairobi Paris Singapore
Taipei Tokyo Toronto

and associated companies in
Berlin Ibadan

Oxford is a trade mark of Oxford University Press

Published in the United States
by Oxford University Press Inc., New York

© British Nuclear Fuels plc, 1996

All rights reserved. No part of this publication may be
reproduced, stored in a retrieval system, or transmitted, in any
form or by any means, without the prior permission in writing of Oxford
University Press. Within the UK, exceptions are allowed in respect of any
fair dealing for the purpose of research or private study, or criticism or
review, as permitted under the Copyright, Designs and Patents Act, 1988, or
in the case of reprographic reproduction in accordance with the terms of
licences issued by the Copyright Licensing Agency. Enquiries concerning
reproduction outside those terms and in other countries should be sent to
the Rights Department, Oxford University Press, at the address above.

This book is sold subject to the condition that it shall not,
by way of trade or otherwise, be lent, re-sold, hired out, or otherwise
circulated without the publisher's prior consent in any form of binding
or cover other than that in which it is published and without a similar
condition including this condition being imposed
on the subsequent purchaser.

A catalogue record for this book is available from the British Library

Library of Congress Cataloging in Publication Data
The nuclear fuel cycle: from ore to wastes / edited by P.D. Wilson
Includes bibliographical references and index
1. Nuclear fuels.   I. Wilson, P.D. (Peter D.)
TK9360.N827 1996      621.48'335–dc20      96-4093
ISBN 0 19 856540 2 (Hbk)

Typeset by Oxford University Press using Synthæsis
Printed in Great Britain by
Biddles Ltd, Guildford & King's Lynn

# Foreword

W. L. Wilkinson, FRS
*British Nuclear Industry Forum, London*

Nuclear power is now established as an important source of electricity, already accounting for some 25–30% of the world's supply and growing. In France, over 75% of the electricity supply is provided by nuclear power. In Europe as a whole, it is the dominant source. It is becoming increasingly important in Japan and Korea. In the United Kingdom, over 20% of our electricity is supplied from nuclear stations.

But nuclear power in general and the nuclear fuel cycle in particular have been controversial issues in recent years. Total agreement is too much to expect, but the purpose of this book is to present the main facts of the nuclear fuel cycle as fairly as possible. Parts of the cycle have already been described in many articles, often however confined to particular areas and addressed to an audience with professional interest. The aim here is to cover the entire range of processes, from ore in the ground to recycled products and wastes ready for final disposal, at a level suited to non-specialists with a general scientific background. The emphasis will be on current practice, with some thoughts on possible developments.

There are in fact two quite distinct cycles, based respectively on uranium and thorium. In the first, uranium-235 undergoes fission on absorbing a neutron, yielding energy and several more neutrons. On average, one of these from each fission goes on to cause a further fission. The rest are absorbed by control mechanisms, by structural, shielding or other materials present, or by the predominant uranium-238. This then forms neptunium-239 which quickly decays to plutonium-239, another fissile nucleus. In favourable conditions the amount of fissile material so bred may slightly exceed that consumed.

Natural thorium has no fissile isotope, but like U-238 it can absorb neutrons and be transmuted by way of an unstable intermediate element to a fissile isotope, in this case U-233. Given an initial fissile charge, a breeding cycle based on this system can be set up, and indeed has considerable support, notably in India with its large deposits of thorium minerals. However, for various reasons it has not been generally favoured elsewhere, and the bulk of the world's effort on fission power is therefore devoted to the uranium–plutonium cycle. So accordingly is this book.

Power reactors may be divided into *thermal* or *fast* types, the distinction lying in the energy spectrum of the neutron flux. Fission in U-235 or Pu-239 is most efficiently induced by thermal neutrons, at energetic equilibrium with their surroundings. In thermal reactors, materials of low atomic number are therefore introduced specifically to take up or *moderate*, by elastic scattering, the surplus energy that neutrons released by fission initially carry. Fast reactors exclude such materials as far as practicable, for reasons explained later.

Some early types of power reactor, still operating, run on uranium of natural composition, with only 0.72% of the fissile isotope. Maintaining reaction with so low a proportion requires a large core and a neutron flux moderated by passage through graphite or heavy water with minimal absorption. However, to reduce the cost per unit of generating capacity, modern designs use uranium artificially enriched in the fissile U-235 isotope so as to sustain reaction in a smaller core, with significant neutron losses in a moderator of purified but otherwise ordinary water. On discharge the fuel still contains a proportion higher than natural, so there are obvious advantages in recovering uranium for re-enrichment, despite complications due to other isotopes then present.

Alternatively, once plutonium has been separated, it may be used as the fissile component of fresh fuel, but there is a complication here too: after the high irradiations needed for efficient power generation, plutonium contains relatively large proportions of the isotopes Pu-240 and Pu-242 that absorb thermal neutrons without themselves undergoing fission. Incidentally, they render the plutonium unattractive for military purposes. In a flux of unmoderated neutrons however, all plutonium isotopes are fissionable, and all uranium isotopes may be either fissioned or converted to plutonium. Reactors operating with such a flux can in principle consume uranium almost totally, in contrast with the 1–2% which is the most (allowing for the proportion in *tails* from enrichment) that thermal reactors can achieve by themselves.

Since plutonium generated during normal operation can serve as a replacement for fissile U-235, it may be asked why the fuel should be discharged and processed rather than kept longer in the reactor. One reason is that in thermal reactors, formation of plutonium lags behind consumption of U-235, while some fission products formed at the same time are *poisons*, strongly absorbing neutrons and so further detracting from reactivity. Fast reactors, it is true, are less sensitive to poisoning and can consume 10–20% of the fuel before discharge. However, although such reactors may generate more plutonium than they consume, much is in outer regions rather than the main power-generating core; moreover, in all reactor systems, the fuel structure is slowly degraded by the neutron flux. It must accordingly be removed from the reactor, of whatever type, while still containing much valuable material, and if this is to be recovered a chemical process is necessary.

This process and ancillary operations account for the back end of the fuel cycle. The front end essentially comprises: mining, separation and purification of uranium; enrichment, where required, to increase the proportion of the fissile U-235 (with correspondingly depleted tails which may serve as fertile material in a subsequent plutonium-driven cycle); and fabrication into fuel elements suited to the chosen type of reactor. Fuel is then irradiated in the reactor for a period limited by structural deterioration, consumption of fissile material or accumulation of neutron-absorbing fission products. On discharge, it is stored or *cooled* to allow decay of the more highly radioactive products and so simplify later operations. For political or economic reasons, discharged fuel may be prepared for disposal intact. Otherwise it is eventually reprocessed to separate it into plutonium, mostly unchanged uranium, and wastes. The wastes, consisting of fission or activation products, structural elements and process residues, are treated for disposal, while recovered uranium and plutonium are stored for future use or re-fabricated into fuel to be fed back into the cycle. A greatly simplified scheme is illustrated in the frontispiece.

Enrichment, and the associated conversion stages needed to bring material into the appropriate form, can be by-passed for those reactors using natural uranium, or where the fissile content of the fuel is to be made up with plutonium. The use of plutonium-enriched fuel (MOX, or mixed oxide) as a routine measure in thermal reactors is a fairly recent proposal, since for most of the industry's lifetime plutonium has been reserved for the fast reactors that can most effectively use it. Consequently the capacity for manufacturing thermal MOX is still quite small. Commercial fast reactors are unlikely to be built until well into the twenty-first century at the earliest, so most of the plutonium separated over the past four decades remains in store.

About half the fuel discharged from reactors around the world is stored with a view to eventual disposal without further treatment. Whether that is preferable to reprocessing, and how it might best be done, are interdependent questions still under debate.

The complete cycle involves a wide range of disciplines in physics, chemistry, engineering and environmental studies, besides the commercial, managerial and administrative skills necessary to a large, complex and heavily-financed industry with world-wide ramifications. This book concentrates on technical aspects, but even so there are more than could be covered adequately by a single author. The various sections are therefore written by specialists in their own fields. For the sake of speed and convenience in assembling the text, most of them have been chosen from the United Kingdom, but that is in no way intended to disparage the distinguished and often vital contributions to principles and practice that have been made elsewhere.

The scope might be criticised for omitting one nuclear power source, fusion—the merging of hydrogen isotopes to form helium and convert surplus mass to energy—widely expected to provide abundant energy in the future without forming the radioactive products that arouse such concern about the fission-based industry. If development is successful this belief may be partly vindicated, since helium is stable and products of activation in the neutron flux will be relatively short-lived. Nevertheless the proviso is vital; assuming technical success on the most promising lines, difficulties in the way of economic application are still more severe than commonly recognised, and to suppose that they could be overcome in less than 30–50 years seems unduly optimistic.

The industry is still evolving, as it must to survive, and any description is liable to be overtaken in some way by events before publication. With respect to installed plant, it should remain substantially accurate for some decades, since lead times before construction and the active lifetime needed to recoup investment are both of this order. Improvements may sometimes be back-fitted, but because of the difficulty and expense particularly in modifying radioactive facilities, this is undertaken only for compelling reasons. The industry is in any case highly conservative, of necessity to satisfy its own safety standards and the requirements of regulatory bodies. The view of the future is however another matter; new ideas could totally transform the prospects, and to that extent authors may well hope for their impressions to be rendered obsolete as soon as possible.

# *Preface*

This book was originally proposed by a senior member of one of the British learned societies, in order to provide factual information in an area clouded by controversy and misunderstanding. The range of subject matter was too broad to be covered by any individual author, so British Nuclear Fuels plc was approached to produce the material.

Since the instigating society decided for internal reasons not to undertake publication itself, commercial publishers were sounded, and the Oxford University Press was eventually chosen. It is a pleasure to acknowledge the help, courtesy and support of the staff, notably Richard Lawrence (Engineering Editor).

Oxford University Press was of course responsible for publishing the three-volume 'Nuclear Power Technology' in 1983. Any direct competition with that formidable text would be presumptuous, and the present work is justified as a more compact source of essential information, brought up to date in the mid-1990s and contemplating possible future developments. The aim has been to describe what is done or might be done in the various segments of the civil nuclear industry and its environment, the principles underlying the operations, and the reasons for them, without going into polemics or the kind of detail that would be required by an actual practitioner. A more basic introductory chapter is included in case of need. Military applications are excluded for obvious reasons, while nuclear fusion is scarcely touched as being still in the experimental stage with little prospect of commercial implementation for at least some decades.

A glance at the list of contributors will show the predominance of authors from BNFL or related organisations. This is solely for practical reasons, and certainly without any intention of promoting the Company line. To whatever extent that the result may fall short of complete objectivity, or has given undue prominence to British practice, I can only beg forbearance with human limitations.

Academic works customarily use SI units for physical quantities. In industry these are often inconvenient and therefore less prevalent; since the present book is meant as a key to industrial practice, practical units are therefore used in preference. The choice should cause no real difficulty, as quantities are given for internal reference or comparison with related documents. Generally they are in rounded numbers so that (for instance) the slight difference between metric and imperial tons may be neglected. Where other documents imply greater precision, readers should remember the convention that quantities of nuclear fuel are specified by the mass of heavy metal (uranium plus plutonium) before irradiation in a reactor, regardless of other chemical elements with which they may be combined or of anything that may have happened since.

The task of producing the book has involved many people apart from the authors. Among my colleagues, special thanks go to Alf Davies for an immense amount of administrative work, to Steve Scott for handling contractual arrangements, to Karen Powell for assistance with the artwork, and to the various experts in particular fields whose comments or supporting information, without earning a place in the list of contributors, have added materially to the accuracy or coverage of the presentation. In particular, Chapter 11 on environmental radioactivity was to have been written by Professor Steve Jones of the Westlakes Research Institute, who indeed provided most of the source material but was unfortunately not in a position to compile it.

*Sellafield, Cumbria, UK*     P.D.W.
*December 1995*


If you would appear to know anything, by far the best way is actually to know it.
*Sir James Frazer, 'The Golden Bough'*

# Contents

| | | |
|---|---|---|
| Contributors | | xv |
| Abbreviations | | xvii |

**1 Basic principles** — 1
*P. D. Wilson*
- 1.1 Introduction — 1
- 1.2 Atomic structure — 1
- 1.3 Reactors — 10
- 1.4 Fuel management — 11
- 1.5 Fuel supply — 12
- 1.6 Waste management — 12
- 1.7 The fuel cycle — 13
- 1.8 Economics — 14
- 1.9 The future — 16
- 1.10 References — 17

**2 Mining and milling uranium ore** — 18
*G. M. Ritcey*
- 2.1 Introduction — 18
- 2.2 Occurrence — 18
- 2.3 Mining practice — 19
- 2.4 Milling — 21
- 2.5 Tailings management — 37
- 2.6 References — 40

**3 Fuel fabrication** — 41
*G. Marsh and H. Eccles*
- 3.1 Introduction — 41
- 3.2 Design considerations — 43
- 3.3 Uranium purification — 44
- 3.4 Conversion — 46
- 3.5 Magnox fuel — 48

|     |     |                                         |     |
| --- | --- | --------------------------------------- | --- |
|     | 3.6 | Oxide                                   | 50  |
|     | 3.7 | Oxide fuel assembly                     | 61  |
|     | 3.8 | Conclusion                              | 65  |
|     | 3.9 | References                              | 66  |
| 4   | **Isotopic enrichment of uranium**      |     | 67  |
|     | *P. C. Upson*                           |     |     |
|     | 4.1 | Introduction                            | 67  |
|     | 4.2 | Gaseous diffusion                       | 69  |
|     | 4.3 | Centrifuge enrichment                   | 71  |
|     | 4.4 | Laser enrichment                        | 74  |
|     | 4.5 | Re-enrichment of reprocessed uranium    | 75  |
|     | 4.6 | Conclusion                              | 76  |
|     | 4.7 | References                              | 77  |
| 5   | **Power reactors**                      |     | 78  |
|     | *K. W. Hesketh*                         |     |     |
|     | 5.1 | Introduction                            | 78  |
|     | 5.2 | Fissions in nature                      | 78  |
|     | 5.3 | Fundamentals of fission reactors        | 79  |
|     | 5.4 | Reactor types                           | 84  |
|     | 5.5 | Future perspective                      | 99  |
|     | 5.6 | References                              | 101 |
| 6   | **Transport and storage of irradiated fuel** |     | 102 |
|     | *D. E. Haslett*                         |     |     |
|     | 6.1 | Introduction                            | 102 |
|     | 6.2 | At the reactor                          | 102 |
|     | 6.3 | Transport                               | 105 |
|     | 6.4 | Interim storage                         | 110 |
|     | 6.5 | References                              | 115 |
| 7   | **Reprocessing irradiated fuel**        |     | 116 |
|     | *I. S. Denniss and A. P. Jeapes*        |     |     |
|     | 7.1 | Introduction                            | 116 |
|     | 7.2 | Head-end processes                      | 117 |
|     | 7.3 | Solvent extraction                      | 123 |
|     | 7.4 | Conclusion                              | 135 |
|     | 7.5 | References                              | 136 |
| 8   | **Recycling uranium and plutonium**     |     | 138 |
|     | *J. P. Patterson and P. Parkes*         |     |     |
|     | 8.1 | Introduction                            | 138 |

|       |                                              |     |
| ----- | -------------------------------------------- | --- |
| 8.2   | Rationale for recycling                      | 139 |
| 8.3   | Brief history of recycling                   | 142 |
| 8.4   | Recycling uranium                            | 143 |
| 8.5   | Recycling plutonium                          | 149 |
| 8.6   | Future trends in recycling                   | 158 |
| 8.7   | References                                   | 160 |

## 9 Waste treatment — 161
*G. V. Hutson*

|       |                          |     |
| ----- | ------------------------ | --- |
| 9.1   | Introduction             | 161 |
| 9.2   | Waste types              | 162 |
| 9.3   | Liquid wastes            | 163 |
| 9.4   | Solid wastes             | 173 |
| 9.5   | Aerial effluent          | 178 |
| 9.6   | Practices at other sites | 181 |
| 9.7   | Conclusion               | 183 |
| 9.8   | References               | 183 |

## 10 Disposal of fuel or solid wastes — 184
*S. Richardson, P. Curd, and E. J. Kelly*

|        |                                       |     |
| ------ | ------------------------------------- | --- |
| 10.1   | Introduction                          | 184 |
| 10.2   | Waste categories                      | 185 |
| 10.3   | Low-level waste                       | 185 |
| 10.4   | Intermediate level waste              | 190 |
| 10.5   | High level waste                      | 195 |
| 10.6   | Direct disposal                       | 197 |
| 10.7   | Review of world-wide disposal systems | 200 |
| 10.8   | Conclusion                            | 204 |
| 10.9   | References                            | 205 |
| 10.10  | Further reading                       | 206 |

## 11 Environmental radioactivity — 207
*P. D. Wilson*

|       |                                             |     |
| ----- | ------------------------------------------- | --- |
| 11.1  | Introduction                                | 207 |
| 11.2  | Atmosphere                                  | 208 |
| 11.3  | Terrestrial environment                     | 209 |
| 11.4  | Seas, rivers, etc.                          | 212 |
| 11.5  | Artificial radionuclides of special interest| 216 |
| 11.6  | Natural radioelements                       | 221 |
| 11.7  | Monitoring and research                     | 223 |
| 11.8  | Conclusion                                  | 227 |
| 11.9  | References                                  | 227 |

## 12 Decommissioning nuclear facilities 229
*S. Buck*
- 12.1 Introduction 229
- 12.2 Principles 229
- 12.3 Timing and strategy 232
- 12.4 Planning 235
- 12.5 Practical techniques 236
- 12.6 Waste management 242
- 12.7 Progress 246
- 12.8 Future prospects 247
- 12.9 Conclusion 249
- 12.10 References 250
- 12.11 Further reading 251

## 13 Management of safety 252
*A. C. Fryer, M. Merry, C. Sunman, and A. E. Waterhouse*
- 13.1 Introduction 252
- 13.2 Risk assessment 255
- 13.3 Criticality control 263
- 13.4 THORP: a specific example 267
- 13.5 References 271

## 14 Future perspectives 272
*P. D. Wilson*
- 14.1 Introduction 272
- 14.2 Aims of the industry 272
- 14.3 Special topics 283
- 14.4 Conclusion 292
- 14.5 References 293

Glossary 295

Appendix 1: elements of the periodic table 307

Appendix 2: decay chains of heavy radionuclides 310

Appendix 3: products from thermal fission of U-235 314

Appendix 4: decay properties of selected radionuclides 317

Index 319

# Contributors

**S. Buck**  Mallard Consultants, Owson Place, Lamplugh, Workington, Cumbria, CA14 4SQ, UK

**P. Curd**  Peter Curd Associates, 9 Elm Road, Faringdon, Oxfordshire, SN7 7EJ, UK

**I. S. Denniss**  BNFL, UK Group, R&D Department, Sellafield, Seascale, Cumbria, CA20 1PG, UK

**H. Eccles**  BNFL, Company Research Laboratory, Springfields, Preston, Lancashire, PR4 0XJ, UK

**A. C. Fryer**  BNFL, UK Group, WR&D, B548, Sellafield, Seascale, Cumbria, CA20 1PG, UK

**D. E. Haslett**  BNFL, THORP Division, Sellafield, Seascale, Cumbria, CA20 1PG, UK

**K. W. Hesketh**  BNFL, Fuel Division, Springfields, Preston, Lancashire, PR4 0XJ, UK

**G. V. Hutson**  BNFL, UK Group, R&D Department, Sellafield, Seascale, Cumbria, CA20 1PG, UK

**A. P. Jeapes**  BNFL, UK Group, R&D Department, Sellafield, Seascale, Cumbria, CA20 1PG, UK

**E. J. Kelly**  BNFL, UK Group, R&D Department, Sellafield, Seascale, Cumbria, CA20 1PG, UK

**G. Marsh**  BNFL, Fuel Division, Springfields, Preston, Lancashire, PR4 0XJ, UK

**M. Merry**  BNFL, THORP Division, Sellafield, Seascale, Cumbria, CA20 1PG, UK

**P. Parkes**  BNFL, UK Group, R&D Department, Sellafield, Seascale, Cumbria, CA20 1PG, UK

**J. P. Patterson**  BNFL, UK Group, R&D Department, Sellafield, Seascale, Cumbria, CA20 1PG, UK

**S. Richardson** BNFL, UK Group, R&D Department, Sellafield, Seascale, Cumbria, CA20 1PG, UK

**G. M. Ritcey** G. M. Ritcey & Associates, 258 Grandview Road, Nepean, Ontario, Canada K2H 8A9.

**C. Sunman** BNFL, UK Group, Plant Safety, B113, Sellafield, Seascale, Cumbria, CA20 1PG, UK

**P. C. Upson** Urenco Limited, 18 Oxford Road, Marlow, Buckinghamshire, SL7 2NL, UK

**A. E. Waterhouse** BNFL, UK Group, Plant Safety, B113, Sellafield, Seascale, Cumbria, CA20 1PG, UK

**W. L. Wilkinson, FRS** British Nuclear Industry Forum, 22 Buckingham Gate, London, SW1E 6LB, UK

**P. D. Wilson** BNFL, UK Group, R&D Department, Sellafield, Seascale, Cumbria, CA20 1PG, UK

# Abbreviations

| | |
|---|---|
| µSv | micro-Sievert |
| ADU | ammonium diuranate, $(NH_4)_2U_2O_7$ |
| AEA | see UKAEA |
| AGR | advanced gas-cooled reactor |
| ALARA | as low as reasonably achievable |
| ANF | Advanced Nuclear Fuels Corporation |
| AUC | ammonium uranyl carbonate |
| AVLIS | atomic vapour laser isotope separation |
| BFS | blast furnace slag |
| BNFL | British Nuclear Fuels plc |
| BVG | Borrowdale Volcanics Group |
| BWR | boiling water reactor |
| CCD | counter-current decant |
| CEA | Commisariat à l'Énergie Atomique (France) |
| CIX | continuous ion exchange |
| CMPO | octyl(phenyl)-$N,N$-diisobutylcarbamoylmethylphosphine oxide |
| COSHH | control of substances hazardous to health |
| DEHPA | diethylhexylphosphoric acid |
| DF | decontamination factor |
| DFR | Dounreay Fast Reactor |
| DoE | (UK) Department of the Environment |
| $D_X$ | distribution ratio of component X in solvent extraction etc. |
| EARP | Enhanced Actinide Removal Plant |
| EU | European Union |
| eV | electron volt |
| FBR | fast breeder reactor |
| FMEA | failure modes effect analysis |
| GCHWR | gas-cooled heavy water reactor |
| GW | gigawatt, $10^9$ watts |
| GWd | gigawatt-day |
| GWd/t | gigawatt-days per tonne |
| HA | highly active |
| $H_2MBP$ | monobutylphosphoric acid |

| | |
|---|---|
| HDBP | dibutylphosphoric acid |
| HEPA | high efficiency particulate-in-air (filter) |
| hex | uranium hexafluoride |
| HLW | high level waste |
| HMIP | Her Majesty's Inspectorate of Pollution |
| HSE | (UK) Health and Safety Executive |
| HTGR | high-temperature gas-cooled reactor |
| IAEA | International Atomic Energy Agency |
| ICRP | International Commission on Radiological Protection |
| IDR | Integrated Dry Route (for $UO_2$ production from $UF_6$) |
| IFP | insoluble fission products |
| IFR | Integral Fast Reactor |
| ILW | intermediate level waste |
| JPDR | Japan Power Demonstration Reactor |
| kPa | kiloPascal |
| LLW | low level waste |
| LMR | liquid metal (cooled) reactor |
| LOCA | loss-of-coolant accident |
| LWR | light water reactor |
| MAFF | Ministry of Agriculture, Fisheries and Food |
| MEB | multi-element bottle |
| MLIS | molecular laser isotope separation |
| MOX | mixed oxide fuel |
| MTHM | metric ton of heavy metal |
| MTR | material testing reactor |
| MW | megawatt |
| NaDBP | sodium dibutylphosphate |
| NII | (UK) Nuclear Installations Inspectorate |
| NRPB | (UK) National Radiation Protection Board |
| OECD | Organisation for Economic Cooperation and Development |
| OECD/NEA | OECD Nuclear Energy Agency |
| OMEGA | Options Making Extra Gains from Actinides |
| OML | oxalate mother liquor |
| OPC | ordinary Portland cement |
| P&T | partition and transmutation (of waste elements) |
| PFA | pulverised fuel ash |
| PFR | prototype fast reactor (successor to DFR) |
| PIE | post-irradiation examination (of fuel elements) |
| POCO | post-operational clean-out |
| PRA | probabilistic risk assessment |
| PWR | pressurised water reactor |
| RWMAC | Radioactive Waste Management Advisory Committee |
| Sv | Sievert |

| | |
|---|---|
| SWU | separative work units |
| TBP | tri-*n*-butyl phosphate |
| TD | theoretical density |
| th | thermal |
| THORP | THermal Oxide Reprocessing Plant (at Sellafield, Cumbria, UK) |
| TOPO | tri-octylphosphine oxide |
| TRU | transuranic (waste) |
| UKAEA | United Kingdom Atomic Energy Authority |
| UOC | uranium ore concentrate |
| VLLW | very low-level waste |
| WAK | Wiederaufarbeitungsanlage Karlsruhe |
| WHIMS | wet high-intensity magnetic separation |

# 1 Basic principles

P. D. Wilson
*BNFL, UK Group, Sellafield, Cumbria*

## 1.1 Introduction

The essentials of the nuclear fuel cycle have been summarised in the Foreword to this book, while details of the individual stages are to be given in the following chapters. Some basic explanations may nevertheless be helpful. All nuclear data are taken from 'Nuclides and Isotopes,' 14th edition, General Electric Company, San Jose, California (1989).

The source of nuclear energy lies in the equivalence of mass and energy according to the Einstein equation $E = mc^2$, where $c$ is the velocity of light. Thus the total annihilation of one gram of matter would produce $9 \times 10^{20}$ ergs ($9 \times 10^{13}$ joules) of energy, enough to heat 215 000 tons of water from freezing to boiling point; the yield from 3.4 grams would melt a square kilometre of ice a metre thick. No process on Earth can totally annihilate ordinary matter, but some at the atomic level can convert a small proportion, about one part in a thousand, which is enough to be useful. The reason lies in the way that the components of atoms bind together.

## 1.2 Atomic structure

Atoms are built from three types of particle: protons, neutrons and electrons. Actually the protons and neutrons are themselves composite, in ways that are of no concern here. Both have very slightly more than unit mass on the atomic scale (about $1.67 \times 10^{-24}$ g). Neutrons are electrically neutral. The protons each have unit positive electrical charge, balanced in the neutral atom by an equal number of negative electrons, each with 1/1836 of a proton mass. The protons and neutrons thus comprise practically all the mass, but are confined to a nucleus occupying a minute volume of the order of $10^{-12}$ cm across. The remaining volume, to a radius typically of the order of $10^{-8}$ cm, is filled by the electrons, which determine the chemical behaviour of the element.

In the simplest atom, that of ordinary hydrogen, the nucleus is a single proton. Other elements have more protons, matching the atomic number, Z, which

2   Basic principles

| GROUP | | | | | | | | | | | | | | | | | |
|---|---|---|---|---|---|---|---|---|---|---|---|---|---|---|---|---|---|
| 1 | | | | | | | | | | | | | | | | | 18 |
| 1 H | 2 | | | | | | | | | | | 13 | 14 | 15 | 16 | 17 | 2 He |
| 3 Li | 4 Be | | | | | | | | | | | 5 B | 6 C | 7 N | 8 O | 9 F | 10 Ne |
| 11 Na | 12 Mg | 3 | 4 | 5 | 6 | 7 | 8 | 9 | 10 | 11 | 12 | 13 Al | 14 Si | 15 P | 16 S | 17 Cl | 18 Ar |
| 19 K | 20 Ca | 21 Sc | 22 Ti | 23 V | 24 Cr | 25 Mn | 26 Fe | 27 Co | 28 Ni | 29 Cu | 30 Zn | 31 Ga | 32 Ge | 33 As | 34 Se | 35 Br | 36 Kr |
| 37 Rb | 38 Sr | 39 Y | 40 Zr | 41 Nb | 42 Mo | 43 Tc | 44 Ru | 45 Rh | 46 Pd | 47 Ag | 48 Cd | 49 In | 50 Sn | 51 Sb | 52 Te | 53 I | 54 Xe |
| 55 Cs | 56 Ba | 57 La | 72 Hf | 73 Ta | 74 W | 75 Re | 76 Os | 77 Ir | 78 Pt | 79 Au | 80 Hg | 81 Tl | 82 Pb | 83 Bi | 84 Po | 85 At | 86 Rn |
| 87 Fr | 88 Ra | 89 Ac | 104 | 105 | 106 | | | | | | | | | | | | |

| Lanthanides (Rare Earths) | 58 Ce | 59 Pr | 60 Nd | 61 Pm | 62 Sm | 63 Eu | 64 Gd | 65 Tb | 66 Dy | 67 Ho | 68 Er | 69 Tm | 70 Yb | 71 Lu |
|---|---|---|---|---|---|---|---|---|---|---|---|---|---|---|
| Actinides | 90 Th | 91 Pa | 92 U | 93 Np | 94 Pu | 95 Am | 96 Cm | 97 Bk | 98 Cf | 99 Es | 100 Fm | 101 Md | 102 No | 103 Lr |

**Fig. 1.1**   The periodic table

identifies the position of the element in the periodic table (Fig. 1.1 and Appendix 1).

With more than one proton, the mutual electrostatic repulsion is overcome by a strong binding force, provided by the neutrons. In the lightest elements the

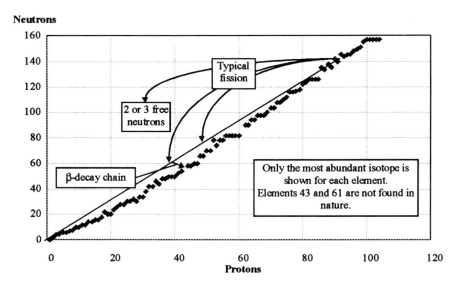

**Fig. 1.2**   Relationship between neutron and proton numbers

numbers are similar, but with increasing total numbers the proportion of neutrons needed for stability gradually rises (Fig. 1.2) although it must never be too great as neutrons themselves are rather unstable in excess. Since chemical properties are determined by the number of electrons, and so of protons, the number of neutrons affects them only in rather subtle ways that depend on the precise atomic mass and can usually be ignored; it often varies somewhat within a given element, giving different *isotopes* (from the Greek meaning the same place). These are distinguished by the mass number, the sum of protons and neutrons (collectively *nucleons*). In notation it is appended or prefixed to the name or symbol of the element, as in strontium-89, Sr-89 or $^{89}$Sr. Hydrogen is unique in having different names for its isotopes H-2 and H-3 which are usually known as deuterium and tritium, D and T respectively.

## 1.2.1 Radioactivity

In a given element, the number of neutrons can vary only so much. Beyond a certain range, and sometimes well within it, a nucleus tends to be more or less unstable. When there are too many, one may be converted to a proton plus an electron; less commonly, when there are too few, the nucleus may emit the positive equivalent of an electron (i.e. a positron) or capture an electron from the surrounding swarm. Any such change is known as a disintegration or decay, and may be repeated several times.

Other decay mechanisms are possible if the nucleus is too big. It may split in two, a process known as fission and considered later. Less drastically, it can eject a helium nucleus—two protons and two neutrons. This then reduces the atomic number by two and the mass number by four. Converting a neutron to a proton and an electron increases the atomic number by one, leaving the mass number unchanged. Either way, one element is changed into another, a process impossible by chemical means and known as transmutation (Fig. 1.3).

Because the mass number changes by 4 or not at all, four decay series are known in the heavy elements, according to whether the nuclei have mass numbers of $4n$, $4n + 1$, $4n + 2$ or $4n + 3$, where $n$ is an integer (whole number); see Appendix 2.

Decay is always to a state of lower energy than the original. When a particle is lost from the nucleus, it carries much of the surplus energy in the form of motion, eventually degraded to heat. A large part often appears as electromagnetic radiation, like radio waves, light or X-rays but more penetrating. The rest, a small proportion determined by the conservation of momentum, may cause the nucleus to recoil out of its crystal lattice position or state of chemical combination.

For historic reasons, dating from before the emitted particles or rays were understood, they are known by letters of the Greek alphabet. The helium nuclei are called alpha ($\alpha$) particles, electrons beta ($\beta$) particles and the pure electromagnetic radiation, gamma ($\gamma$) rays. If electrons and positrons have to be distinguished, they are symbolised by $\beta^-$ and $\beta^+$ respectively. The process of disintegration in

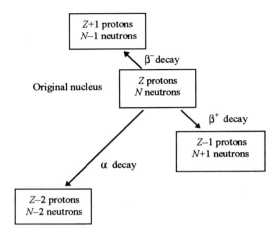

**Fig. 1.3** Radioactive decay

general is called radioactivity, and isotopes subject to it are called radioisotopes or radionuclides (because it is a property of the nucleus).

The energy of these particles and rays is enough to dislodge electrons from other materials that they meet, leaving positive ions, so they are collectively known as ionising radiation. Particularly in biological systems, the ions can cause damaging reactions, and substances emitting ionising radiation have therefore to be carefully controlled. Fortunately the radiation can be contained: alpha particles, which deposit their energy very quickly and so are the most damaging within their range, are stopped by a thin film of plastic or by the dead surface of human skin, beta particles by a few millimetres of aluminium or about a centimetre of water, gamma rays by a heavier barrier for instance of lead, steel or concrete.

The principal sources of radiation on Earth are:

- minerals containing uranium and thorium, present from the earliest times and very widely distributed in small but highly variable amounts, together with their decay products (*daughters*) including the radon that is by far the greatest single cause of exposure to a population spending most of its time in poorly ventilated buildings;
- other radionuclides, also present from the formation of the Earth or derived by decay, such as potassium-40 which is a natural component of the sea and of all living things;
- cosmic radiation (from the sun or other sources in space) and the particles, rays or radionuclides formed from it by collisions in the atmosphere;
- medical X-rays.

The first three are completely natural, and all life on Earth has developed against their background. X-rays are of course artificial, and the slight risk associated with them is considered justified by the benefits. Whether the very much smaller amount caused by nuclear power generation is justified in the same way is much debated.

Industrial operations utilising radioactivity, such as checking welded structures for cracks or bubbles, commonly employ artificial radioisotopes obtained by exposing stable elements to an intense flux of neutrons. Thus cobalt-59 is converted to cobalt-60, which decays to nickel-60 emitting a beta particle and a particularly penetrating gamma ray. Such sources have the advantage over X-ray equipment of being very compact and needing no power supply; on the other hand, they cannot be switched off and have to be kept in heavily shielded containers. Other artificial radionuclides are used as tracers, for instance in medical investigations, or to treat cancers by locally destroying their tissue. All such uses again carry risks justified by the benefits.

The decay of a particular nucleus is a matter of chance. It may happen in the next hour, or not for another year. Since atomic units are so small by familiar standards, however, even traces of matter contain many millions of atoms, and in such numbers the overall behaviour is statistically predictable. There is a certain likelihood that any particular type of nucleus will decay within (say) one second. Thus the rate of decay is proportional both to that likelihood and to the number present, and indeed being readily detected it is the most commonly-used measure of how many there are. Provided that there are enough to smooth out random variations, a half of any amount will disintegrate in a particular interval of time known as the half-life. Actual values range from small fractions of a second to billions of years. Large values, which cannot be measured directly, are calculated from rates of decay observed over shorter times.

It follows from the pattern of decay that it can never be known with total certainty to be complete. After one half-life the rate of decay, as well as the amount remaining, are both halved, and so on until it is a matter of pure chance whether or when the last few individual nuclei disintegrate. By that time however the radioactivity is not practically detectable against the natural background; it would be like trying to see the faintest stars in broad daylight. After say ten or twenty half-lives, when there will be only a thousandth or a millionth left, decay may be considered effectively complete.

## 1.2.2 Nuclear energy

The atomic unit of mass is defined as one-twelfth of the mass of one carbon-12 atom, which has six protons and six neutrons. On this scale every isotopic mass is thus very close to the mass number. The isotopic masses of the lightest and heaviest elements, and of protons and neutrons themselves, however, are slightly greater than the mass numbers, while those in between are slightly less. The difference is expressed in terms of the *packing fraction*, a measure of the mass per

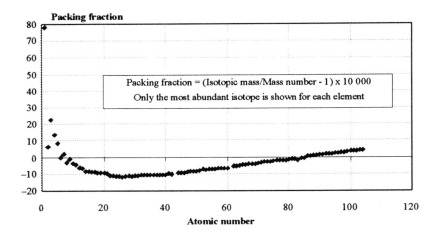

**Fig. 1.4** Variation of packing fraction with atomic number

nucleon, which as can be seen from Fig. 1.4 has a shallow minimum in the range of atomic numbers from 20 to 40 (calcium to zirconium).

The reason is that in combining together to form nuclei, protons and neutrons lose a small fraction of their mass in the form of energy. The loss is greatest for elements in the region around atomic number 30. Accordingly energy can be released by combining light nuclei into heavier ones (fusion), and indeed the conversion of hydrogen to helium is the main reaction powering stars such as the Sun. The heaviest elements were formed against the gradient in the extreme conditions of exploding stars, and some of their energy can be regained by splitting such atoms into two very roughly equal parts (fission, see Fig. 1.2).

Both fission and fusion have been used to generate energy artificially. In principle fusion could be an almost unlimited source of power, but under controlled conditions has so far been sustained for periods no longer than a second or two, returning only a fraction of the energy needed to start it. Other difficulties promise to be extremely troublesome even if these are solved. All of them may possibly be overcome in time, but for the foreseeable future, nuclear power means fission.

### 1.2.3 Conditions for fission

Although splitting a heavy nucleus such as uranium is favoured by the energy balance, and can occur spontaneously, it is very much more probable if provoked by some other disturbance of the structure. This is most readily caused by the absorption of a neutron, which being uncharged is not repelled either by the concentration of positive charge at the core of an atom or by the surrounding electrons. Since fission produces typically two or three free neutrons besides the main fragments, one of them can go on to cause a further fission, and so on. Then a chain reaction occurs.

The speed of the neutron and the structure of the particular heavy nucleus are both important. If the neutron is slow, it may induce fission in one type of nucleus but not in another. If it is fast, it spends too little time near any nucleus for capture to be likely, but when this does occur the additional energy is enough to disrupt either kind. Uranium as found in nature consists mainly of two isotopes, 0.72% U-235 which is of the first kind, and 99.27% of the second. Thus slow neutrons, especially with certain particular energies at which 'resonances' occur, are likely to cause fission in U-235 but not in U-238, while fast neutrons are much less likely to cause fission at all but can disrupt U-238. (The percentages given here are by numbers of atoms; the same proportion of U-235 may be represented as 0.711% by weight).

When released by fission, neutrons are fast, and can be slowed (*moderated*) most readily by elastic collisions with particles of similar mass, according to the usual laws of motion. Nuclei of ordinary hydrogen, having almost exactly the same mass, are very effective but can sometimes absorb the neutron inelastically. This forms the second isotope of hydrogen, deuterium. Further absorption is very much less likely, so deuterium oxide ('heavy water') is sometimes used instead of the ordinary variety. Graphite is another common choice. When fully moderated, neutrons have energies corresponding to the temperature of their surroundings, and are therefore called *thermal*.

### 1.2.4 Transmutation

Although U-238 is not fissioned by thermal neutrons, it can absorb one to form an unstable nucleus of U-239. This decays, by successive beta emissions, first to neptunium-239 and then to plutonium-239, which like U-235 can be split (is *fissile*) at thermal energies. Instead of fissioning, however, it may simply absorb neutrons to form higher isotopes Pu-240, Pu-241 and Pu-242 (Fig. 1.5). Unlike the other plutonium isotopes which are alpha-emitters, Pu-241 decays by beta-emission to americium-241, which can lose an alpha particle to form neptunium-237. Other sequences of neutron absorption and beta decay lead to the higher elements curium, berkelium and californium in decreasing amounts. Neptunium and these elements beyond plutonium are often called the *minor actinides*, minor because there is much less of them than of uranium or plutonium, actinides because (like uranium and plutonium themselves) they are chemically members of a sub-series starting with actinium, analogous to the lanthanides, as shown in the periodic table.

Some of these transmuted products can be fissioned by thermal neutrons; for instance, the odd- but not the even-numbered plutonium isotopes. All however can be fissioned by fast neutrons.

In a power reactor, the formation of fissile plutonium compensates to some extent for the consumption of U-235, and towards the end of the fuel's residence

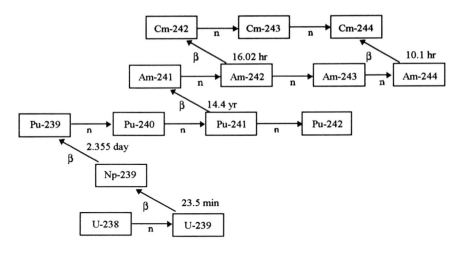

**Fig. 1.5** Typical transmutation reactions

the plutonium may supply a substantial proportion of the power. At higher neutron energies where fission releases more free neutrons, the amount of new fissile material may exceed that consumed, and the system is then said to breed. Hence comes the term *fast breeder reactor*, rather misleading because it is the neutrons that are fast, not the breeding. Incidentally, a fast reactor need not breed at all if that is not required, as at present when stocks of plutonium are already ample for the time being.

Besides uranium, thorium could also be used as a nuclear fuel. Although not fissile itself, it can be transmuted by way of protoactinium-233 into uranium-233 which is very much so, and given a start with some other fissile material, a breeding cycle similar to that with uranium-238 and plutonium can be set up. India, with large deposits of thorium minerals, has based part of its nuclear power development programme on this fact. However, the intermediate protoactinium-233 has a less convenient half-life (27 days) than neptunium-239, and the process chemistry, more troublesome than with uranium and plutonium, has never reached an industrial scale. Even so, Germany also had a project to develop thorium reactors, but eventually abandoned it on the grounds that the U-233 produced would be particularly suitable for proliferating nuclear weapons.

### 1.2.5 Fission products

As mentioned earlier, light nuclei generally contain roughly equal numbers of protons and neutrons, but as the atomic number increases, the proportion of neutrons to protons gradually rises. For uranium it is about 1.56, for elements in the region of atomic numbers 40 to 60 it is some 1.3 to 1.4. Thus on fission, even with a

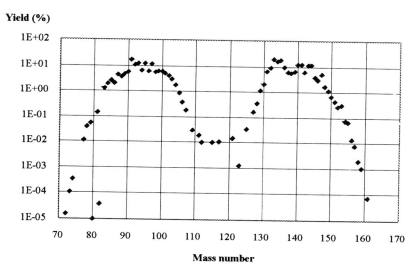

**Fig. 1.6** Variation of fission yield with mass number

few free neutrons released, the main fragments still have too many to be stable. These fragments or fission products therefore decay by successive emissions of beta particles.

The primary fission products have typically about ten neutrons too many for the protons (Fig. 1.2). They decay rapidly, with each beta-decay converting a neutron to a proton, and after perhaps five transitions many soon achieve a stable balance. Some however reach a stage one or two steps short of complete stability and thereafter decay much more slowly, with half-lives of a few years or decades. In a few instances, notably iodine-129 and technetium-99, half-lives are very much longer. Their radioactivity lies behind much of the anxiety about nuclear power.

The range of fission products is enormous, covering about a third of all elements from around copper to the mid-lanthanides, but the most common are clustered around two broad peaks with mass numbers about 95 and 135 respectively; symmetrical fission, yielding products of nearly equal mass, is far less frequent, see Fig. 1.6. Chemically, the more abundant after a short period of decay range roughly from krypton to palladium, and from iodine to europium. Some have neutron-absorbing properties that are important in the operation of a reactor, for instance limiting the effective life of fuel. Half-lives range from fractions of a second to millions of years; the shorter they are, the more intense the radioactivity, which overall therefore decays in a complex manner with time according to the currently predominant species.

Because of the decaying fission products, reactor fuel still generates heat at a substantial rate even after the fission reaction has been stopped, since heat is the ultimate form to which all energy is naturally degraded. Provision must therefore be made for cooling when a reactor is shut down or fuel unloaded.

## 1.3 Reactors

The purpose of a nuclear power reactor is to supply energy in a steady and controlled fashion to an electric generator. There have been proposals to utilise fission heat directly, but current power reactors merely replace coal, oil or gas burners by raising steam which drives turbo-alternators in conventional fashion.

Ways of meeting this simple description vary greatly in more or less substantial detail, but in thermal reactors, the essentials are:

- the fissile material, generally uranium metal or oxide, sometimes with plutonium added;
- cladding, a metal shell around the fissile material to protect it from corrosion and to prevent fission products from escaping;
- a moderator, composed of light elements to slow down the neutrons from their initially high speeds;
- a coolant, to carry the heat generated to the steam generator or directly to the turbine;
- a control mechanism, moving neutron-absorbing materials into or out of the reaction zone so as to adjust the rate of fission or stop it altogether;
- shielding to protect operators and the public from radiation;
- a containing shell to prevent radioactive material from escaping in case of excessive internal pressures.

The crucial difference in a fast reactor is that the moderator is omitted.

To transfer heat from the nuclear fuel to the steam-raising equipment, fast reactors with their high energy density use liquid sodium as coolant; thermal reactors may use water or a gas such as carbon dioxide. To slow down the neutrons without absorbing too many of them, thermal reactors also need a moderator, currently graphite, water or heavy water (deuterium oxide). The most common types, the pressurised or boiling water reactors, use ordinary water as both moderator and coolant. Reactors are often designated according to the combination.

The fissile material, cladding, and sometimes a part of a graphite moderator are closely combined; together with any necessary appendages, they form the fuel as customarily understood. It may be changed many times during the lifetime of a reactor, and so is manufactured as discrete, easily-handled units. To prolong the time between fuel changes, or to allow operation with less fuel than would otherwise be needed, the proportion of U-235 to U-238 is usually increased or *enriched* above the natural 0.72%. This allows the reactor core to be smaller than with natural uranium, so the structure is cheaper to build. Reactors for submarines, where space is limited and refuelling extremely inconvenient, need very high enrichment.

During the time that fuel spends in a reactor:

- a part of the original U-235 or plutonium is gradually consumed;
- some of the U-238 is transmuted into plutonium, although in thermal reactors only partly making up for the loss of initial fissile content;
- fission products build up, some of them heavily absorbing neutrons.

The net effect is to reduce the ease with which the nuclear reaction can be maintained, and eventually the fuel has to be changed. In thermal reactors, this is necessary when at most a few per cent of the uranium has been consumed. Fast reactors may achieve 10–20% burn-up per cycle.

## 1.4 Fuel management

When the fuel has been discharged, there are two alternative ways of dealing with it:

- direct disposal—discarding it completely as waste, after a period of storage to allow heat-generating fission products to decay;
- chemically treating it (*reprocessing*) to separate remaining uranium and plutonium from the fission products and minor actinides which are discarded.

Either way, strict precautions must be taken against the escape of potentially dangerous fission products. Some countries, notably the USA, are committed to the first course, while others including Britain, France and Japan prefer the second, with a view to using again the recovered uranium and plutonium. Over the whole world, roughly equal quantities of fuel follow each path.

For the present, recycled uranium and plutonium are intended to be used in thermal reactors. However, these cannot effectively utilise the even-numbered isotopes of plutonium. Fast reactors can. In theory, given several rounds of recycling, they could eventually extract all the energy obtainable from uranium including the predominant U-238, directly or by way of plutonium. In contrast, thermal systems are limited to about 2% (isotopically enriched fuel in the reactor may reach a higher percentage, but there is also the remaining uranium-238 to be taken into account). Although fast reactors have not yet been built commercially because they are likely to be more expensive than thermal stations, they would thus conserve uranium stocks against the time when fresh supplies are expected to become scarce. Even before then, they would make the most complete use of material for which the heaviest environmental penalty has been paid at the mine. Ironically, they would also be the most straightforward means of consuming plutonium stocks, notably surplus military material of which 200 tons or so are available from dismantled weapons.

## 1.5 Fuel supply

If fuel is not to be made from recycled material, fresh uranium must be mined and is indeed the predominant source. Uranium is naturally present in most rocks but usually in mere traces. Even valuable deposits generally contain only small proportions, sometimes associated with other ores which are the main object of mining operations. They are a long way from fuel manufacturing centres, and the ore is locally processed (*milled*) to concentrate the valuable minerals and avoid having to transport all the worthless material.

At the fuel manufacturing plant the concentrated ore is further purified. If, as is usual, the uranium is to be made into fuel with an increased proportion of U-235, it is then converted to uranium hexafluoride which can easily be vaporised, since the current enrichment processes (diffusion or high-speed centrifuge) both need a gaseous feed. Eventually it is converted to metallic rods, or more often oxide pellets, and sealed into cladding tubes of a suitable metal—a magnesium alloy for metal fuel, stainless steel or a zirconium alloy for oxide. Because oxide conducts heat poorly, the pellets are narrow to avoid unduly high internal temperatures; individual packed tubes or *pins* thus contain less fuel than is convenient to handle, and they are assembled into clusters that may contain over two hundred.

When fuel is to contain plutonium, this may be added before or after the conversion to oxide, and particular care is needed to ensure that it is uniformly mixed at the finest level. Otherwise it might form local hot spots during irradiation in the reactor, and subsequently lead to highly-radioactive insoluble particles on reprocessing. For thermal reactors the proportion is likely to be about 6%; for fast reactors, 20% to 30% is usual.

## 1.6 Waste management

Every industrial operation produces waste. If discharged nuclear fuel is discarded directly, the whole is considered to be waste; otherwise the principal wastes arising from it are the separated fission products, the remains of the cladding, and various solids, liquids or gases that have become more or less contaminated through contact with radioactive substances in the course of processing. Eventually the reactors and process buildings, together with the equipment that they contain, have to be decommissioned and themselves become waste or recycled scrap.

The object of radioactive waste management is

(a) to concentrate the radioactive material as far as possible into a small volume that can be isolated indefinitely from human contact;

(b) where streams such as the water from fuel storage ponds are too bulky for anything but release into the environment, to remove from them all radioactivity or harmful material that poses a significant risk.

This is not the place to go into the arguments about what should be considered

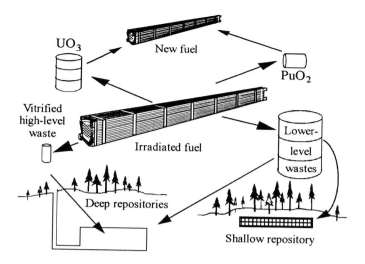

**Fig. 1.7** Closing the fuel cycle

significant, and *zero discharge* is of course impossible, since no separation is ever completely clean-cut. The general principle is that risks should be no greater than people are willing to accept in other areas.

So far, disposal has been completed only for liquids, gases and low-level solids, all containing small proportions of radioactive material. The main bulk of fission products, separated during reprocessing, at first generates a great deal of heat through radioactive decay. That heat has to be safely dispersed. For a time the material is kept in concentrated solution, then formed into glass blocks that are still too hot for disposal which is expected to take place after about fifty years. A similar delay is of course also necessary for intact fuel if that is to be discarded. Intermediate-level waste, such as the remains of fuel cladding, is formed for instance with cement into blocks that are intended to be buried deep underground, although possible burial sites are still being investigated.

Fuel manufacture also produces wastes, notably the decay products which in freshly-mined ore are in equilibrium with the parent uranium. Their total radioactivity is less intense than that of the fission products separated during reprocessing, but still considerable, and similar principles apply to their management.

## 1.7 The fuel cycle

Where fuel discharged from a reactor is reprocessed and the uranium or plutonium returned for further use, at least some of the material follows a closed loop or *cycle* (see Fig. 1.7). Reprocessing for this purpose is often described as *closing the back end of the fuel cycle*. By analogy, the option of discarding discharged fuel completely is often called the *once-through cycle*, despite the contradiction in terms. The third possible practice, prolonged storage with provision for eventually retrieving the fuel, is not a permanent option but a means of buying time pending a decision some decades in the future.

The genuine fuel cycle comprises:

- mining and milling the ore;
- purifying the ore concentrate, enriching the U-235 content if necessary and manufacturing fuel;
- utilising the fuel in reactors of various kinds;
- reprocessing the discharged fuel to separate uranium and plutonium from waste;
- returning the uranium and plutonium for further use;
- disposing of wastes.

The various aspects are described in more detail in the chapters that follow.

Incidentally, the costs of electricity produced by a modern nuclear station—estimated, since much depends on the particular accounting rules assumed—may be broken down roughly as 88% associated with the reactor itself, 9% with the fuel supply (the so-called front end of the cycle), and 3% with the back end.

## 1.8 Economics

Although the emphasis throughout this book is primarily technical, something needs to be said about the economics of nuclear power—not too much, as the subject has been aptly described as 'boringly complicated'. A comparison as source of energy with fossil hydrocarbons, particularly coal, is probably the most fruitful way to proceed.

The comparison is hampered by two basic difficulties:

- the wide differences in performance between the nuclear industries of different countries, and indeed in some instances between individual plants within a single country;
- the incompatible cost structures, with capital representing a far greater proportion of the total in nuclear than in fossil-fired generation, possibly as high as 70%.

The contrasting experiences of Britain and France are nevertheless instructive.

France generates over three quarters of its electricity in nuclear stations and claims it to be the cheapest in Europe, apart from some hydroelectric sources in particularly favoured locations.[1a] Some 4–5% French electical production is exported to Britain. There the proportion of nuclear generation is 23%[1b] although the installed capacity is only some 14% of the total. This disparity between capacity and output in fact reflects the cost structure, since a nuclear generating station, once built, is cheaper to run than an equivalent burning coal, and is therefore used in supplying the base load—that part of the total consumption which is always in demand, regardless of time or season. Nevertheless, because of the capital and other fixed charges, the overall cost in Britain is usually taken to be greater than with coal.

Part of the difference between the two countries undoubtedly lies in the sheer size of the French programme, which permits sequence construction where lessons learned in one project can be applied directly to the next; construction times and costs are thus progressively reduced. This possibility is due in turn largely to consistent Government backing, on the strategic grounds of overcoming France's relative lack of indigenous energy resources. There is little coal or oil, coal mining has virtually or completely ceased on the grounds of cost, the Lacq gas field is approaching exhaustion, while hydroelectricity accounts for only a fraction of requirements and is already fully utilised. With much of the balance made up in the early 1970s by imported oil, France suffered severely from the sudden price increases and threats to supply, and is determined not to be caught in the same way again. Since one ton of uranium is equivalent to around twenty thousand of coal even in thermal systems, it can be stockpiled more easily and effectively than fossil fuels. French electricity generation is therefore based predominantly on nuclear reactors with a systematic construction programme and a few standard types. Meanwhile the British nuclear history has been characterised by a succession of one-off designs, by irregular orders leading to a dissipation of experience, and after early enthusiasm by chronic political indecision.

Although a nuclear reactor is relatively cheap to operate, reaching that stage takes five to ten years, occasionally more as at Dungeness B where industrial problems disrupted the schedule. During that time the investment earns nothing while interest charges are accumulating. Thus the actual costs are very sensitive to unforeseen delays, and depend heavily on somewhat unpredictable interest rates. The wide differences in economic performance between various nuclear utilities in the USA must in part be due to the particular opportunities for protracted and obstructive litigation in that country, in one instance forcing closure almost immediately after a much-delayed start to commissioning (see Chapter 12, Section 3.3).

Any new project must include a relative assessment of capital cost and eventual earnings. Because much of the expenditure arises early and the earnings come only after it is complete, the two cannot be directly compared. To judge the present value of future money, it is *discounted* to an extent related to expected interest rates but generally rather larger, reflecting the desire in financial institutions for early returns. A further complication is the need to allow for eventual decommissioning, which in a nuclear station would probably be completed in stages during a period of perhaps fifty to a hundred and fifty years after closure. Because of the uncertainties in this task (uncertainties which have actually diminished over the last few years), it is customary to apply a low or zero discount rate to the necessary financial provision, although the logic appears questionable. Thus the provision must be larger than if the rate were the same as assumed for future earnings. An element of wishful thinking may be suspected in the choice of rates for any particular calculation on either side of the discussion.

With a fossil-powered station, a much larger proportion of the total cost is incurred for the fuel itself only a short time before the earnings that cover it, and a correspondingly smaller proportion is subject to the vagaries of interest and discount rates. Moreover decommissioning can proceed as soon as power generation has ceased. Overall the uncertainties are very much less.

A survey suggests that on reasonable accounting suppositions, the financial costs of coal-fired and nuclear electricity from modern stations are quite similar.[2] The balance depends on the assumed discount rate and on the predicted price of coal. Finance, however, is not the only consideration. Security of supply has already been mentioned. Further issues are the future effects of radioactive wastes on the one hand and of carbon dioxide emissions on the other (desulphurisation of flue gases is assumed to be mandatory); these are different in kind and therefore hard to evaluate comparatively. Then there are the risks of serious accidents, or a diversion of civil nuclear material to military or terrorist use, risks which cannot be dismissed completely although often poorly understood and arguably much exaggerated in some quarters.

None of these can be quantified at all accurately. Nevertheless attempts have been made to cost such considerations as an addition to the price of electricity. For the UK, a recent estimate of these additions yields values lower for nuclear generation than for any other source except possibly gas.[1c]

In summary, nuclear power evidently can be economically competitive with fossil sources, given good management and consistent political support. Tangible factors that favour it are low interest rates, a firm programme with regular repeat orders for stations of a given design, high prices for coal etc., and expeditious construction. Against it are the concern about radioactive wastes in comparison with the environmental detriment due for instance to emissions of carbon dioxide and ash dust, anxieties about a possible misuse of material, the risk of another Chernobyl however remote the likelihood, and above all, the demand for short-term profit.

## 1.9 The future

The technology of nuclear power is well established; even the most controversial aspect—the final disposal of wastes—could arguably be implemented immediately, given the political will, with better provision for the safety of present and future generations than is demanded of other industries. Nevertheless improvements are desirable throughout, whether in technical performance or economics, and development continues. Whether it will lead to the acceptance of new reactor types, new reprocessing methods, and new ways of managing wastes, or merely variants of those already well known, remains to be seen. Some of the possibilities that can already be envisaged are considered in the concluding chapter.

## 1.10 References

1 D. Rooke, I. Fells and J. Horlock (eds). *Energy for the future*. E and F N Spon for The Royal Society (1995).
  a R. Carle. Chapter 3 French energy policy.
  b I. Fells and J. Horlock. Chapter 11 Energy strategy.
  c D. Pearce, Chapter 2 Costing the environmental damage from energy production.
2 T. Price. *Political electricity*, Chapter 6 Nuclear power economics. Oxford University Press (1990).

# 2 Mining and milling uranium ore

G. M. Ritcey
*G. M. Ritcey and Associates, Ontario, Canada*

## 2.1 Introduction

Whole books have been written about every aspect of uranium mining and milling, so this chapter presents a selection of material particularly related to the author's experience, largely in milling. Much of it is drawn from his own and other contributions to the comprehensive 'Uranium Extraction Technology' published by the IAEA, Vienna.[1]

## 2.2 Occurrence

Of the roughly 200 known uranium minerals, only the few in Table 2.1 are mined commercially.

Deposits, in decreasing order of importance, may be unconformity-related, sandstone, quartz-pebble conglomerate, veins, breccia complex, intrusive, phosphorite, collapse breccia pipe, volcanic, surficial, metasomatite, metamorphite, lignite, black shale etc. Their complexity and association with other oxides or sulphides present many problems in analysis, processing, recovering by-products and minimising environmental impact.

The value of a deposit depends on:

- quantity;
- geology, grade and mineralogy;
- the mineralogy of associated metals and worthless material;
- whether access is by underground or open-pit mining;
- the number and costs of process stages required;
- the cost of safely impounding residues.

Some deposits that would be uneconomic if worked for uranium alone are nevertheless valuable on account of other metals recovered at the same time, whether as principal or subsidiary products.

**Table 2.1** Commonly mined uranium minerals

| Type | Mineral | Formula |
|---|---|---|
| Oxides | Uraninite | |
| | Pitchblende | $(UIV_{1-x}UVI_x)O_{2+x}$ |
| | Gummite | |
| | Becquerelite | $7UO_3.11H_2O$ |
| Mixed oxides | Brannerite | $(U,Ca,Fe,Th,Y)(Ti,Fe)_2O_6$ |
| | Davidite | $(Fe,Co,U,Ca,Zr,Th)_6(Ti,Fe,V,Cr)_{15}(O,OH)_{36}$ |
| | Pyrochlore | $(Na,Ca,Fe,U,Sb,Pb,Th,Zr,Ce,Y)_2(Nb,Ta,Ti,Sn,Fe)_2O_6(O,OH,F)$ |
| Silicates | Coffinite | $U(SiO_4)_{1-x}(OH)_{4x}$ |
| | Uranophane | $Ca(UO_2)_2(SiO_3)_2(OH)_2.5H_2O$ |
| | Uranothorite | $(Th,U)SiO_4$ |
| Phosphates | Autunite | $Ca(UO_2)_2(PO_4)_2.10$ to $12\ H_2O$ |
| | Torbernite | $Cu(UO_2)_2(PO4)_2.12H_2O$ |
| Vanadates | Carnotite | $K_2(UO_2)_2(VO_4)_2.1$ to $3H_2O$ |
| | Tyuyamunite | $Ca(UO_2)_2(VO_4)2.5$ to $8H_2O$ |
| Carboniferous | Thucholite | Uraninite complex with carboniferous material |

## 2.3 Mining practice

The depth of mineralisation in relation to local geology decides between underground and open-pit mining. Figure 2.1 shows diagrammatically the features and nomenclature of both practices.

### 2.3.1 Open-pit mining

Where practicable, this may have the advantages of higher productivity, higher recovery, easier dewatering, safer working conditions and usually lower costs than the alternative. The environmental impact is however greater both during and after the operating period.

Controlled blasting breaks the rock into *benches* 5–20 m high, driven progressively into the face (Fig. 2.2). On closure, the pit may be back-filled with waste or flooded, subject to the approval of several regulatory agencies. Management of tailings and mine closure is a topic in itself.[2]

### 2.3.2 Underground mining

Below about 200 m, underground mining is generally advantageous. Methods differ for sedimentary and vein-type ores.

Sedimentary ores are mined by the *room and pillar* method, with stopes 2 m or more high. The amount mined, about 50%, depends on the integrity of the walls.

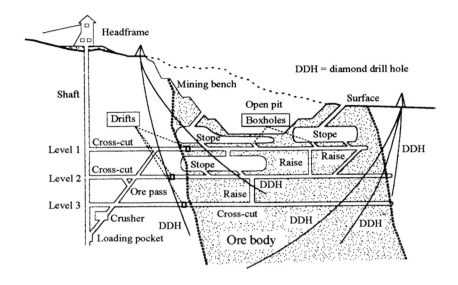

**Fig. 2.1** Open-pit and underground mining: vertical section and nomenclature

Vein-type ores are usually won by the cut and fill method, where a vein is cut into horizontal slices 3–4 m wide. When a slice has been mined, it is back-filled with sand from the mill before the next slice is started. Before transfer to the surface, ore is crushed underground to 200–300 mm size.

### 2.3.3 Control

Radioactivity is a guide to all stages from prospecting to ore concentration. The location, extent and grade of a deposit are indicated by airborne, surface and drill-hole measurements, although petrographic and chemical examinations of drill cores are also necessary to assess whether it is worth mining.

During mining, measurements at the wall face and in blast holes continue the grade control. Extracted material is likewise classified as product or waste. Sometimes trucks carrying broken ore pass under a radiometric bridge to estimate the content. (Such estimates, though inevitably less accurate than is possible after milling, are nevertheless very close.) Ore is then stockpiled by grade near the mill.

### 2.3.4 Risks

To the hazards of noise, dust, vibration, rock falls, chemicals and explosives, common to all mining operations, uranium mining adds those of radiation, both external and internal.

**Fig. 2.2** Uranium ore mine, Saskatchewan

External exposure is mainly due to gamma-emitting nuclides in the uranium decay series. The risk of internal exposure comes partly from radon, which can migrate through rock, be inhaled and deposit its daughter elements, partly from radionuclides in dust that can settle in the lungs. Workers must therefore wear dust-filter masks and some radiation-recording device to ensure compliance with limits set by the International Committee for Radiological Protection (ICRP).

Radon is also a problem in tailings beds unless they are kept wet or adequately covered.[2] In any type of mining operation, all aqueous and gaseous effluents must be continuously monitored.

## 2.4 Milling

### 2.4.1 Outline

The unit operations are much the same as for other ores (Fig. 2.3) with some differences in implementation. The choice depends on the mineralogy of ore and gangue, the process of dissolution, the desired product and the need for environmental precautions.

The ore is first crushed and ground to a suitable size. Roasting may be necessary to destroy organic carbon which could interfere with purification. The ground

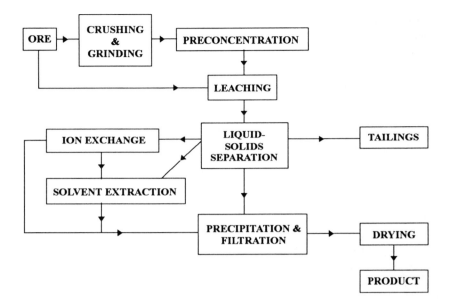

**Fig. 2.3**   Schematic flowsheet for processing uranium ores

ore is leached, usually with sulphuric acid and an oxidant to convert U(IV) to the more soluble U(VI);[3] where the gangue material is basic, however, leaching may be with a solution of sodium carbonate and bicarbonate to avoid excessive consumption of acid.

The solution is usually separated from the solid residue and purified by ion exchange, solvent extraction or both in that order (Eluex process). To avoid the expense of filtration, resin-in-pulp and solvent-in-pulp methods are beginning to be considered. At some stage, by-products may be separated for recovery.

The final product is usually precipitated and calcined.

The various effluents are directed to a tailings impoundment where as far as practicable, water is separated and returned to the mill. Eventually the area is covered and reclaimed.[2]

### 2.4.2 Comminution (size reduction)

Uranium ores differ from others in their requirement for size reduction only in that the minerals are susceptible to leaching and so need only be exposed, not completely liberated from the gangue. The nature of the rock and the required degree of crushing or grinding determine the type of equipment to be used and its energy requirement.

## Primary crushing

The size of rock fragments depends on the mining technique and on the geological and mineralogical disposition of the ore. Veins extending to depth are attacked underground in patterns of drilling and blasting that produce small fragments, usually crushed further before transport to the surface. Massive ore bodies, characteristically of lower grade, are extracted by the more economical open stoping method with less fragmentation, and so need more comminution.

The choice of crushing equipment depends on the location and throughput. The low-profile jaw crusher is probably more common underground; in open-pit mines, the higher throughput of the gyratory type usually outweighs the need for a more massive and expensive concrete base to withstand the impact.

## Fine crushing

Primary crushing yields fragments nominally up to 6 inches maximum dimension; later rod milling can accommodate pieces of ½ to ¾ inch. Fine crushing reduces the size by a factor not greater than three or four, so two stages are needed. Cone crushers are used, the second in closed circuit providing for recycle if necessary.

Efficient dust collection is particularly important with radioactive material. Typically a fine crushing plant of 5000 tpd capacity would require 40 000 cfm collection capacity, allowing 5000 cfm per dust generator (crusher or screen) and 1500 cfm per transfer point. Depending on the dust type and the required efficiency, wet or dry cyclones, electrostatic precipitators or bag filters are used to separate the dust.

## Fine grinding

Efficient leaching commonly requires ore to be reduced to a size range 6% +297 microns, 50% −74 microns. For decades, uranium mills followed the conventional practice of a single-stage open-circuit rod mill followed by single- or multiple-stage closed-circuit ball milling, capable of treating up to 10 000 tons per day.

However, the wear and stresses induced by the tumbling steel rods and balls have prompted a move towards autogenous grinding in which the work is done by the ore itself. Two stages of secondary crushing can then be eliminated. Where this is not entirely effective, a few large steel balls, up to 4% of the ore weight, may be added to give semi-autogenous grinding (SAG); this has become the most popular type of new installation in all metal-mining operations.

Output from the grinder is usually classified, commonly by hydraulic cyclones, and the excessively coarse fraction returned for further comminution.

### 2.4.3 Beneficiation

Generally the minimum $U_3O_8$ content for economical recovery is 0.07%, unless other metals are also recovered (as from the gold ores of South Africa), or the ore is amenable to heap or *in situ* leaching. Profitable working of very low-grade ores may depend on the possibility of pre-concentration. Many methods have been investigated for over 20 years, but with limited success because of the fine dissemination of uranium minerals.

The purposes of beneficiation are to:

- enhance feed grades for subsequent treatment;
- remove minerals that are likely to prove deleterious in the uranium leaching or recovery stages;
- provide clean tailings that can be rejected without causing environmental concerns.

Ore particles may be segregated according to radioactivity, size, shape, density or surface characteristics.

### Radiometric sorting

Originally developed in Canada in the late 1940s, and taken up throughout the world, this is an efficient, inexpensive process usually applied after primary crushing. There are three conditions for success:

- sufficient heterogeneity of uranium mineral;
- coarse fragmentation (usually 200–250 mm);
- secular equilibrium between uranium and radium.

### Photometric sorting

In the early 1970s the UKAEA developed an optical system which differentiated between opaque uranium-bearing particles and translucent quartz. Scanning laser sorters, developed by Ore Sorters, have been operated in South Africa.

### Separation by size or shape

Separation by size, while partially successful, has not proved economic. Ores in which uranium is finely disseminated are unsuited to physical separation processes. In Witwatersrand however sorting by 3 and 19 mm screens has permitted 60% recovery in 30% of the weight.

### Separation by gravity

Concentration by gravity is limited to situations where the uranium minerals are coarse and capable of resisting breakage and sliming, or association with other minerals or gangue. Uraninite and pitchblende qualify, but usually the ore is too finely ground.

Spiral concentrators, in which a combination of drag, gravity, frictional and centrifugal forces drive the heaviest particles rapidly to the bottom, have been used successfully. Reichert cones alone, although useful for gold and pyrite, concentrate uranium only slightly, but combined with a magnetic separator and shaking tables have recovered 60% of uranothorianite from copper tailings.

The commonest mechanical and gravity separators have been the jig and shaking tables, used respectively for the 25 mm–75 micron and 3000–15 micron ranges of pitchblende ore, in Canada (Port Radium) from the 1930s to 1960, also in South Africa and China.

Separation in a heavy medium, which requires a density difference of at least 0.1 unit between the ore and the gangue, has been applied successfully in Witwatersrand, Australia and Canada.

**Magnetic separation**

The magnetic characteristics of ore and associated material are usually too similar to provide a basis for separation. However, wet high-intensity magnetic separation (WHIMS) has recovered 84% uranium from Witwatersrand residues of 106–322 micron. It has also separated pyrite from Canadian, Australian and Chinese ores to prevent generation of acid in the tailings empoundment.

**Flotation**

Flotation of uranium minerals has never really been very successful, but some ores have responded to 70–90% uranium recovery with good concentration ratios. A problem has been the amount of uranium remaining in the tailings. Nevertheless flotation has been successfully applied to the removal of undesirable material such as sulphides from the uranium.

**Roasting**

Occasionally the ore has to be roasted to improve the recovery of uranium. The results can include:

- solubilisation of vanadium (and possibly other metals that could be by-products);
- elimination of sulphides;
- oxidation of uranium;
- elimination of carbonaceous material;
- dehydration of clays with consequent improvement in liquid-solids separation.

The stability of uranium oxides subjected to heat is as follows:

- $UO_3$ is stable in air to 450–600 °C;
- above 600 °C, $UO_3$ begins to convert to $U_3O_8$;
- in the temperature range 650–900 °C, $U_3O_8$ is the stable oxide in air;
- above 900 °C, $U_3O_8$ starts decomposing to $UO_2$.

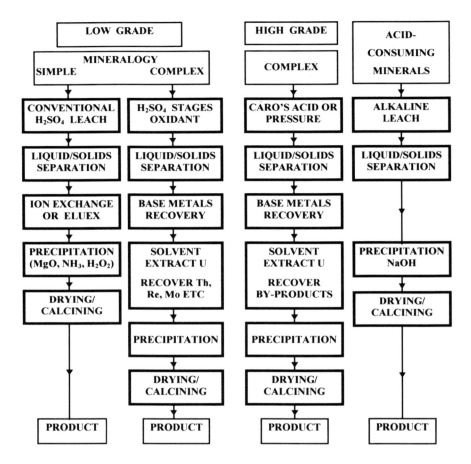

**Fig. 2.4** Process options for various types of ore

Also, above 600 °C, uranyl silicates are formed and are difficult to solubilise.

Carbon is removed from lignite and asphaltic ores at about 400 °C; sulphides may also be eliminated; and clays are dehydrated at 300–600 °C.

Although not commonly applied, roasting has always improved economics. Multiple-hearth roasters or rotary kilns are normally used.

### 2.4.4 Leaching

Depending on the nature of both ore and gangue minerals, leaching may be by sulphuric acid or a mixture of sodium carbonate and bicarbonate (Fig. 2.4). Where either could be effective, the choice is based on energy requirements, economics and environmental impact.

Leaching can be performed in stirred tanks, columns (pachucas), pressure vessels, heaps or *in situ*. The choice depends on the ore grade, the complexity of mineralisation, reagent costs, the time required to solubilise the values, and the overall capital and operating costs.

**Chemistry**

Uranium occurs naturally as hexavalent and tetravalent oxides. The former is readily solubilised, forming uncharged or anionic complexes with sulphate or carbonate ions in acid or alkaline leaching methods respectively.

$$UO_3 + 2H^+ \rightarrow UO_2^{2+} + H_2O \tag{2.1}$$

$$UO_2^{2+} + 3SO_4^{2-} \rightarrow UO_2(SO_4)_3^{4-} \tag{2.2}$$

$$UO_2^{2+} + 3CO_3^{2-} \rightarrow UO_2(CO_3)_3^{4-} \tag{2.3}$$

In the alkaline process, the hydrogen ion for eqn (2.1) is supplied by bicarbonate.
Uranium dioxide is not solubilised quickly and requires oxidation, e.g.

$$UO_2 \rightarrow UO_2^{2+} + 2e^- \tag{2.4}$$

The oxidation has been studied extensively and in an acidic medium is electrochemical. Ferric iron is an effective oxidant and its presence in solution enhances the efficiency of the process when manganese dioxide or sodium chlorate is added.

$$UO_2 + 2Fe^{3+} \rightarrow UO_2^{2+} + 2Fe^{2+} \tag{2.5}$$

Dissolution is fastest at about pH 2, and will not occur until conversion from the tetravalent to the hexavalent state is complete.

Heap leaching, or *in situ* leaching underground, can be assisted by the bacterium *Thiobacillus ferrooxidans*. This oxidises divalent iron to the trivalent form which then acts on the uranium mineral as above.

In alkaline leaching, U(VI) forms soluble carbonates (eqn 2.3). Much of the gangue material remains undissolved. Unless bicarbonate is also present to maintain solubility, hydroxyl ion is formed and precipitates the solubilised uranium as diuranate:

$$UO_3 + 3Na_2CO_3 + H_2O \rightarrow Na_4[UO_2(CO_3)_3] + 2NaOH \tag{2.6}$$

$$2Na_4[UO_2(CO_3)_3] + 6NaOH \rightarrow Na_2U_2O_7 + 6Na_2CO_3 + 3H_2O \tag{2.7}$$

$$UO_3 + Na_2CO_3 + 2NaHCO_3 \rightarrow Na_4[UO_2(CO_3)_3] + H_2O \tag{2.8}$$

Again, as in acid leaching, an oxidant is required, usually air or oxygen. The order of effectiveness in alkaline media is

$$O_2 < Cu(NH_3)_4^{2+} < H_2O_2 < Fe(CN)_6^{3-}$$

**Mineralogy**

The most important primary minerals are the oxides, pitchblende and uraninite, and the silicate coffinite. They may be gradually oxidised to the secondary minerals also listed in Table 2.1. Their properties, and those of the gangue, determine the degree of grinding necessary, the potential for physical beneficiation, the type and consumption of lixiviant, the ease of separating leachate from residual pulp, and the composition of the solution.

Ores may be classified as containing:

(a) uraninite, pitchblende or coffinite, with much of the uranium quadrivalent and needing oxidation;

(b) secondary minerals such as gummite, becquerelite, autunite, saleeite, sklodowskite, torbernite, uranophane, carnotite etc., with uranium already hexavalent;

(c) multiple oxides including niobates, tantalates, titanates, zircon etc.

Only type (a) minerals respond to alkaline leaching. Both (a) and (b) minerals dissolve in sulphuric acid at pH 1.8–2 in about 12 hours at 40 °C, although the presence of phosphate or vanadate may need higher acidity; phosphate (which precipitates uranium at low acidity) also complexes ferric ion and so hampers oxidation.

Type (c) minerals, of which brannerite is the most important, require severe leaching conditions with high temperature and acidity, with heavy consumption of reagent.

**Acid leaching—atmospheric**

Acid leaching is the most common, conducted in a cascade of agitated tanks or in pachucas; they may account for 25–33% of the mill's electrical energy consumption. Pulp density is usually 50–55% solids, determined by the physical properties as affecting the liquid-solids separation, as well as the solution composition. The form and content of silica can be troublesome in the subsequent solvent extraction processing, and thus the pulp density may be dictated by this aspect as well.

The concentration of acid must be enough to attack the uranium minerals with minimal attack on the gangue. At the end of leaching there should be sufficient free acid to prevent precipitation in the washing circuit.

Acidity may be maintained by fixed additions, or controlled automatically by pH or conductivity measurements. Oxidant, commonly sodium chlorate or pyrolucite, is added to maintain the redox potential in the range 475–525 mV relative to a saturated calomel electrode; in one Australian plant Caro's acid, $H_2SO_5$, is used with the advantages of economy (despite its higher intrinsic cost), improved control and virtually eliminating manganese from the process.

Some refractory Nigerian ores have been finely ground, mixed dry in a revolving drum with a small amount of concentrated sulphuric acid (pugging), cured at 65–100 °C for 12–24 hours and leached with water.

### Acid leaching—pressurised

Some complex, refractory ores, uneconomic to process by conventional atmospheric leaching, can be broken down under increased oxygen pressure. Pyrite is converted to sulphuric acid and ferrous sulphate,

$$2FeS_2 + 7O_2 + 2H_2O \rightarrow 2H_2SO_4 + 2FeSO_4 \quad (2.9)$$

and the latter subsequently to ferric ion which is an ideal oxidant for U(IV). In the most recent application (Key Lake, Saskatchewan), a two-hour atmospheric leach at 50–70 °C is followed by a second stage for 3–4 hours in a series of ten autoclaves at 70 °C, 650 kPa and pH 1.0. Overall recovery is 99.5%, 35% in the first stage.[4] Nevertheless, at the nearby Midwest Lake with similar mineralogy, extraction for 4 hours with Caro's acid and sodium chlorate at 60 °C gives comparable results more economically under atmospheric pressure. No plant is yet in operation.

### Alkaline leaching—atmospheric

Few plants use this process. One of the first (Eldorado Beaverlodge in Canada) operated from 1953 with frequent modifications until 1982, and none has been built since then.

The ore, containing 0.5% each of calcite and pyrite, was ground in a solution of 10 g/l sodium bicarbonate and 40 g/l sodium carbonate. The carbonate and sulphide fractions were separated by flotation, from which the rougher tails were thickened to 55% solids and further carbonated with flue gas to increase the sodium bicarbonate to 20 g/l. Leaching for 96 hours with a sparge of pure oxygen recovered 92–94% of the uranium.[5]

### Alkaline leaching—pressurised

Leaching in stainless steel autoclaves has been used in many alkaline circuits. In the most recent (Lodeve, France[6]) there are two stages of leaching with mechanical agitation and injection of oxygen at 600 kPa.

### *In situ* and in-place (heap) leaching [7,8]

Nearly all *in situ* operations are in sandstone aquifers, less than 300 m below the surface, confined by shale or mudstone of low permeability, and containing deposits that could not be mined economically. Leaching solutions, either acid or alkaline and usually with added oxidant, are forced through vertical injection wells into the deposit, recovered and brought to the surface for processing. The dilute uranium solution is generally concentrated by ion exchange before precipitation from the eluate. The main concern is the risk that leach solution might contaminate ground water, during or after operation.

In-place or heap leaching has been used for many years, particularly in mined-out stopes, to recover remaining low-grade uranium. The rock is broken, acid leach solution containing bacteria to aid dissolution is percolated through the heap, and the resulting uranium solution processed by ion exchange, with or without solvent extraction to follow.

### 2.4.5 Solids-liquid separation

This is one of the most important operations in the mill process, but also the most expensive, so that careful choice and design are essential. In the past both counter-current decant (CCD) thickeners and multiple drum filters have been used. Development has aimed to cut costs by minimising requirements for wash water, reducing the necessary area, improving the clarity of solutions and reducing maintenance.[9] The result has been the development of high-rate thickeners and horizontal belt filters.

#### High-rate thickeners
With the aim of reducing the required settling area, the feed launder above the slurry-liquid interface in a CCD thickener has been replaced by a horizontal pipe beneath the interface, so that solids need no longer settle through the free liquid zone.

Flocculents at about 50 g/t are used in both CCD and high-rate thickeners, but more effectively in the latter as there is less opportunity for disruption of the floc. The new design reduces the settling area and capital cost by factors of about 5 and 2 respectively; it is best used for pulps with over 15% solids, while the conventional type is best with 10–15%.

#### Cyclones
These centrifugal devices give an incomplete phase separation and are mainly used to classify wet or dry solids into fine and coarse fractions, supporting thickeners and filters to increase the overall capacity. They are small, and a circuit may contain 20 units in series.

#### Horizontal belt filters
Conventional filters of the disc, plate and frame, pressure and vacuum drum types are expensive, energy-intensive, have small surface areas and need considerable maintenance. The concept of feeding leach pulp on to a moving horizontal belt filter, through which the solution drains under gravity, was developed over several years, and many have been installed since 1973.

## 2.4.6 Solution purification

The clarified leach solution may be purified in several ways depending on its concentration and composition, and on the desired product purity. The first two criteria depend on the mineralogy of the ore and on the leaching medium. Among the alternative schemes are:

- direct precipitation;
- two-stage precipitation;
- solvent extraction, stripping and precipitation;
- ion exchange, elution and precipitation;
- ion exchange, elution, purification of eluate, precipitation of product;
- pre-treatment to remove impurities or by-products, ion exchange, elution and precipitation;
- ion exchange, elution, solvent extraction, stripping and precipitation;
- solvent extraction, selective scrubbing to remove impurities or by-products, stripping of uranium and precipitation.

As noted on Fig. 2.4, ion exchange, solvent extraction or both are suited to conventionally acid-leached low-grade ores. Where they contain nickel, cobalt, arsenic etc., solvent extraction with possible recovery of by-products is indicated. The same applies to high-grade complex ores, although they may need a more aggressive leach with Caro's acid or under pressure. Alkali leachates contain very little gangue material and may be directly precipitated. Future chloride leachates may best be treated by solvent extraction to recover uranium and by-products other than radium, which would be separated by ion exchange.

### Ion exchange

The first commercial application was by West Rand Consolidated Mines Ltd in 1952; in some cases it was combined with solvent extraction (Eluex process) or replaced by it. Batch processing in fixed-bed systems was at first universal, but continuous ion exchange (CIX) is now accepted and its application to unclarified solutions is the most recent technological development.

Uranium in either sulphate or carbonate solution exists as anionic complexes with two or three divalent anions attached to the divalent uranyl cation, and as such is strongly sorbed by strong- or weak-base resins. In practice the only carbonate liquors normally treated by ion exchange are the dilute solutions from *in situ* leaching.

Ion exchange serves both to concentrate and to purify the product. Ideally, in the presence of other anions, the resin should absorb uranyl complexes selectively, rapidly, reversibly and to a high loading.

Most systems operate at a pH between 1.5 and 2, except where there is a large excess of iron (mole ratio >15:1) when it must be at 1.5. Acid at higher concentrations decreases the loading by competing for resin sites.

The loading attainable with fresh resin normally decreases rapidly during the first few cycles of operation, and then more slowly. This is due partly to structural changes in the resin, partly to poisoning by irreversible absorption of impurities. Common poisons include molybdenum, polythionates, silica and sulphur; others are titanium, zirconium, thorium and organic materials. Various treatments with caustic soda have been used to remove the common poisons, while 6M sulphuric acid has successfully removed the second group.

Batch systems that have been used include conventional fixed-bed types and so-called moving-bed systems in which the loaded resin is moved to a second column for elution.[10] Several continuous systems and resin-in-pulp units have been installed around the world.

Fixed-bed systems, installed in the 1950s and in some instances still working, comprised three columns piped and valved so that any two could operate in series on sorption while the third was eluted—an automatic operation controlled by timers and volumetric counters. They were designed to treat clarified acid leach liquors containing 0.6–1.0 g $U_3O_8$/l. Resin attrition losses were typically less than 5% per year.

The moving-bed Porter–Arden system, used at several locations in Canada and the USA, comprises two sets of 3 sorption columns linked to one backwash column (where foreign matter is removed) and one set of three elution columns. By an automated combination of liquor diversions with resin transfers, the saturated leading batch from the sorption chain is back-washed, to become the trailing batch in elution at the next move, while the regenerated leading resin from the elution chain becomes the trailing batch in sorption. Losses by attrition are typically around 10% per annum.

Continuous ion-exchange represents the current state of the technology. Many systems have been devised to move the resin countercurrently to the liquid stream through the different phases of the process, as described by Merritt;[11] a summary by Himsley[10] in 1986 gives supplementary information and references. The advantage of continuous ion exchange is that solids, perhaps up to 15%, can be tolerated, thus eliminating expensive liquid-solids separation. The resin bed is fluidised in some of the designs by an upflow of feed solution; these include the NIMCIX, the USBM-MCIX column, the Himsley and CANMET columns of Canada, and the multiple tank system of Porter. By contrast is the pulsed downflow design of Chem-Seps, often called the Higgins Loop CIX system. Depending on the mode of operation, attrition losses range from 3% to 70% per year.

The basket resin-in-pulp (RIP) system was first operated commercially in the USA in 1955. The resin was contained in baskets suspended in the desanded leach pulp, and by absorbing uranium before the solid-liquid separation, avoided the

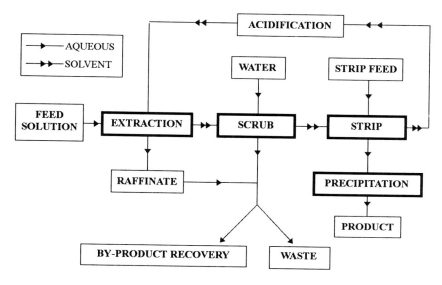

**Fig. 2.5** Purification by solvent extraction

loss liable to occur at that stage. Nevertheless none of the six installations is now operating.

## Solvent extraction[11-15]

The atomic energy industry was the prime user of solvent extraction in the separation and recovery of metals until the first large copper plants came into production in the late 1960s. Application to uranium ore processing began in 1955. A schematic flowsheet is shown in Fig. 2.5. The metal-bearing process solution flows countercurrently to the organic solvent in a mixing device where the metal is transferred from the aqueous to the organic phase, usually over several stages of contact for maximum recovery. Following coalescence of the two phases, the depleted aqueous phase or raffinate may be treated to recover other values, recycled to another point in the overall process or discharged to waste.

The loaded solvent is usually contacted with a scrub solution to remove impurities (iron, silica, aluminium etc., mostly due to entrainment of feed liquor in the solvent) which would otherwise promote emulsification and crud formation, besides contaminating the product. Then it is stripped with a suitable aqueous solution, possibly pre-equilibrated with sulphuric acid, and returned to the extraction circuit.

### Extractants

These may be acidic, basic or neutral. The first used, in the DAPEX process, was di-2-ethylhexyl phosphoric acid (DEHPA) which is not very selective and needs close control of pH to limit the co-extraction of impurities. Since 1957 secondary or particularly tertiary amines (AMEX process) have been more popular.

$$2(R_3NH)_2SO_4 + UO_2SO_4 \rightarrow (R_3NH)_4UO_2(SO_4)_3 \qquad (2.10)$$

The order of effectiveness in extraction from chloride or sulphate media is

quaternary > tertiary > secondary > primary.

They selectively extract uranium in the presence of iron, thorium, phosphate or rare earths, although vanadium, molybdenum and zirconium may be co-extracted and require selective stripping to remove them.

Amines can be used in sulphate or chloride media without degradation, but in the presence of nitrates or other oxidants (such as excess sodium chlorate in the leach solution) they are degraded very quickly.

Among neutral extractants, tri-*n*-butyl phosphate (TBP) has been used very extensively in the nuclear industry, but mostly to refine the mill precipitate or in post-irradiation reprocessing (Chapter 7): the only mill to use it (in extraction from nitric acid) is Palabora in Africa. Trioctyl phosphine oxide (TOPO) has however been used to recover by-product uranium from phosphoric acid liquors.

**Diluents**

A diluent is often necessary to improve the physical properties of the solvent or limit its extractive power, and is then generally the major part of the solvent phase. It may be of the kerosene type, or with great benefit tailor-made with a blend of aliphatic, naphthenic and aromatic constituents to optimise extraction chemistry, phase separation and flash point—usually over 120 °F (49 °C).

**Modifiers**

A third component of the solvent may be necessary to improve phase separation, increase the solubility of the extracted metal complex and so prevent third-phase formation, or reduce crud formation. The choice is often TBP in the DEHPA system, or a long-chain alcohol such as *iso*decanol or tridecanol with amines.

**Stripping**

Stripping can often be accomplished by a change in pH. Sodium carbonate was used at first, but a solution of ammonia and ammonium sulphate is now the most popular, despite the difficulty of controlling pH to prevent premature precipitation with consequent crud formation and solvent losses. When molybdenum has been co-extracted, sodium chloride may be used to strip the uranium alone before a second strip with carbonate to remove the molybdenum.

Other stripping agents include sodium sulphate and ammonium carbonate. The latter has the advantage of leading simply, by way of the tricarbonate, to an ammonium uranyl carbonate product that can readily be converted to $UO_2$, $UO_3$, $U_3O_8$ or $UF_4$.

**Contacting equipment**

Solvent extraction is usually performed in continuous countercurrent mode, either in discrete stages (mixer-settlers) or differential contactors (columns), depending on the hydraulic properties of the two phases and the kinetics of the transfer mechanisms. Mixer-settlers are most commonly used, but early designs generated a finer dispersion than was necessary for effective mass transfer and so tended to form emulsions and cruds; subsequent designs, such as the Krebs or the Combined Mixer Settler (CMS) of Davy, have overcome this problem. Where kinetics are fast, pulsed sieve-plate columns may be used.

**Solvent losses**

Solvent losses, whether by solubility, volatility, entrainment (including emulsion and crud formation) or degradation, are important environmentally as well as economically. The issues are discussed elsewhere.[13,16]

### 2.4.7 Product recovery

The mill product, commonly uranium diuranate (*yellowcake*), is formed by precipitation. It must be highly purified, settle and filter quickly, dry easily and handle well.

The choice of precipitant—hydrogen peroxide, ammonia, magnesia, magnesium hydroxide, or sodium hydroxide—depends on the purity of the feed solution, the product specification, cost, and environmental considerations. Although much studied, this stage of the milling process can still be among the most troublesome; the precise conditions required are specific to each site and have to be optimised experimentally.

**Precipitation from acid solution**

Leach liquor, if not otherwise purified, is usually part-neutralised with lime to pH 3–4 as a preliminary treatment to precipitate iron and other impurities. Precipitation of uranium, however pre-treated, is often with magnesia, but ammonia or ammonium hydroxide may be used, and peroxide is sometimes mandatory to prevent co-precipitation of other metals or the release of ammonaical pollutants.

Magnesia is added in suspension to reach a pH of 7.0–7.5:

$$UO_2SO_4 + MgO + xH_2O \rightarrow UO_3 \cdot xH_2O + MgSO_4 \ (x = 1\text{--}2) \quad (2.11)$$

Ammonia, in solution or more usually as gas mixed with three volumes of air, is applied at 30–50 °C to a final pH of 7–8.

$$2H_2UO_2(SO_4)_2 + 10NH_3 + 3H_2O \rightarrow (NH_4)_2U_2O_7 + 4(NH_4)_2SO_4 \quad (2.12)$$

70% hydrogen peroxide may be added in 35–100% excess to uranium solution at about 25 °C and pH 3.5, maintained by addition of ammonia or sodium hydroxide to neutralise the acid formed.

$$UO_2^{2+} + H_2O_2 + 2H_2O \rightarrow UO_4.2H_2O + 2H^+ \quad (2.13)$$

**Precipitation from alkaline solution**

Alkaline leach liquors are often, but not always, pure enough for direct precipitation with sodium hydroxide, a process performed at 50–80 °C over a period of 6–12 hours.

$$2Na_4UO_2(CO_3)_3 + 6NaOH \rightarrow Na_2U_2O_7 + 6Na_2CO_3 + 3H_2O \quad (2.14)$$

A portion of the precipitate is normally recycled to nucleate the next batch.

With ion-exchanged or solvent-extracted liquors, the retention time is shortened to 2–4 hours. The final pH is 12.

**Ammonium uranyl tricarbonate system**

The salt ammonium uranyl tricarbonate (AUC) can be produced from crude uranium hydroxides, peroxides, uranates or uranyl salts by dissolution in dilute ammonium carbonate solution to form the tricarbonate complex, filtration to remove residues, and addition of more ammonium carbonate. AUC crystallises and can be filtered off; in well-controlled operations, enough of the impurities remain in solution for the product to meet nuclear grade specifications.

### 2.4.8 Solids-liquid separation

The precipitated product is dewatered by thickening or centrifugation, and may then be dried directly or filtered and washed on drum or belt filters. Filtration may prove difficult in the presence of excessive impurity or mother liquor, or if the precipitate is slimy or very fine; the conditions of precipitation may then have to be modified.

The uranium precipitate is washed and dewatered at the same time, batchwise in plate and frame presses or continuously in thickeners, rotary vacuum drum and disc filters, belt filters or centrifuges. Washing may be, for instance, with ammonium sulphate solution or process water, in a single stage or countercurrent. The pH may need adjustment to maximise decontamination with minimal loss of uranium.

In the USA, initial dewatering was usually by single-stage thickening, otherwise by two-stage CCD thickening, but centrifugation is now the most popular practice, in combination with a single-stage dewatering thickener. This eliminates most entrained impurities.

### 2.4.9 Drying or calcination

The wet concentrate is dried or calcined and packaged in steel drums with a polyethylene liner. The ammonium diuranate product is often dried or spray-dried at

150–250 °C, whereas sodium diuranate is commonly calcined at about 400 °C. Some mills calcine ammonium diuranate or peroxide products, eliminating water completely and driving off ammonia; sintering however must be avoided as it reduces subsequent solubility in nitric acid. Driers may be of the single or multiple hearth, drum, belt, screw or radiant-heat types.

## 2.5 Tailings management[2]

This has changed more than any other aspect of the process. In the past it was disregarded, but is now an important contributor to decommissioning costs, while environmental considerations often require modifications to the process, such as the elimination of ammonia and nitric acid from many circuits.

### 2.5.1 Tailings and waste rock characteristics

The particle size of tailings, the type of material (e.g. slimes or clay), porosity and permeability all affect the water retention, and so the dewatering needed to consolidate the mass and return water to the plant. The mineral content, particularly sulphides, and any neutralising agents still present will affect the composition of effluents, the disposal technique and the design of impoundment.

### 2.5.2 Toxicity of effluents

Even when the discharge of effluents to the impoundment area is properly planned and engineered with the best practical technology, seepage and run-off can occur with the risk of affecting the aquatic environment. Reagents should therefore be evaluated with that in mind. On the other hand, levels of pollutants actually leaving the impoundment may be reduced by sorption processes within the tailings. For instance, Ra-226 is retarded in its mobility by precipitation on surfaces such as gypsum, so that its rate of passage through the tailings impoundment is orders of magnitude slower than that of the acid.

### 2.5.3 Design of overall treatment system

The four main considerations are the site, its preparation and design, the physical properties of the tailings, and their chemical preparation in the mill. Problems and costs can be minimised by designing the mine and mill complex as a whole, taking account of all the chemical, physical and possibly biological interactions in the overall process.

Possible sources of pollution are mine water, mining waste rock, overburden from an open pit, tailings slurry, iron and aluminium hydroxide sludges (possibly containing arsenic), gypsum sludges, mill process residues, neutralised mill effluents, gaseous effluents, organic constituents (such as solvents, diesel oil, flotation agents, humic acids, bacteria, algae, and fungi) and barium-radium precipitates.

### Site selection options

Each site will have advantages and disadvantages, depending on the type of milling operations. Criteria include capacity, availability, hydrology, initial cost, ease of operation, geotechnical and geological considerations, freeboard to accommodate a 200-year flood, and total engineering design.

Wastes may be discharged to a valley dam, a ring dike, a mine pit, a specially-dug pit, an underground mine, a deep lake or ocean. Each has advantages and constraints, and the selection must be environmentally approved.

### Disposal

The management system includes treatment in the mill, slurry thickening, transport, impoundment, water recycle, the treatment and disposal site, effluent treatment, evaporation, and eventually restoration of the site. Solids may be impounded on land by random discharge over the whole site, usually as a thin slurry from a single point; by subaerial discharge in which a thicker slurry with 30% or more solids is spread in a systematic pattern, so that in any area each layer has time to dry out before the next is applied; or after thickening to some 70% solids which leads to very little subsequent seepage and migration of contaminants. Alternatively they may be used as backfill in the mine; discharged to deep lakes or the sea; or processed to recover secondary metals before disposal. Because uranium tailings are easily dewatered, the trend is towards the subaerial or possibly the thickened-discharge method.

### Water management

This is vital to every aspect of the mining complex. The control plan must cover:

- collection and treatment of run-off from the mine area;
- diversion of run-off from surrounding areas;
- recycling of process water.

Details are specific to the site.

## 2.5.4 Weathering and migration

Weathering will be governed by climatic conditions, particularly the occurrence of acid rain, and the mineralogy of the tailings, combined with the effects of process residues and possibly bacteria. In time it may cause many changes in chemical and physical characteristics.

Upon close-out the weathering process, although unchanged externally, may be considerably different within the impoundment. Chemical processes may alter; some effects—precipitation, sorption, redissolution, and permeability to seepage with consequent leaching of radionuclides or other contaminants—may increase or diminish with time. The nature of the tailings and waste rock themselves, the surrounding rock and sediments, and the integrity of the retaining structure, are all important, besides the global cycles of oxygen, carbon dioxide, nitrogen and sulphur.

### Effluent treatment

Waste water (both the input to the impoundment and the effluent from it) may need treatment to control organic content, pH, suspended solids, toxic materials, colour and volatile components. According to the particular circumstances, the process must be chosen from the many available, including coagulation, sedimentation, neutralisation, sand filtration (e.g. to remove radium-barium sulphate), carbon filtration, in-line filters, coalescers, centrifuges, ion exchange, precipitation and biological treatment.

### Decommissioning, reclamation and covers

Restoration aims to:

- stabilise the surface against erosion by wind or water;
- contain contaminants such as radon, other radionuclides, sulphuric acid and metals;
- provide an aesthetic appearance.

Few of the possible covering methods meet all these objectives. The most effective are growing vegetation, covering with soil, rock etc., or forming a crust from fine-grained minerals by chemical means.

Costs vary with the type of tailings, the disposal method, the location and weathering conditions.

### Site monitoring

A well-designed environmental monitoring programme will determine:

- present levels of toxicants in the effluents;
- their environmental impact at the mine site and various other locations;
- pollution-abatement measures necessary to comply with standards.

The particular mineralogy and process conditions will of course dictate the specific requirements, but common parameters are dissolved and suspended solids, dissolved oxygen, biological oxygen demand, temperature, pH, conductivity, cations and anions. Once the total process is understood, a model can be derived and applied to the control of both existing and new disposal areas.

The author has established a very useful testing and research procedure, involving simulated and accelerated weathering, to assess long-term effects on material going to the tailings area.[2] Inexpensive tests provide information necessary for impoundment design, disposal type, vegetation and rehabilitation.

## 2.6 References

1. Uranium extraction technolog, IAEA Technical Report Series 359, Vienna (1993).
2. G. M. Ritcey. *Tailings management—problems and solutions in the mining industry*, Elsevier (1989).
3. K. E. Haque and G. M. Ritcey. Comparative efficiency of selected oxidants in the acid leaching of uranium ores. *CIM Bulletin*, May 1982, p. 127.
4. W. Floter. The Key Lake project. In *Proceedings of IAEA Technical Committee meeting 25–28 November 1985*, Vienna. IAEA-TC-453, 5/16, pp. 79–91 (1987).
5. D. G. Feasby. Eldorado Nuclear Limited—milling practice in Canada. In *CIM Special Volume 16*, pp. 300–2.
6. G. Lyaudet, P. Michel, J. Moret and J. M. Winter. A new unit for purification of uranium solution in the Lodeve mill. In *Proceedings of Technical Committee meeting, 25–28 November 1985*, Vienna. IAEA-TC-453, 5/13, pp. 55–67 (1987).
7. *In situ* leaching of uranium: technical, environmental and economic aspects. In *Proceedings of a Technical Committee meeting organised by the IAEA, Vienna, 3–6 November 1987*. IAEA-TECDOC-492.
8. *Uranium in situ leaching. Proceedings of a Technical Committee meeting, Vienna, 5–8 October 1992*. IAEA-TECDOC-720.
9. *Uranium extraction technology—current practice and new developments in ore processing*. Joint report by OECD Nuclear Energy Agency and IAEA, IAEA, Vienna, 1989.
10. A. Himsley. Application of ion exchange to uranium recovery, in *Ion exchange in the nuclear fuel cycle* IAEA, TECDOC-365, pp53–84 (1986).
11. R. C. Merritt. In *The extraction metallurgy of uranium,* Colorado Schools of Mines Research Institute. pp. 137–220 (1971).
12. J. W. Clegg and D. D. Foley. *Uranium ore processing*. Addison Wesley, Reading, Massachusetts (1958).
13. G. M. Ritcey and A. W. Ashbrook, *Solvent extraction—principles and applications to process metallurgy*. Vols 1 & 2, Elsevier, Amsterdam (1978, 1982).
14. P. A. Schweitzer (ed.) *Separation techniques for chemical engineers*. 2nd. edn. McGraw-Hill (1988).
15. T. C. Lo, M. H. I. Baird and C. Hanson (eds) *Handbook of solvent extraction*. Wiley, New York (1983).
16. G. M. Ritcey and E. W. Wong. Methodology for crud characterisation. In *Proceedings of International Solvent Extraction Conference, ISEC '88*, Moscow, p. 116.

# 3 Fuel fabrication

G. Marsh and H. Eccles
*BNFL, Springfields, Lancashire*

## 3.1 Introduction

Nuclear fuel comprises two essential components:
- the reactive substance, generally uranium or one of its compounds, possibly with plutonium or other constituents also present;
- a metal cladding to protect the fuel substance from its chemical environment, and to prevent fission products from escaping.

There are also appendages or fittings to hold composite structures together, locate the fuel in the reactor or aid loading and unloading. Various types are illustrated in Fig. 3.1.

The two principal types of fuel in current use are based respectively on uranium metal and uranium oxide. (Thorium oxide can also be used but is not widely adopted, despite considerable interest on the part of the Indian nuclear industry where relative abundance and a desire for technological independence tend to counterbalance its disadvantages.) Metal can be used in the form of massive rods, which are conveniently handled individually. The oxide has a much lower conductivity; it is therefore formed into thinner rods or *pins*, and to avoid unnecessary handling operations within the reactor these are built up into clusters, assemblies, bundles, elements or sub-assemblies (the terminology varies according to the type of reactor, but with much the same meaning) which may each contain several hundred.

Fuel may be further distinguished according to whether the uranium has the natural U-235 content of 0.71% by weight, or has this proportion enhanced artificially to increase the nuclear reactivity, or is supplemented by plutonium for the same purpose. Considerations peculiar to plutonium-enriched fuel are covered in Chapter 8, and the process of enhancing or 'enriching' the U-235 content in Chapter 4.

The essential stages of fuel manufacture are
- purification of uranium;
- if enrichment is needed, conversion to hexafluoride, and enhancement of U-235 content;

## 42  Fuel fabrication

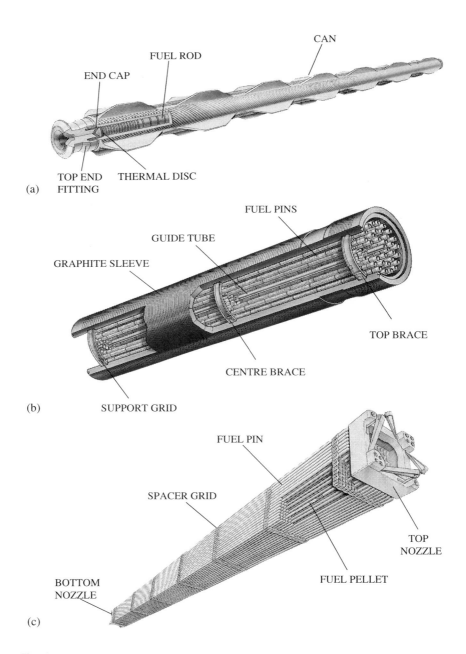

**Fig. 3.1**  Fuel elements— (a) Magnox, (b) AGR and (c) PWR

- conversion to the required chemical form, usually metal or dioxide;
- formation into the required shape;
- application of cladding;
- building into clusters (for oxide);
- addition of fittings or appendages.

## 3.2 Design considerations

### 3.2.1 Metal fuel

Metal fuel (Fig. 3.1) is now used in the first generation of British power reactors, and was used in two similar stations in Japan and Italy. It comprises an alloy of natural uranium metal with traces of iron and aluminium, added to help form the required grain structure, contained within a magnesium-based cladding alloy that has a low capture cross section for thermal neutrons. This alloy, Magnesium Alloy AL80 (containing 0.8% aluminium, 0.002–0.05% beryllium, 0.008% calcium and 0.006% iron), was specifically developed to resist the oxidising effects of the carbon dioxide coolant in the reactor—hence the name Magnox, applied not only to the alloy but also to the fuel as a whole and to the reactors that use it.

As with any type of fuel, the physical properties of the materials used place limits on the performance that can be obtained. In the case of Magnox fuel there are two principal considerations:

- the cladding must be kept below 450 °C to avoid excessive oxidation;
- at 662 °C the uranium metal undergoes a change from the $\alpha$- to the $\beta$-phase accompanied by a significant expansion, so to avoid undue stress, this temperature must not be exceeded anywhere within the body of the fuel rod.

Elements typically contain about 12 kg uranium as a rod 1.1 inches in diameter and 38–42 inches long (28 × 965–1070 mm). In service, they are located within cylindrical channels in the graphite moderator and exchange heat with the coolant gas by way of heat exchange fins machined into the clad. On some designs, the cooling fins are also integral with braces which provide mechanical support to prevent bowing in service, besides improving the overall heat transfer characteristics of the fuel element by inducing turbulent gas flow around the fins.

Magnox fuel is of natural U-235 content, but to extend the time between reloads, a slight degree of enrichment has been suggested.

Experience of gas-cooled, metal-fuelled reactors is not confined to the UK Magnox type. In particular, the French developed and operated commercial gas-cooled reactors fuelled by uranium metal in the form of tubular elements with graphite cores. The cladding was of magnesium alloyed with 6% zirconium, again machined into chevron cooling and straight bracing fins.

A quite different type of metal fuel, alloyed with zirconium for high-temperature performance, has been used in experimental fast reactors, but has not so far entered the commercial cycle and seems unlikely to do so for some time.

### 3.2.2 Oxide

Because of the temperature limitations on uranium metal fuel, the thermodynamic efficiency of a generating station using it is limited. With the low proportion of fissile U-235, large arrays are needed to maintain the necessary power, and construction costs are accordingly high; moreover, the amount of energy that can be extracted between reloads is limited. To overcome these drawbacks, oxide fuel with about 4% U-235 may be used. Magnox alloy is then an unsuitable cladding metal, and a zirconium alloy (*Zircaloy*) or stainless steel is used instead. Such fuel assemblies are used for both water-cooled reactors of the PWR, BWR or VVER type, and gas-cooled AGR reactors. Essentially similar fuel but with uranium oxide of natural U-235 level is also used in the Canadian-designed CANDU reactors.

Because the pins are so narrow, thermal stresses could induce bowing which might cause contact and local over-heating, while the flow of coolant could induce vibrations with possible fatigue damage. The elements are therefore held together by structures of various designs, with grids distributed along the length to maintain the spacing.

Fuel elements for Western light water reactors are generally square in section; Russian VVER designs are hexagonal and RBMK cylindrical. AGR elements are also cylindrical and contained in a graphite sleeve which forms a replaceable component of the moderator.

## 3.3 Uranium purification

Uranium is received at the manufacturing plant in the form of uranium ore concentrate (UOC, or *yellowcake* in Canadian terminology). Although in the mining and milling operations the uranium has been separated from the majority of other elements and radionuclides (such as thorium and radium) present in the ore, it still contains impurities detrimental to its final use and needs further purification to reach *nuclear grade*. Specifically it must be separated from

- elements such as boron and cadmium with high capture cross sections for thermal neutrons;
- those forming volatile fluorides (e.g. molybdenum, vanadium, tungsten and chromium) and therefore liable to contaminate the uranium hexafluoride produced at a later stage;
- others with chemical properties similar to uranium, e.g. thorium.

UOC is refined by either the *dry* or the *wet* process. The former, used by Allied

Chemicals (USA), is based on the anhydrous fluoride volatility process developed from a conceptual design by the Argonne National Laboratory;[1] the uranium feed is hydrofluorinated directly, wastes are solid and the uranium hexafluoride product is purified by distillation.

Apart from the Japanese PNC process briefly described in section 4.2,[2] all commercial UOC refiners use the classical wet process (developed by the USAEC) or an approximation to it, so only this route will be described in any detail.

A liquid-liquid extraction (solvent extraction) circuit is used. Extractants such as substituted phosphine oxides, phosphonates, phosphinic acids, sulphoxides, ketones and ethers have been evaluated or used in pilot-plant studies, but a diluted solution of tri-$n$-butyl phosphate (TBP) was selected as the preferred solvent in the mid-1950s. Apart from some minor variations such as the choice of diluent, TBP concentration and contactor design, the liquid–liquid technology employed by all UOC refineries is very similar. The conditions employed by BNFL's refinery at Springfields (near Preston in Lancashire) are typical.

All refineries in the United Kingdom, USA, France and Canada dissolve the UOC directly, or after calcination, in nitric acid:[3-5]

$$(NH_4)_2U_2O_7 + 6HNO_3 \rightarrow 2UO_2(NO_3)_2 + 2NH_4NO_3 + 3H_2O \quad (3.1)$$

$$U_3O_8 + 8HNO_3 \rightarrow 3UO_2(NO_3)_2 + 2NO_2 + 4H_2O \quad (3.2)$$

UOC is fed at a controlled rate into a multi-stage continuous dissolver co-currently with 50% nitric acid at 90 °C. Continuous on-line measurement of uranium concentration and free acidity in the product allows automatic control to yield a slurry of 450 g/l uranium and 80 g/l free nitric acid.

After cooling, insoluble material such as sand or silica is removed by filtration. The filtrate combined with washings contains approximately 350 g/l uranium, together with soluble impurities from which it is separated by continuous counter-current solvent extraction. The contactors are mixer-settlers using 20% (by volume) TBP in *odourless* kerosene as solvent; each unit comprises eight stages each of extraction and scrub followed by twelve of strip. The product contains 110 g/l uranium and is stored to await the results of quality control analysis.

The solvent extraction process depends on association between the uranyl and nitrate ions to produce a neutral complex, and solvation of the metal by the extractant:

$$UO_2^{2+}{}_{(aq)} + 2NO_3^-{}_{(aq)} + 2TBP_{(org)} \leftrightarrow UO_2(NO_3)_2 \cdot 2TBP_{(org)} \quad (3.3)$$

The extracted complex is a well-defined compound with a sharp melting point of 6 °C.

Neglecting activity coefficients, the equilibrium constant may be written as

$$K = [UO_2(NO_3)_2.2TBP]_{org}/([UO_2^{2+}]_{aq}[NO_3^-]_{aq}^2[TBP]_{org}^2)$$

which clearly shows the dependence of uranium extraction on the concentrations of uranium and nitrate ions in the aqueous phase, and of free TBP in the solvent. Reaction 3.3 is exothermic to the right, so that temperature is also significant, with a rise decreasing the extraction of uranium.

## 3.4 Conversion

As in the purification of uranium, so in converting it to the hexafluoride, similar process technology is employed by the various UOC refineries, with differences chiefly in the choice of contactor. Processes at BNFL's Springfields works are described as representative.

### 3.4.1 Production of uranium dioxide

Purified uranyl nitrate solution is evaporated in a battery of four-effect climbing-film evaporators to a molten salt containing 1100 g/l uranium. Condensate is used to pre-heat the incoming feed and then recycled to the solvent extraction circuit. The molten uranyl nitrate is denitrated in a fluidised-bed reactor at a temperature of 300 °C, maintained by a combination of internal heating elements and external jackets, forming uranium trioxide, oxides of nitrogen, oxygen and steam.

$$UO_2(NO_3)_2.xH_2O \rightarrow UO_3 + NO_2 + NO + O_2 + xH_2O \tag{3.4}$$

The uranium trioxide overflows continuously from the denitrator and is transported pneumatically to storage hoppers. Fluidising air and exhaust gases are filtered and directed to an acid recovery plant, where nitric acid is reconstituted by condensation and continuous counter-current absorption for recycling to the dissolution stage; this achieves the double benefit of reducing cost and lowering the nitrate level in effluent.

The uranium trioxide at this stage consists of hard spherical particles in which surface area and internal porosity are both low. To increase them, and so ensure rapid diffusion of gases in subsequent reactions, the product is hydrated in a stainless steel trough through which it is carried with a metered proportion of hydrating fluid by an interrupted-screw conveyor. The temperature is controlled so as to form a dry, free-flowing dihydrate with low density and high surface area.

The dihydrate is screw-fed into a stainless steel rotary kiln, about 10 m long by 1.2 m diameter, counter-currently with metered quantities of hydrogen. Precise control of conditions is necessary to ensure complete reduction without loss of reactivity.

$$UO_3 + H_2 \rightarrow UO_2 + H_2O. \tag{3.5}$$

The kiln rotates at 3 rev/minute with an internal temperature of 500 °C (maintained by three independently controlled electric muff heaters and forced-air cooling of barrel sections as appropriate), and a system of flights and dam rings within it assures the necessary gas-solid contact and residence time. The product is transported pneumatically on nitrogen to the feed hoppers of the hydrofluorination kiln; exhaust gases containing nitrogen, steam and surplus hydrogen are filtered and vented to atmosphere by way of a water-cooled condenser and caustic scrubber.

### 3.4.2 Hydrofluorination

The dioxide is hydrofluorinated with controlled quantities of anhydrous hydrofluoric acid in an Inconel rotary kiln, about 16.7 m long by 1.2 m diameter, where dam rings and flights similar to those in the reduction kiln provide the residence time and counter-current gas contact required to ensure complete conversion to uranium tetrafluoride:

$$UO_2 + 4HF \rightarrow UF_4 + 2H_2O \qquad (3.6)$$

Just as in the reduction kiln, the temperature in the reacting zones is carefully controlled at about 450 °C to prevent thermal damage to the basic particle, causing deactivation and incomplete conversion or in the most extreme instance sintering of the bulk material.

Exhaust gases containing steam, nitrogen and the small amount of unused hydrogen fluoride are filtered, condensed and finally scrubbed before discharge to the atmosphere. After satisfactory assay, the uranium tetrafluoride product is fed to the production of either hexafluoride or uranium metal.

In parenthesis, PNC (Japan) produces uranium tetrafluoride by a different route. Uranium ore concentrate is dissolved in sulphuric or hydrochloric acid, and the resulting uranyl salt solution reduced electrolytically to the tetravalent state, from which hydrated $UF_4$ is precipitated by aqueous hydrofluoric acid at 95 °C.

### 3.4.3 Production of uranium hexafluoride

Uranium hexafluoride (*hex*) is produced by the reaction of the tetrafluoride with elemental fluorine, generated in cells derived from a basic design developed by ICI. The molten salt KF.2HF is electrolysed at 80 °C between amorphous carbon electrodes in mild steel cells. The HF content of the electrolyte is maintained at the desired 42% by weight through automatic addition of anhydrous hydrofluoric acid. HF in the fluorine product was formerly removed to less than 0.2% wt/vol by absorption in sodium fluoride. Hydrogen evolved at the cathode is scrubbed with caustic potash solution before discharge to atmosphere by way of a flame trap.

Uranium tetrafluoride is fed at a controlled rate into an inert bed of calcium

**Table 3.1** Melting and boiling points of some volatile fluorides

| Element | Fluoride | Melting point, °C | Boiling point, °C |
|---|---|---|---|
| Uranium | $UF_6$ | 64 (triple point) | 56 (sublimation) |
| Vanadium | $VF_5$ | 19 | 48 |
| Molybdenum | $MoF_6$ | 17.5 | 35 |
| Tungsten | $WF_6$ | 2.5 | 17.5 |
| Antimony | $SbF_5$ | 7 | 150 |
| Chromium | $CrF_4$ | – | 300 |
| Hydrogen | $HF$ | –83 | 20 |
| Fluorine | $F_2$ | –218 | –188 |

fluoride maintained at 450–475 °C in a Monel reactor and fluidised with a mixture of nitrogen and fluorine at reduced pressure. Reaction is instantaneous,

$$UF_4 + F_2 \rightarrow UF_6 \qquad (3.7)$$

and the gaseous uranium hexafluoride is first filtered to remove entrained solids, then condensed. The incondensable gases containing nitrogen and unused fluorine are recycled.

The solid uranium hexafluoride in the condenser is melted and run off under gravity into transfer cylinders, where a *gassing back* operation is performed to eliminate contamination by light gases. Condensers operate as cooling and liquefying units in turn, using a fluorocarbon liquid at –40 °C when on primary or secondary cooling duty or fluorocarbon vapour at about 105 °C for liquefaction of hexafluoride.

If the UOC had not been purified, certain metals besides uranium would also produce volatile fluorides and contaminate the product; see Table 3.1.

## 3.5 Magnox fuel

All Magnox fuel is produced at the BNFL's Springfields site. As with most industrial-scale metallurgical process, the terms used to describe the various stages from metal production to the finished, machined component, are those originally used for iron and steel processes.

### 3.5.1 Uranium metal production

The uranium metal alloy is produced as a billet with a nominal mass of 350 kg, by utilising a highly exothermic, thermite-type reaction between uranium tetrafluoride ($UF_4$, a dense green powder) and magnesium metal. The $UF_4$ is intimately mixed with coarse chips of magnesium and compacted into pellets of approximately 3 kg in weight which are loaded into a batch reaction vessel along with an appropriate quantity of alloying aluminium as metal coupons.

The metal-forming reaction is carried out within a graphite containment held in a sealed stainless steel pressure vessel under an inert atmosphere of argon, which is admitted after air and most of the absorbed moisture have been evacuated from the vessel and its contents. The reaction is started simply by external heating to 650 °C and is thereafter self-sustaining. It reaches a maximum temperature of over 1500 °C, necessary to reduce the viscosity of the liquid products and so allow efficient separation.

Magnesium fluoride ($MgF_2$) is a by-product and separates from the resultant molten uranium metal as a lighter liquid slag. The overall reaction is

$$UF_4 + 2Mg \rightarrow U + 2MgF_2 \qquad (3.8)$$

After cooling, the uranium billet is removed from the graphite crucible container and cleaned; the solid magnesium fluoride slag is crushed and leached with acid to remove any residual uranium before final disposal.

### 3.5.2 Uranium rod production

A casting technique has proved reliable and provides a rod suitable for the downstream heat-treatment and machining stages necessary to meet both mechanical and metallurgical requirements. Fresh billet stock and recycled metal scrap from later manufacturing stages are melted in a large, thick-walled graphite crucible by means of a vacuum induction furnace. After being brought to 1500 °C over a 2 hour period, and allowed to dwell for 15–20 minutes to ensure complete melting and absorption of a certain carbon level, the molten metal is poured through a tapping hole in the base of the crucible into a mould assembly.

The rod moulds are made of thick-wall spheroidal graphite (SG) cast iron, spray-coated internally with powdered alumina which prevents reaction with the molten uranium. After pouring, the mould and contents are cooled to around 250 °C under an argon blanket before final cooling in air, disassembly and extraction of the rods.

### 3.5.3 Heat treatment and finishing of rods

Rods as cast have a strongly textured grain structure which would cause anisotropic grain growth in the uranium metal during irradiation within the reactor. Also the orthorhombic α-phase of uranium has a negative coefficient of thermal expansion along one axis and highly positive along the other two, so that alignment of the crystals is particularly undesirable. To prevent this, a finer, less textured grain structure is induced. Rods are heated to 660–700 °C so that the metal is transformed to the β-phase, and then rapidly quenched by means of water sprays directed at the rod surface as it leaves the heating zone. The presence of the iron and aluminium alloying elements added earlier in the manufacturing process helps achieve the preferred grain structure in that intermetallic precipitates, which formed on casting, serve as grain nucleation centres. The complete heat treatment and grain structure modification process is referred to as beta-quenching.

After heat treatment, the rods are machined on automatic lathes to final dimensions and have circumferential anti-ratcheting grooves cut at regular intervals along their lengths. They are inspected and cleaned before insertion into the Magnox cladding.

### 3.5.4 Magnox cladding and fuel element assembly

Various designs of Magnox can are manufactured on high-speed conventional machining tools which take an extruded former through to the finished component complete with integral heat-exchange fins and threaded end profiles. Spacing ribs are commonly formed outside the helical fins, and stiffening braces may be welded on afterwards. Designs currently in use have evolved progressively through operational experience, particularly of vibration and bowing of the element during service in early reactors.

The uranium rod, and ceramic end discs to protect the end welds from the full fuel temperature, are inserted into the can, threaded Magnox end caps are fitted, and the remaining space between rod and cladding is purged and filled with helium. The end caps are keyed, locked and welded into position, then tested for helium leaks with a mass spectrometer. The welding process is very carefully controlled and the result inspected by X-radiography.

The fuel element is then externally pressurised to force the clad down plastically on to the uranium rod and into the anti-ratcheting grooves, both to ensure good thermal contact and to lock the components so that they expand and contract together in service. Depending on the destined location of the fuel element in the reactor core, the external pressure is applied either by hydraulic oil at 230 °C or by gas at 550 °C, with the latter resulting in a much coarser grain in the Magnox cladding.

After final cleaning, the conical end fittings are attached and the fuel element is then ready for use at the reactor.

## 3.6 Oxide

The performance expected of modern commercial fuel types places stringent demands on the quality and integrity of fuel assemblies. Typical international specifications may be quoted, but vary in response to the requirements of the reactor operator and the licensing authority of the country in which the reactor is situated. Often, the licensing authority is a government body charged with ensuring all aspects of nuclear safety. The design of the fuel, its quality of manufacture and its operational reliability are then primary considerations, and assurance of meeting specification is therefore an integral part of the manufacturing process from feed material to end-product.

Uranium dioxide fuel is a ceramic material contained within, and supported by, metallic structural components. The processes involved in making the ceramic

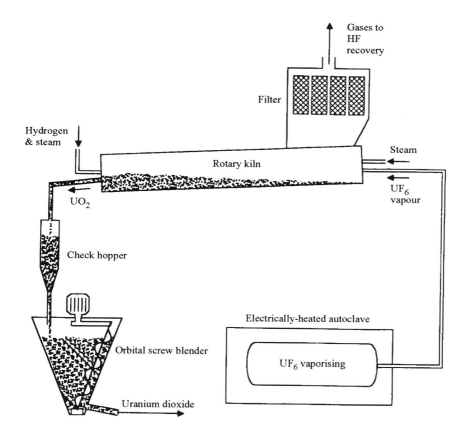

**Fig. 3.2** Integrated Dry Route (IDR) for UO2 production

and incorporating the component into a suitable engineered structure have much in common with similar operations in other industries. The four main stages are preparation of UO$_2$ powder, pellet manufacturing, rod manufacture and, finally, fuel assembly and despatch.

## 3.6.1 Enriched uranium hexafluoride conversion to UO$_2$

Apart from the Canadian-designed CANDU, all reactors utilising uranium oxide fuel need a proportion of U-235 higher than natural, and current enrichment processes operate on the volatile uranium hexafluoride (Chapter 4). This is therefore the feed material for the following stages.

Enriched UF$_6$ (*hex*) is now a commodity product traded around the world. Electric power utilities that require the fuel are able to purchase the services of converting UOC to $UF_6$ and of enrichment so that hex of specific U-235 content, manufactured to internationally-recognised quality standards, is delivered to the fuel fabricator on a schedule enabling manufacturing lead-times and investment in enriched uranium stocks to be optimised for maximum economy.

Originally, all processes for the conversion of $UF_6$ to a ceramic-grade, dry $UO_2$ powder were based on a production route which involved:

- hydrolysis of the $UF_6$ in water;
- additional stages to produce a pure uranium compound;
- finally a high-temperature conversion of the uranium compound to $UO_2$ with the necessary physical properties to be used as a precursor for ceramic pellet manufacture.

Wet processes are still in general use today, but developments by many fuel fabricators have concentrated on dry routes involving fewer stages. Both wet and dry processes have their advantages and disadvantages, and several are in current commercial use throughout the world.

Uranium dioxide is also an intermediate in some $UOC$–$UF_6$ processing routes, but is then of natural enrichment and not generally considered a ceramic-grade product.

### Dry conversion routes to ceramic-grade $UO_2$

In dry conversion of $UF_6$ to $UO_2$, the hexafluoride is usually decomposed by steam to solid $UO_2F_2$

$$UF_6 + 2H_2O \rightarrow UO_2F_2 + 4HF \tag{3.9}$$

which is then reduced to $UO_2$ in a multi-stage method using fluidised bed technology, a two-stage method using a rotary kiln or a single stage method using flame reaction technology. Only gas-phase or gas-solid reactions are involved hence the term *dry* route.

Several variations of conversion processes using fluidised bed technology are cited in the literature,[6] but follow a common theme involving:

(a) Reaction with superheated steam to form $UO_2F_2$ particles;
(b) Then pyrohydrolysis of the $UO_2F_2$ with steam and hydrogen,

$$UO_2F_2 + H_2O \rightarrow UO_3 + 2HF \tag{3.10}$$

$$UO_3 + H_2 \rightarrow UO_2 + H_2O \tag{3.11}$$

(c) Additional pyrohydrolysis in a separate fluidised bed to reduce the fluoride content further, followed by conditioning and stabilisation of the powder using an air/nitrogen gas mixture which has the effect of increasing the O/U ratio so that the powder remains stable in air.

**Table 3.2** Typical properties of ceramic-grade UO2 powder produced by various routes

|  | IDR | ADU | AUC |
|---|---|---|---|
| **Physical properties** | | | |
| Specific surface area ($m^2/g$) | 2.5–3.0 | 2.5–6.0 | 5.0–6.0 |
| Pour density ($g/cm^3$) | 0.7 | 1.5 | 2.0–2.3 |
| Tap density ($g/cm^3$) | 1.65 | 2.4–2.8 | 2.6–3.0 |
| Mean particle size (μm) | 2.4 | 0.4–1.0 | 8 |
| Powder morphology | Dendritic platelet | Irregular spheroid | Large porous agglomerate |
| **Chemical properties** | | | |
| O/U ratio | 2.05 | 2.03–2.17 | 2.06–2.16 |
| Fluorine (μg/g $UO_2$) | <25 | 30–50 | 30–70 |
| Carbon (μg/g $UO_2$) | 20 | 40–200 | 120 |
| Iron (μg/g $UO_2$) | 10 | 70 | 10–20 |
| Boron (μg/g $UO_2$) | <0.05 | 0.2 | 0.1 |

The BNFL Integrated Dry Route Kiln process,[7] which was first operated commercially in 1969, and has been adopted in France and in the USA, converts $UF_6$ to $UO_2$ in a single rotary kiln which operates at a relatively high temperature. Solid $UF_6$ within a transport cylinder is vaporised by heat and metered into the kiln where it undergoes the overall reactions 3.9 to 3.11 as it passes down the kiln. Figure 3.2 is a schematic diagram of an IDR kiln and shows the arrangement of both the co-current and countercurrent gas flows and the off-gas filtration system. Table 3.2 lists the principal physical and chemical properties of the powder product and typical values obtained with this and other processes. Such values are usually incorporated into specification documents and are routinely checked to ensure that the product is fully suitable for onward processing to pellets.

The IDR process was developed by BNFL in preference to other dry routes because of its relative simplicity, low environmental impact and desirable sintering properties in the $UO_2$ product. It is suitable for manufacturing all types of fuel pellet, in particular for AGR fuel where pellets have to attain more than 97.2% of theoretical density (TD, 10.96 $g/cm^3$), i.e. >10.65 $g/cm^3$. Where a lower pellet density is required, as for PWR fuel, IDR $UO_2$ powder is usually used in conjunction with a fugitive pore-forming additive. BNFL has a proprietary compound for this purpose, called CONPOR. Other fuel fabricators use similar compounds or larger quantities of $U_3O_8$ which achieves a similar density-depressing effect.

The French Commisariat à l'Énergie Atomique (CEA) also has a kiln conversion process operated by FBFC at Pierrelatte. This process[8] is a variant on the IDR introducing the reactant gases in the filter chamber which is stated to be a separate vessel. Products of Reaction 3.9 are conveyed by screw into a closely-coupled rotary kiln fitted with deep baffle plates and having a countercurrent steam supply

## 54  Fuel fabrication

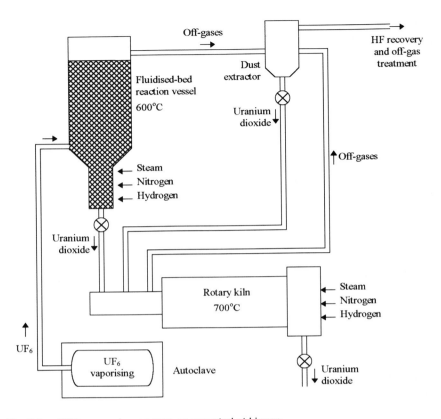

**Fig. 3.3**  ANF conversion process as operated at Lingen

to the centre of the kiln. The kiln has two heated zones which maintain temperatures of 600–760 °C and 760–800 °C respectively.

Advanced Nuclear Fuels Corporation (ANF) at Richland, Washington, USA, operates a hybrid process based on a conversion method developed and patented by Siemens Power Corporation.[9] It has recently been brought into operation at the Lingen, Germany, site of ANF GmbH. The process, shown schematically in Fig. 3.3, involves a fluidised bed reactor vessel and a rotary kiln. Within the first reaction vessel, $UF_6$ and steam are injected through a concentric nozzle and form uranyl fluoride, which is held in a fluidised state by nitrogen and hydrogen flows and maintained at 600 °C to allow partial pyrohydrolysis and reduction to $UO_2$. The powder is then passed to a rotary kiln which is heated to 700 °C, and the pyrohydrolysis and reduction reactions are completed. The properties of the $UO_2$ powder product, a direct result of process variables such as the temperature and residence time in the fluidised bed vessel, are very different from those in the IDR process and are described in some detail in the patent literature.

The General Electric Company flame reaction process is a single-stage reduction and hydrolysis within an active flame of excess hydrogen and oxygen:

$$UF_6 + 3H_2 + O_2 \rightarrow UO_2 + 6HF \tag{3.12}$$

Advantages claimed over other production routes include simplicity of equipment, the ease with which a continuous process can be engineered, and the prevention of reaction products from building up at the reactant inlet owing to the flame being displaced from the end of the jet.

### Wet-chemical conversion routes to ceramic-grade $UO_2$

Two wet-chemical routes are in commercial operation and are known by the name of the intermediate uranium compounds, respectively ammonium diuranate (ADU route) and ammonium uranyl carbonate (AUC route).

As well as serving in the main production line, both routes are also used as the final stage in residue recovery processes, i.e. as a means of recycling contaminated residues. These are normally dissolved in nitric acid, purified by solvent extraction and utilised as a dilute nitrate solution.

### Ammonium diuranate (ADU) route

The ADU route involves several stages comprising, as a minimum, hydrolysis of the $UF_6$, filtration of a precipitated ADU, and furnace treatment to $UO_2$.

Normally, gaseous $UF_6$ is hydrolysed by reaction with a dilute solution of ammonium hydroxide within a stirred reaction vessel. A fine precipitate of yellow ammonium diuranate is formed with a controlled particle size and well-characterised filterability. The solids are filtered, washed with dilute ammonium hydroxide solution to remove as much as possible of the entrained fluoride, dried to remove moisture and excess ammonia, and passed to a pyrohydrolysis and reduction furnace to complete defluorination and conversion to $UO_2$. After furnace treatment, the powder product is normally milled to break down large physical agglomerates, and sieved.

The physical properties of the $UO_2$ product are very dependent upon the actual conditions used and particularly the particle size distribution of the precipitate formed in the first stage. This also has a significant effect on the sintering characteristics of the pellets produced at later stages. Typically, pellets produced from ADU-derived $UO_2$ have a sinter density just above 95% TD which allows some scope for the addition of $U_3O_8$ as will be described later.

Overall, the chemical reactions for the ADU-$UO_2$ process can be represented as:

$$UF_6 + 2H_2O \rightarrow UO_2F_2 + 4HF \tag{3.9}$$

$$2UO_2F_2 + 8HF + 14NH_4OH \rightarrow (NH_4)_2U_2O_7\downarrow + 12NH_4F + 11H_2O \tag{3.13}$$

## 56  Fuel fabrication

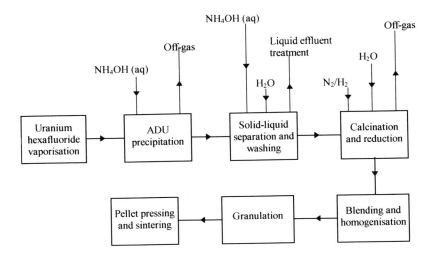

**Fig. 3.4** Ammonium diuranate (ADU) process

$$(NH_4)_2U_2O_7 + 2H_2 \rightarrow 2UO_2 + 2NH_3 + 3H_2O \quad (3.14)$$

Where uranyl nitrate solution is used, nitrate replaces fluoride as anion in Reaction 3.13.

Figure 3.4 is a schematic representation of the process stages involved. The treatment of ammonia-containing effluents, and avoidance of criticality in the volumes of liquors involved, are important aspects that need to be effectively managed.

Table 3.2 includes typical values of the physical and chemical properties of the $UO_2$ product of the ADU route.

### Ammonium uranyl carbonate (AUC) route

The AUC route has found widespread application in some countries such as Germany and Sweden, and the technology has been successfully exported elsewhere. Developed originally as a commercial process by Nukem GmbH and RBU (Reaktor Brennelement Union, now part of the Siemens group of companies), the AUC process has the main advantage over the ADU or other wet routes that it produces a free-flowing, granular $UO_2$ of uniform particle size, which does not require precompaction and granulation before pressing (see later Section 3.6.2—$UO_2$ powder to pellets). Like the ADU process, it can also accommodate recycled residues as an additional nitrate stream.

Hex, $UF_6$, vaporised by heat from the transport cylinder, is directed to a precipitation vessel containing a solution of ammonium hydroxide and ammonium carbonate. Uranium is precipitated as ammonium uranyl carbonate,

$$UF_6 + 5H_2O + 10NH_3 + 3CO_2 \rightarrow (NH_4)_4UO_2(CO_3)_3\downarrow + 6NH_4F \quad (3.15)$$

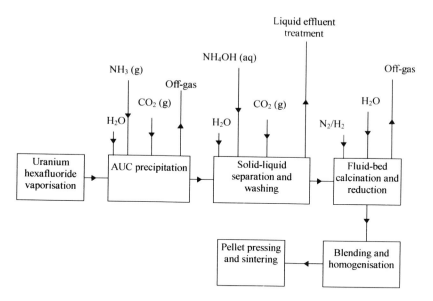

**Fig. 3.5** Ammonium uranyl carbonate (AUC) process

As with the ADU process, the precipitation conditions determine the characteristics of the resultant $UO_2$; the best results are obtained with a large crystallite size which is only produced at high solution concentrations, also desirable to minimise the amount remaining dissolved. To avoid dilution, additional ammonia and carbon dioxide are both fed in gaseous form. From the precipitation vessel, the solids are transferred to a filter where they are washed with clean ammonium carbonate solution to remove most of the entrained liquor and dissolved fluoride. The solids are dried and transferred to a fluidised-bed reduction furnace, supplied with steam and hydrogen gas at 650 °C; further fluoride is removed as HF, and the ammonium uranyl carbonate is first decomposed and then immediately reduced to $UO_2$ by the hydrogen:

$$(NH_4)_4UO_2(CO_3)_3 + H_2 \rightarrow UO_2 + 3CO_2 + 4NH_3 + 3H_2O \quad (3.16)$$

Figure 3.5 represents the process stages involved, and typical values for the chemical and physical properties of the $UO_2$ product are given in Table 3.2.

### 3.6.2 $UO_2$ powder to pellets

All commercial fuel fabricators use uniaxial dry pressing to form $UO_2$ powder into the required pellet shape. To permit fast and reproducible pressing, the feed must flow well and be consistent in physical properties, besides having the required chemical and isotopic composition. To meet these conditions, powder produced by any but the AUC route requires some pre-treatment; most raw powders flow poorly, pack loosely and erratically, and would require excessive tool movement if loaded directly into the die. Pressing is therefore generally a two-stage process, with the powder first compacted into granules which serve as feed to the pelleting press.

**Powder blending and press feed preparation**

To maximise use of uranium and minimise waste, recovered and recycled uranium oxide is added to the main bulk, after purification if necessary. Clean sintered pellets rejected on the grounds of defects such as cracks or chips are usually oxidised in air to form $U_3O_8$ powder which after sieving, homogenisation, analysis etc. is suitable for direct addition.

A pore-forming compound is also added at this stage when, as for PWR and BWR fuels, the final pellet density is to be less than would be obtained by sintering the blended powders alone. The pore-former is usually an organic compound that decomposes early in the sintering process leaving a void.

The whole homogenised blend is finally analysed and tested to ensure that chemical and physical properties meet specific requirements and that pellets produced from it are fully satisfactory. From this point material usually achieves a *batch* identity that stays with it through to pin manufacture and assembly.

Final preparation of the press feed depends on whether or not an organic binder is added. In the past a binder has been used extensively throughout the fuel manufacturing industry, principally because it produces a strong *green* pellet (as pressed but before sintering—in this context, the term is unrelated to colour) without relying on the cohesion of the $UO_2$ powder. This is important, since the green pellet strength directly affects the amount of damage that a pellet will suffer during handling, and hence the rate of rejection.

Binder route processes are, however, being rapidly replaced by binderless granulation processes to reduce the number of process steps, the amount of residue material requiring recovery and so the overall process costs. Subsequent pellet manufacturing stages are also somewhat simplified by the absence of a debonding process to remove the organic binding agent.

The production of AGR fuel, at the Springfields plant of BNFL, is one example where a binder route is still in use in 1995, although due to be superseded. The steps are:

- micronising the powder by jet milling;
- slurrying the micronised powder with a solution of the organic binding agent;
- spray-drying of the resultant slurry in a fluidising vessel to produce spherical, hollow granules;
- sieve-classifying the granules to remove fines and oversize material.

Other fuel manufacturers use a small amount of an organic polymer binder dissolved in water to achieve the same end without the need for a separate debonding stage.

Production of press feed granules by a binderless route simply relies on the consolidation strength of the $UO_2$ powder to produce a compact body strong enough to fracture into granules when forced through a coarse mesh. For powders produced from ADU or by most dry routes, the pressure to achieve this is of the order of 0.75 tonne/cm$^2$, usually applied within a large-diameter press die in a *slugging* press, capable of containing enough low-density powder to produce a compact several millimetres in depth.

When this is broken up, the initial granules have a very wide size distribution which needs to be reduced for the sake of consistent flow-rate. Some 0.2% of lubricant, such as powdered zinc stearate or Acrawax (a wholly organic stearate that produces no zinc fume), is commonly added at the same time in a process termed conditioning.

## Pellet pressing

$UO_2$ granules prepared as described above, or blended and homogenised granular powder derived from AUC, are pressed into the shape of the final fuel pellet by profiled top and bottom punches sliding within an essentially cylindrical tool die. Fuel fabricators use different types of press, at different speeds and often with very different control philosophies. The most common is the rotary pelleting press, similar to those used in the pharmaceutical industry for tablet-making.

Such a press typically has 16 tool positions and can produce up to 400 pellets per minute. It can be used with a variety of tooling dependent upon the type of pellet. Tungsten carbide, a ceramic material with improved wear resistance, commonly replaces the traditional tool steels as material for the die and wearing parts of the punches.

Designs of pellet, and hence of tooling, vary in detail according to the reactor type and licensing requirements, but are of two generic forms, solid or annular. Solid pellets are used in most PWR, BWR and CANDU reactors, annular in AGRs and VVERs. Annular pellets require the use of a retractable, central core-pin within the tool assembly and slightly increase the complexity of operation.

Pelleting pressures are generally in the range 3–4 tonne/cm$^2$ and the resultant green pellet density is approximately 5.5–6.0 g/cm$^3$, the practical limit that can be achieved by pressure alone. Friction between powder particles at these densities is very high, while higher pressures risk damage to the tooling and increase wear. Further densification, and conversion to a true ceramic form, has therefore to be achieved by subsequently heating the pellet to a sintering temperature. Friction is reduced by a lubricant in the granules, or for AUC-derived $UO_2$, by a thin film of a liquid lubricant applied to the wall of the press die during each pressing cycle.

The main control parameters for the pressing operation are the depth to which the die is filled, the position of the pellet within it (avoiding upper sections slightly tapered to ease ejection), intensity and duration of applied pressure, the balance of hold-down and upward forces as the pellet is ejected from the die, and the gradual removal of the hold-down force to prevent sudden and damaging relaxation of the green pellet which can give rise to so-called *capping* problems—the presence of circumferential cracks at one or both ends of the pellet. Control methods differ, but each type of $UO_2$ powder has its own specific requirements that fuel fabricators have come to recognise.

For ADU-and AUC-derived $UO_2$ powders, green and final sinter densities of the pellet are closely linked. Hence tight control limits must be imposed and the product frequently sampled to ensure that the pelleting process stays within the control parameters. There is a correlation between the amount of physical shrinkage during subsequent sintering and the diameter and density of the green pellet; since the diameter of the final sintered pellet must be within very tight limits, provision is made for any variation, usually by having press tooling to cover a range of green pellet diameters.

For IDR-derived $UO_2$ and other dry-route powders, the final sintered density is much less variable. Thus the shrinkage depends only on the green density of the pellet and the press tooling requirements for this type of powder are reduced.

**Pellet sintering**

Sintering the fuel pellets significantly increases their density and converts the $UO_2$ from green compressed powder to a true ceramic which has all the microstructural and physical properties necessary for service at high temperatures in a reactor core. Small voids are removed as the particles shrink and coalesce into a crystalline grain structure. The overall shape of the pellet is however retained.

Sintering involves solid-state diffusion processes which occur at practical rates only at high temperatures, in this instance of the order of 1750 °C ($UO_2$ melts at 2850 °C). Various types of furnace are used, but all are electrically heated and have high-purity, refractory alumina brickwork in the high temperature zones. Pellets are held in molybdenum metal trays or boats.

To prevent oxidation of the $UO_2$, a reducing atmosphere of hydrogen, hydrogen and nitrogen or cracked ammonia is used. Some moisture within the sintering furnace is also desirable to prevent degradation of the alumina refractories, but oxidation of the refractory metal heating elements has to be avoided.

Since sintering is a very energy-intensive process, developments have largely been aimed at reducing duration and temperature. The Nikusi technique developed by Siemens is an example, with oxidation by carbon dioxide at temperatures between 900 and 1100 °C followed by reduction annealing in a hydrogen-rich atmosphere at the same temperature.[10] The lattice diffusion constant of uranium is high in the oxidising environment, and sintering is much quicker than under reducing conditions. Operation of a furnace with two distinct atmosphere conditions however makes the process equipment complex, and although the technique was used for some time in Germany, it has now been replaced by more conventional techniques.

### Pellet finishing

The diameter precision demanded by most fuel design specifications can be achieved only by a final grinding, chiefly because during sintering the pellet shrinks more at the centre than at the ends, where the full pressure has been applied without friction losses. A centreless-grinding technique is universally used to produce a parallel-sided pellet with a tolerance of a few microns. Anti-ratcheting grooves are formed in the centres of about one third of them.

Wet grinding is customary, followed by a wash with clean water, and the removed material is recycled. If a silicon carbide grinding wheel has been used, the uranium oxide must be dissolved and purified before processing by the ADU or AUC route. If the wheel is of diamond, the residues usually require only oxidation to $U_3O_8$ before addition to a powder blend.

## 3.7 Oxide fuel assembly

The detailed operations involved at the oxide fuel assembly stage depend on the design, and in turn on the type of reactor, the fuel vendor and the operating utility. Changes in PWR fuel are introduced mainly to improve performance or safety, and product designs are therefore recognised by the licensing authorities as an important factor in predictable reactor operation. This section will nevertheless be concerned only with generic types.

The assembly of AGR, PWR and BWR fuel is described in some detail; procedures for CANDU, RBMK and VVER types may be presumed to follow similar lines, and only the form of the structure is covered. Details of many different designs are given in *Nuclear Engineering International, 40(494)*, (September 1995) pp. 26–35.

### 3.7.1 AGR

For all the UK advanced gas-cooled reactors, the fuel elements are of a similar design and the sequence of manufacturing operations the same. There are two distinct stages—manufacture of fuel pins containing the $UO_2$ pellets, followed by assembly into elements.

### Pin manufacture

$UO_2$ pellets are assembled into stacks for insertion into the stainless steel cladding. They are first air-dried, then arranged (along with a uranium-free ceramic insulating pellet at each end of the stack) according to an interval pattern of grooved pellets determined by the customer, checked for stack length and integrity and then weighed. The stack is loaded into a pin by a ram device, checked again for length inside the can, then prepared for gas-filling and insertion of the end-cap. Preparation comprises cleaning the internal bore of the can end, purging with helium and then insertion of ceramic discs and stainless steel end pieces.

The end caps are attached to the can by resistance welding which produces a series of tiny overlapping spot welds providing the primary seal. Afterwards the pins are tested for helium leaks by a mass spectrometer, and pass to extension piece welding and edge-welding stations. Finally they are pressurised hydraulically to approximately 17500 psi, so that the can is plastically deformed into the anti-ratcheting grooves to minimise relative motion of pellets and clad during service in the reactor and to prevent large gaps from appearing after periods of thermal cycling.

Operations are however soon to be slightly modified when new facilities are brought on line. The biggest single change will be that the main seal will be welded by an automatic technique with lasers as the energy source.

After pressurising, the pins are annealed by heating for a set period of time to a temperature of 930 °C in a tube furnace containing hydrogen, and then cooled slowly. This operation removes cold-working stresses introduced mainly during the pressurising stage and reduces the risk of cracking in service. After annealing, it is quite normal for the pins to require some straightening and this is done carefully to avoid reintroducing undue stresses.

Immediately before assembly into fuel elements, the pins are electrolytically cleaned, dried and inspected both externally and internally. The internal inspection is achieved by a gamma-scanning device which checks enrichment and gaps between pellets.

**Fuel element assembly**

The detailed design of an AGR fuel element assembly varies only slightly according to the reactor being fuelled. Typically it consists of a skeletal structure supporting 36 fuel pins inside a single or double graphite sleeve, see Fig. 3.1. Two end support grids and a central bracing grid are aligned to allow easy insertion of the individual fuel pins. Construction involves a number of complicated automatic assembly and welding operations, before the pins are inserted and fixed into position by locating one end of each into the bottom grid support and rolling over the extension piece so that it is forced against the outside edge of the support grid.

Finally the finished assemblies are tested by applying an internal pressure to demonstrate that bursting pressure exceeds a specified value, then packed in a polyethylene sleeve and loaded into a transit container.

Before loading, assemblies are combined at the reactor site into stacks of eight held together by a tie rod made from a nickel-based superalloy.

### 3.7.2 PWR

These fuel elements have a square cross-section and are the full height of the reactor core. Pins, also of full length, are usually arranged in a $17 \times 17$ or $14 \times 14$ array along with control rod guide tubes, or thimbles, which are fixed to top and bottom nozzles—see Fig. 3.1. Spring clip grid assemblies are fastened to the guide

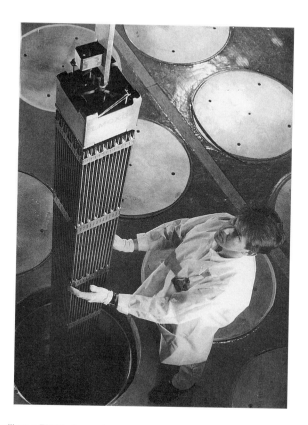

**Fig. 3.6** Handling a PWR element

thimbles at intervals along the height of the fuel assembly to provide support for the fuel pins. The ends of these are not fixed to either the bottom or the top nozzle, but are free to expand axially to prevent bowing.

Again a measured stack of dried $UO_2$ pellets is assembled and weighed. Each stack is inserted into a Zircaloy-4 tube, already cap-welded at one end with a plenum support tube inserted. A plenum spring is inserted and after cleaning of the open end, the tube is evacuated, flooded with helium gas to a specified pressure, plugged with the end cap and welded. After checking for helium leaks, pins are inspected by X-radiography and gamma scanning.

Those which pass all the inspection stages are assembled into the cage structure of the fuel element, built up on the control rod guide tubes and spacer grids. The fuel rods are loaded into this skeleton structure, the top and bottom nozzles are placed in position and the required hold down force on the assembly is applied through springs on the top nozzle.

The finished element is washed, usually in two stages, the second in distilled water. Afterwards the assembly is dried in hot air and stored for despatch to the reactor.

### 3.7.3 BWR

Although BWR and PWR fuel elements are both square in section and conceptually similar, differences in the cooling and reactivity control systems for the two reactor types (Chapter 5) affect the detailed design of the fuel and the choice of materials. In turn, these affect the manufacturing operations.

Fuel pin manufacture follows similar steps to those described for PWR above, the principal difference being that the zirconium cladding alloy is Zircaloy 2 and much thicker. The fuel pellets are also larger in diameter (approximately 10.6 mm compared with about 8.2 mm for PWR).

However, the chief differences are in the assembly of an element. The grid spacers are attached at prescribed intervals along a central full-length rod which does not contain fuel but in service is filled with cooling water. In some BWR designs, there are several such water rods and the spacer grids are attached to each, forming the cage into which fuel pins are inserted. Some fuel pins can also double as tie-rods and provide structural support to the assembly after they have been fixed to the top and bottom end fittings. Typically, an $8 \times 8$ fuel pin assembly will have 8 pins, located towards the corners of the assembly, which double as tie bars. The upper end fitting normally incorporates a handle device which applies hold-down force to the element once inside the reactor. After inspection and cleaning of the fuel element, it is fitted with a Zircaloy shroud to confine the flow of water.

### 3.7.4 CANDU[11]

The CANDU fuel *bundle*, as the assembly is called, has progressively evolved from a design with 7 elements through 19 and 28 to 37, and plans are very advanced for the commercial introduction of a 43-element version 50 cm long by 10.27 cm in diameter, the same overall dimensions as for the 37-element type. In this the individual elements are 13.1 mm in diameter and have an internal coating, usually of a graphite compound, between the Zircaloy clad and the pellets. The bundle has a central Zircaloy spacing grid and flat Zircaloy braces spot-welded to both end caps of each element. These also have external Zircaloy bearing pads which help maintain inter-element spacing, centralise the bundle in the reactor channel, and reduce friction during loading and discharge.

The uranium dioxide pellets within the elements are of natural enrichment but otherwise made by the standard route.

### 3.7.5 RBMK[12]

In a typical RBMK fuel assembly, two bundles are stacked vertically, each with 18

**Table 3.3** Characteristics of VVER fuel assemblies

|  | VVER 440 | VVER 1000 |
|---|---|---|
| No. of pins | 120 | 312 |
| Rows of pins | 6 | 10 |
| Assembly length | 3.2 m | 4.6 m |
| Assembly width | 144 mm | 234 mm |
| Pin length | 2.536 m | 4.020 m (max) |
| Pin diameter | 9.0 mm | 9.1 mm |

fuel rods 3.64 m long by 13.6 mm diameter. The assembly is just over 10 m high by 79 mm in diameter in the active part. Cladding is of zirconium alloyed with 1% niobium, and 0.86 mm thick. Spacer grids are of a chromium-nickel corrosion-resistant steel.

Pellets are 15 mm long by 11.5 mm in diameter, with two enrichments of 2.0 and 2.4% in the RBMK-1000 or a single level of 2.0% in the RBMK-1500.

### 3.7.6 VVER[12]

The two main types, the VVER 440 and the newer VVER 1000, are both pressurised water reactors using enriched $UO_2$ fuel. The assemblies are hexagonal in cross section, with details as in Table 3.3.

Since the break-up of the Soviet Union, some VVER plants outside Russia have invited refuelling business from foreign suppliers, and this has sometimes led to minor changes such as to cladding material or grid design within the same basic dimensions.

## 3.8 Conclusion

The improvements to fuel designs, manufacturing processes and in-service performance, currently being implemented, illustrate the importance attached to optimising the efficiency of generating nuclear electricity. They will undoubtedly continue and be strongly influenced by both economic and safety factors.

The former are concerned with the overall fuel cycle, not just the cost of fabrication but also of storage or reprocessing and recycling after discharge from the reactor. Reprocessing costs are influenced by the burn-up achieved in service as well as by the mechanical design of the fuel assembly. Further, the development of improved fuels utilising fissile plutonium in the form of MOX will continue and can be expected to feature prominently in the plans of the larger utilities. Obvious economic benefits stem from the use of plutonium, recycling of which will also reduce stockpiles of this politically-sensitive material. Fuel designs which optimise its utilisation and can be easily reprocessed are being developed in both France and the UK.

Safety must of course always be paramount among considerations of nuclear fuel. It encompasses all aspects of the cycle and in particular detail the design of fuel and the reactor operating envelope—the range of conditions that in appropriate combinations have been thoroughly checked and declared satisfactory. The trend towards increased safety influences, in turn, developments in the fundamental properties of the materials used and the drive towards very high quality in fabrication processes. Materials research is currently leading, for instance, to fuels with reduced interaction between pellet and clad during temperature cycling, and to the application of new technologies to pin welding.

In the wider scene, the introduction of new reactor systems with integrally safe construction, or with closely coupled reprocessing facilities, will no doubt both influence the relevant fuel designs. The future of the fabrication industry can therefore be seen to hold many interesting developments, each with its own particular challenges and opportunities.

## 3.9 References

1. J. C. Bishop and B. J. Hansen. Fluoride volatility process operating experience. *Uranium '82, 12th Annual Hydromet Meeting,* Toronto, Canada (August 1982).
2. S. Takada, T. Amanuna, and G. Fukuda. Uranium processing pilot mill at the Ningo-Toge mine; recovery of uranium from its ores and other sources. *Proceedings of IAEA Symposium,* São Paulo, Brazil, pp. 97–109 (1971).
3. J. Delannoy and R. Faron. Conversion of concentrates containing uranium and uranium hexafluoride. *Uranium '82, 12th Annual Hydromet Meeting,* Toronto, Canada (August 1982).
4. H. Page, L. P. Shortis, and J. A. Dukes. The processing of uranium ore concentrates and recycle residues to purified uranyl nitrate solutions at Springfields. *Transactions of the Institution of Chemical Engineers,* **38**, pp. 184–96 (1960).
5. R. C. Alexander, L. P. Shortis, and C. J. Turner. The second uranium plant at Springfields. *Transactions of the Institution of Chemical Engineers,* **38**, pp. 177–83 (1960).
6. Mitsubishi Kinzoku Kabushika Kaisha. UK Patent 2,178,418 (June 1986).
7. BNFL. UK Patent 1,320,137 (October 1969); UK Patent 1,341,379 (May 1976).
8. UK Patent 1,548,300 (May 1989).
9. US Patent 4,830,841 (May 1989).
10. W. Dorr and H. Assmann, Sintering of $UO_2$ at low temperatures. *Proceedings of the 4th internaional meeting on modern ceramics technologies, CIMTEC,* Saint-Vincent, Italy (May 1979).
11. *Nuclear Engineering International,* **34(425)**, p. 32 (December 1989).
12. Fuel review. *Nuclear Engineering International Special Publication* (1992).

# 4 Isotopic enrichment of uranium

P. C. Upson
*Urenco Limited, Marlow, Buckinghamshire*

## 4.1 Introduction

Uranium, as it occurs in nature, comprises two main isotopes, U-235 and U-238 in the proportions 0.711% and 99.28% respectively by weight. Only U-235 is fissionable by thermal neutrons, and whilst the Magnox and CANDU reactors were designed to use natural uranium, all others require the concentration of U-235 to be increased to around 3–5%. This is enrichment, and although many possible processes could be used only two, gaseous diffusion and high-speed gas centrifugation, have been employed commercially to date.

Early interest was of course focused on enrichment to much higher levels, when isotope separation became a necessary and key stage of the development programme for nuclear weapons (the Manhattan project). Early in the Second World War, parallel development was carried out on several different processes, including gas centrifuges and gaseous diffusion, but also on other techniques such as electromagnetic separation, thermal diffusion and chemical processes. In fact, the electromagnetic method was used early in the US programme, but as it proved suitable only for very small throughputs, it has not been used commercially. It operates on the same principle as a mass spectrometer; a stream of uranium ions, passed through a transverse magnetic field, separates into different isotopes as the ions follow trajectories depending on their charge-to-mass ratio. The process is now used only for small quantities of special isotopes for medical or other research purposes.

Of all the techniques developed in this early period, while the priority was still to produce very high levels of enrichment for nuclear weapons, only gaseous diffusion and the gas centrifuge were considered to have large scale potential. Owing to significant technical and engineering problems with the latter, it was considered the less likely to achieve success on a short time-scale and work on centrifuges terminated in 1944; gaseous diffusion was adopted in the US, the UK and Russia in the immediate post-war period.

In the 1950s, uranium enrichment was dominated by the three large US diffusion plants at Portsmouth Ohio, Paducah Kentucky, and Oak Ridge, Tennessee,

**Table 4.1** World enrichment plants in late 1980s

| Location | Owner | Type | Status | Capacity, MSWU/y |
|---|---|---|---|---|
| Argentina | CNEA | Diffusion | Operating | 0.02 |
| Brazil | Nuclebras | Jet Nozzle* | Under construction | 0.3 |
| China | | | | |
|   Lanzhou | | Diffusion | Operating | 0.08 |
| France | | | | |
|   Pierrelatte | CEA | Diffusion | Operating | 0.3 |
|   Tricastin | Eurodif (Marketed by COGEMA) | Diffusion | Operating | 10.8 |
| FRG | | | | |
|   Karlsruhe | Steag | Jet Nozzle* | Pilot | 0.05 |
|   Gronau | Urenco | Centrifuge | Operating | 0.2 (1.0 by 1992) |
| Japan | | | | |
|   Ningyo–Toge | PNC | Centrifuge | Pilot | 0.05 |
| | | Centrifuge | Under construction | 0.2 |
|   Shimokita | JNFI | Centrifuge | Planned | 1.5 |
| Netherlands | | | | |
|   Almelo | Urenco | Centrifuge | Operating | 1.0 (1.5 by 1991) |
| S. Africa | | | | |
|   Valindaba | UCOR | Helikon* | Under construction | 0.3 |
| UK | | | | |
|   Capenhurst | BNFL | Diffusion | Being decommissioned | – |
| | Urenco | Centrifuge | Operating | 0.7 (1.5 by 1993) |
| US | | | | |
|   Oak Ridge | DOE | Diffusion | Shut–down | 7.7 |
|   Paducah | DOE | Diffusion | Operating) | 19.5 |
|   Portsmouth | DOE | Diffusion | Operating) | |
| USSR | | | | |
|   Siberia | Techsnabexport | Centrifuge plus some remaining Diffusion | Operating | 7 to 10 (3 generally offered to world market) |

\* Two aerodynamic processes have been taken to demonstration plant level; the jet nozzle (developed in Germany and demonstrated in Brazil) and the Helikon process developed in South Africa. In both cases, $UF_6$ diluted in a light carrier gas is compressed, expanded to very high speeds and forced to turn through a very sharp radius, through either a nozzle or a vortex tube. The lighter isotope will thus be deflected further. It is possible that there is development potential in these processes, but as they have not been taken to large scale use nor are now likely to be, they are not described further in this chapter.

with a much smaller UK plant at Capenhurst, Cheshire. Oak Ridge has been shut down, but despite their enormous power consumption which leads to high operating costs, the other two US plants remain in operation producing low-enriched uranium for civil reactor fuel, and in the early 1970s a new plant was built at Pierrelatte in France for the civil market. The Capenhurst diffusion plant was however shut down in 1981 and replaced by centrifuge plants (at Capenhurst, Almelo in Holland and Gronau in Germany). It appears that a similar switch to centrifuges has been made in Russia from the 1960s onwards.

In uranium enrichment by any method it is usual to measure the plant capacity or output in Separative Work Units (SWU or kgSW) rather than the weight of enriched uranium product, since the added value depends on the concentration. The detailed derivation is given in many of the standard references,[1] but an SWU is, essentially, a measure of the work performed in enriching a given amount of uranium from the feed to the product concentration.

Table 4.1 lists the known enrichment facilities in the late 1980s.

If diffusion is considered to be the first-generation technology and centrifuges the second, work has been undertaken over the past decade on what is seen as the most likely third generation, laser isotope separation. This process remains a long way from commercial operation. These three technology generations are reviewed in more detail below.

## 4.2 Gaseous diffusion

The gaseous diffusion process was the first to be used on a large scale for uranium enrichment, largely because in the 1940s and 1950s it required only one novel plant item (the membrane); gas compressors and heat exchange technology were well understood. The technical basis of the process has been discussed by Tait[1] and London.[2]

The process is based on the simple principle that in a mixture of gases at thermal equilibrium, the average kinetic energy of all the molecules is the same. The lighter molecules therefore travel faster than the heavier ones and will thus hit the container walls more frequently. If the wall is a membrane containing holes of a size to accommodate individual molecules but not bulk flow, then the escaping gas will contain a higher concentration of the lighter molecules than the remainder.

The uranium compound used is hexafluoride, chosen because it is gaseous at temperatures around 60 °C and because fluorine has only the one isotope. The design and operating characteristics of the membrane remain classified and cannot therefore be described here. However it is evident that since the mean free path of the molecules to be separated is only a few tens of nanometres, the membrane pores must also be on this scale for efficient operation.

In addition, the small difference in molecular weights (349 against 352) gives a maximum stage separation factor of only 1.0043—the square root of the ratio. The mechanics of diffusion through such fine-pored membranes favours the use of low gas pressures, while the low separation factor requires a very large number of stages even to reach the modest levels of enrichment required for nuclear power reactors. This combination leads to large interstage flows at low pressures—a relatively inefficient process best handled by axial-flow compressors.

70  *Isotopic enrichment of uranium*

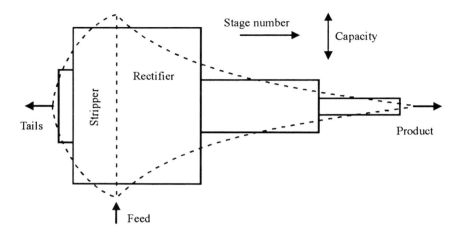

**Fig. 4.1**   Ideal and practical cascade outlines

### 4.2.1 Process description

A gaseous diffusion plant demands a large capital investment because each of the very many operating stages requires a series of physically large process items.

The basic unit of the process is one separation stage. A stage consists of the diffuser containing the permeable membrane, a compressor, a gas cooler and the associated compressor drive motor and flow control valves.

To achieve the required levels of enrichment with such small individual separation factors, the stages must be connected in series, known as a cascade. The enriched output from one stage becomes the feed to the next upstream unit, with the depleted output recycled to the previous downstream unit. In an ideal system, each stage would be individually sized to match the flow-rate at that point in the cascade, and the number of stages would be determined by the product enrichment required.

The calculations for an ideal cascade have been derived elsewhere,[3] and lead to the shape represented by the pecked outline in Fig. 4.1, with the largest capacity at the feed point. As noted above however this ideal cascade would have different molar flow rates at each stage and would be impracticable as a commercial plant, where some standardisation or simplification of design is necessary. To achieve this, a *squared-off* cascade design is used to approximate to the ideal (full outline in Fig. 4.1) whereby each *squared* section is made up of identical units, giving easier construction and reduced cost, whilst retaining much of the efficiency of the ideal.

Operational experience of the US diffusion plant is described in Reference 3, and Fig. 4.2 showing the essentials of a single stage is based on details in that paper. As noted above, the component parts need to be very large to achieve the

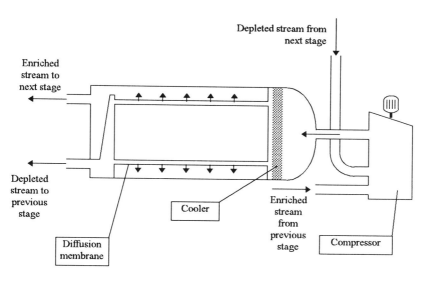

**Fig. 4.2** Essentials of a diffusion cascade stage

enrichment levels required. The diffuser unit (also known as a converter) is approximately 3.5 metres in diameter and 6 metres long. To circulate the gas and to provide the pressure differential across the membrane requires compressors, each weighing around 16 tonnes and driven by a 1600 kW motor.

These stage units, typically in groups of eight, are mounted in cells which then form the smallest parts of a cascade that can be isolated for off-line maintenance. The major problems in diffusion plant operation are to sustain highly efficient and stable performance. Besides routine periods of plant maintenance, procedures have been established to isolate cells where key components have failed in service and replace them expeditiously.

Despite significant design improvements during more than 40 years of operation, culminating in the French plant at Tricastin, which began operation in the early 1980s, the process is still very expensive in capital investment and energy consumption. It also necessitates very large process inventories, making it slow to respond to adjustments (and not ideal for recycled uranium, where the contaminants remain in the plant for many months). It is therefore considered most unlikely that any further diffusion plant capacity will ever be built.

## 4.3 Centrifuge enrichment

The technical basis of centrifuge enrichment is described in some detail by Heriot.[4] Again it exploits the small difference in mass between the isotopes in uranium hexafluoride vapour. The separation effect is proportional to the absolute

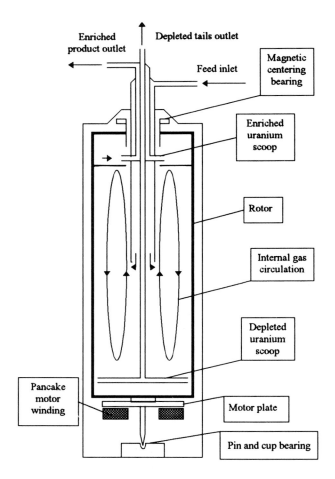

**Fig. 4.3** Zippe centrifuge (schematic)

mass difference, $\delta m$, rather than the square root of the relative difference, $\delta m/m$, as in diffusion.

In the spinning rotor, the isotopes tend to separate, the heavier towards the periphery and the lighter towards the centre. The separation factor depends strongly on the peripheral velocity. In a simple centrifuge, although significantly greater than in a diffusion stage, it is still very small. It can be greatly improved by a countercurrent circulation within the centrifuge, generated mechanically, by thermal convection, or both; the gas at the periphery, depleted in the lighter isotope, is swept to one end and the enriched gas in the central zone to the other. The axial concentration difference so generated is a factor larger than the elementary radial separation factor, and is proportional to the rotor length.

### 4.3.1 Technical development

Early work on enrichment was hampered by the limitations of rotor bearings and structural materials. However, by the early 1960s, Zippe[5] published details of an improved design with a lower bearing comprising a central pivot sitting in an oil filled cup, and a non-contact upper bearing formed by centralising magnets. A schematic diagram is shown in Fig. 4.3. Following from the early Zippe design, and taking advantage of new, stronger materials, longer and faster centrifuges have made enrichment by this means economically viable.

In theory, the maximum output of a centrifuge is proportional to its length and to the fourth power of its peripheral velocity. In practice the exponent is lower, depending on the gas dynamics within the rotor. The velocity depends mainly on the rotor materials. The length depends not only on the materials but also on the dynamics of the rotor, and for a given velocity and constraint on flexural resonance is proportional to diameter. Substantial development is therefore required to support a decision on how much longer or faster to go in order to increase output; much depends on the selection of materials and the ability to *productionise* the centrifuge design for building many thousands to the same standards.

For plants in the UK, Holland and Germany, Urenco has followed a continuous development programme, with significant improvements in centrifuge efficiency and cost. As new materials permit new centrifuge designs, programmes of design studies, materials testing, manufacture and spin-testing of components confirm the parameters. Manufacturing routes must then be qualified by the production of typically 100 to 200 qualification centrifuges; full production follows at rates of several thousand machines per year. In this way Urenco has been able over the past 25 years to bring into operation a family of centrifuge models with significant improvements in both the machine output and plant economics. Power consumption is now less than 5% of that needed for an equivalent diffusion plant.

The lead time for installation can be short because of the small module size, and a plant can be available about two years after the investment decision. This has allowed Urenco to pursue a programme of capacity increases to match firm forward orders, with the added advantage that improved technology can be introduced as it becomes available. Consequently, in the early 1990s Urenco has in simultaneous operation five generations of machines, and technical developments have been identified which will allow at least one further improved generation.

Elsewhere, the US programme of centrifuge development, at its peak in the 1970s with centrifuges on a much larger scale than those in Europe (almost certainly reflecting the larger capacity demand in the US), was stopped in the mid-1980s. In Japan, development has progressed on similar lines to those in Europe with the plant at Rokkasho-mura already having reached the stage of initial cascade operation, and planned expansion to full capacity over the next decade.

In the past few years[6] it has also become known that the USSR had a major programme of centrifuge development with its diffusion plants replaced by successive expansions of centrifuge capacity. However, the Russian development is believed to have progressed no further than the second-generation Urenco machine, reflecting the different economic pressures in Russia, where lower centrifuge-manufacturing costs (for manpower), and perhaps lack of competitive pressure, have not led to the same pressure for economic improvements.

Centrifuge technology became classified in 1960, so it is not possible to discuss later technical improvements in detail. Roughly, since 1975, specific capital costs of Urenco centrifuges have been halved in each decade and their output doubled every five years.[4]

### 4.3.2 Plant description

The general principles of a large plant have been given elsewhere.[4] In summary, the uranium hexafluoride is vaporised from the international transport container in a feed autoclave and distributed to the centrifuge cascade at sub-atmospheric pressure. The enriched and depleted streams are then condensed as solid in cold traps or desublimers. When one is full, the gas stream is directed to an adjacent unit, and the desublimer heated to transfer the product to an international container. The collection stations are simple collecting vessels chilled to condense the material (by air for product, water for depleted tails).

The gas-flow and separation characteristics of the centrifuges allow highly efficient cascades to be designed even for relatively small tranches of separative work capacity. A plant can therefore be extended in modular form and in each module the cascades can be closely approximated to the ideal shape for any enrichment concentration. This is a considerable advantage over gaseous diffusion plant where the design is less ideal to reduce capital costs, and plant economics dictate large tranches of capacity per unit of plant brought on line.

## 4.4 Laser enrichment

The separation of isotopes using tunable lasers has been known for some time and essentially since such lasers have been commercially available, it has been possible, for example, to separate calcium isotopes by photo-ionisation.

The potential for laser separation processes to enrich uranium has been known for some years, (e.g. patents in the early 1970s[8]), and development since then has led to two possible laser routes, the Atomic route (AVLIS, or SILVA in France) and the Molecular route (MLIS).

In the AVLIS process (Atomic Vapour Laser Isotope Separation), uranium metal is evaporated at around 3000 °C in a vacuum chamber by surface melting using an electron beam. The resulting vapour is irradiated with laser light tuned to very specific wavelengths, selectively ionising U-235 which can then be extracted either electrostatically or electromagnetically.

MLIS (Molecular Laser Isotope Separation) uses uranium hexafluoride vapour as the process medium. In a two-step procedure, the $^{235}UF_6$ molecule is first excited by an infra-red laser and then dissociated by an ultra-violet laser to give $^{235}UF_5$ particles and fluorine gas.

Both processes have been the subject of very significant development programmes in USA, France, UK, Germany and Japan (and although less detail is known, also in China and the former USSR). All however now concentrate exclusively on AVLIS except in Japan, where both routes continue to be developed in parallel with a view to a decision between them in the late 1990s. In the USA at the Lawrence Livermore Laboratory, approximately 1 tonne of uranium has been evaporated.

It is agreed that either process is technically feasible; in all continuing programmes the work is aimed at optimisation and assessing the economics for future use. Recent reviews in the USA[9] vary from suggested operating costs of around $40 per kgSW (significantly below that of diffusion) and possible deployment in the late 1990s, to an apparently more realistic view that AVLIS may just be competitive in future with Urenco's centrifuge technology[10] but would not be ready for deployment until 2010 and would require a development programme of around $1000M to reach that stage.[11]

This remains under review in most commercial enrichment organisations, but needs considerable technical and development work to establish whether AVLIS has any economic or technical advantages over existing processes. The lifetime and reliability of laser systems require proving, the maintenance requirements and costs of the lasers and uranium evaporation systems in an industrial environment remain uncertain, and the process changes the existing fuel cycle to one based on uranium metal. Demonstrating that it is acceptable in all these respects, a prerequisite for industrial deployment, will take considerable time and money to justify the new investment in plant and equipment necessary for such a fundamental change of process. Such studies, even if successful, are unlikely to be complete before the early decades of the next century.

## 4.5 Re-enrichment of reprocessed uranium

Uranium recovered from the reprocessing of irradiated fuel differs from natural uranium in containing several additional isotopes. The two of particular importance are U-232 and U-236.

Uranium-232 is produced by neutron capture and decay during and after irradiation. Owing to the very high-energy gamma radiation emitted by one of its daughter isotopes, Tl-208, it can cause gamma dose-rates from the feed containers an order of magnitude greater than from natural feed, increasing as the daughter grows in. This leads to the need for additional shielding at the feed station and a strong incentive to re-enrich as soon as possible after recovery.

Uranium-236 is a neutron poison and so requires additional enrichment in U-235 to compensate, placing an economic penalty on re-enrichment.

Both isotopes naturally tend to follow the enriched product stream rather than the tails. Once they are taken into account there is no technical barrier to using either the diffusion or centrifuge process, but the recycled material must be segregated from natural stock if the product is to be acceptable to customers on the once-through fuel cycle.

With diffusion technology, the plant inventory is high (typically 250 kgU for every tonne SW/yr capacity) and so clearing the contamination would take several months. The gas centrifuge on the other hand has a low inventory (0.1 kgU per tonne SW/yr) which would allow the contamination to be cleared in less than six hours; moreover, as the plant is divided into a number of small cascades each physically separated from the others, recycled uranium can be confined to a small part.

From a theoretical point of view, the laser process should be ideal for re-enrichment as it would selectively enrich the U-235 and ignore contaminants. However, the problems of handling such isotopes in metallic vapour form, and dealing with the subsequent plant contamination, remain unconsidered.

Re-enrichment has already been demonstrated in Urenco plants with both ex-Magnox and ex-Oxide material, and a study is currently under way in conjunction with the French to review the economics of a centrifuge plant to be built alongside the French diffusion plant so that re-enrichment can be accommodated there.

## 4.6 Conclusion

Uranium enrichment technology and plant are well understood after more than 40 years' experience. The early work in the 1940s and 1950s concentrated on the gaseous diffusion technology which although still employed in the USA and France, is recognised to be both capital- and energy-intensive. It is most unlikely that any new commercial plant would be constructed with this technology.

Work in the 1960s and 1970s has led to the much more energy-efficient centrifuge process and the plants in Holland and the UK have close to 20 years operating experience; the technology can now be considered well into maturity. Developments have continued through the '70s and '80s and are expected to bring further improvements in efficiency. This technology would almost certainly be the only choice for any new enrichment plant to be built in the next ten or possibly twenty years.

Studies over the past decade have indicated that the atomic vapour laser enrichment process can be made to work, but needs considerable further effort (and finance) before being considered a serious commercial option, even given major improvements in the reliability and power of lasers. Whether it will ever reach beyond its current infancy to become a mature technology remains to be seen.

## 4.7 References

1 J. H. Tait. Uranium enrichment. In *Nuclear Power Technology*, Vol.2, (ed. W Marshall), Clarendon Press, Oxford (1983).
2 H. London (ed.), *Separation of isotopes*. Newnes (1961).
3 S. Villani (ed.). *Uranium enrichment* Spinger-Verlag, New York (1979).
4 I. Heriot. *Uranium enrichment by gas centrifuges* EEC publication EUR 1146EN (1988).
5 G. Zippe. *The development of short bowl centrifuges*. Report ORO315. University of Virginia (1960).
6 V. Mikhailov. Russian fuel cycle industry—capabilities and aspirations. in *17th International Symposium* held by Uranium Institute (September 1992)
7 A. Johnson. Advanced enrichment techniques. *In Pacific Basin Nuclear Conference*, p. 430 (1987).
8 R. H. Levy and G. S. Janes (USA), US Patent No 3,772,515; I. Nebenzahl and R. Levin (Israel), German Patent application No 2,312,194.
9 *The DOE Avlis programme: an industry assessment.* Edison Electric Institute, Nuclear Report NFC-91–001.
10 K. Schneider. LIS: The View from Urenco, ASANER'94, (22–25 March 1994).
11 Nuclear Fuel Special on USEC, *Nucleonics Week*. Outlook on USEC (30 September 1993). Inside N.R.C. (4 October 1993). Nuclear Fuel (11 October 1993).

# 5 Power reactors

K. W. Hesketh
*BNFL, Fuel Division, Springfield*

## 5.1 Introduction

The nuclear fission reactor is the central element of the fuel cycle, generating the energy that is the primary aim of the industry. It is also the source of fission and transmutation products presenting opportunities and problems in succeeding stages. Various designs have been developed for central power generation and for more specialised applications; the purpose of this chapter is to describe the principles and main practical features of the more important types.

## 5.2 Fissions in nature

### 5.2.1 The fission process

Uranium provides the bulk of the energy in most reactors and is the key to the fuel cycle. Like other heavy elements, it originated in the synthetic reactions brought about upon relatively light nuclei by intense neutron fluxes during the explosion of stars preceding our own, and some of the surplus energy it then gained is released in the fission process.

In 1939 L. Meitner, O. Hahn and F. Strassemann[1] discovered that irradiation of uranium with neutrons, besides forming new heavy atoms as expected, also produced a range of elements with nuclei whose masses were only about half that of uranium. Lise Meitner and Otto Frisch[2] suggested that this was due to the uranium nuclei splitting into two roughly equal fragments. This was nuclear fission.

The stability of atomic nuclei depends on the balance between the strong nuclear force, which attracts the constituent protons and neutrons, and the repulsive electric force between the like-charged protons. The latter force, effective at longer ranges, tends to be favoured in large nuclei such as those of uranium, making them inherently unstable. One manifestation of this instability is that a uranium nucleus can split into two fragments, either spontaneously or more readily after absorbing a neutron. The two daughter nuclei or fission products fly apart owing to the electric repulsion between them, carrying a portion of the stored energy from the original heavy nucleus, ultimately converted to heat.

The energy released from each fission event is very large on the atomic scale, about 200 million electron-volts (eV where 1 eV = $1.6 \times 10^{-19}$ Joules). In comparison the energy released in chemical reactions is typically of the order of 1 eV. Hence uranium constitutes an extremely compact fuel.

### 5.2.2 Natural fission

Small amounts of uranium occur naturally in most rocks, soils and building materials. It consists mainly of 99.3% U-238 and 0.7% U-235, where the identifying numbers of the isotopes denote the total of protons and neutrons in the nuclei. Both are alpha-active, decaying with half-lives of 4.5 and 0.7 billion years respectively, and a small fraction undergoes spontaneous fission.

Uranium-235 nuclei are particularly liable to fission if bombarded with slow neutrons. The impact of cosmic rays on the Earth's atmosphere forms showers of secondary particles, including neutrons. When one of these neutrons collides with a uranium nucleus, there is a good chance that it will cause fission. As with spontaneous fission, this process takes place all around us.

Each fission produces several more neutrons, which under favourable conditions can go on to cause further fissions, setting up a chain reaction. Owing to the relative half-lives, the proportion of U-235 is now too small to sustain such a chain without artificial help, but in the distant past this was not so, and fission chain reactions took place in some uranium ore bodies, notably at Oklo in Gabon (West Equatorial Africa). These natural fission reactors, which pre-dated the first man-made reactor by about two billion years, are described in Section 5.4.1.

## 5.3 Fundamentals of fission reactors

### 5.3.1 The fission chain reaction

Although each of the neutrons released by each fission could in principle go on to cause another, this does not occur in practice since a proportion depending on circumstances escapes from the reaction zone or is absorbed by non-fissile material. The chain reaction can therefore take one of three forms, depending on the average number of neutrons from each fission that are effective in propagating the chain ($k_{eff}$).

- If $k_{eff}$ exceeds one, as may occur with a very high concentration of fissile atoms, the fission rate increases exponentially until slowed by losses. This is the principle of the fission bomb, where a compact core of fissile material (plutonium or uranium highly enriched in U-235) is assembled quickly by an explosive mechanism. The value of $k_{eff}$ is substantially greater than one and this, coupled with the very short time between generations (less than 0.01 microsecond), causes the fissile atoms to release their energy explosively.
- With $k_{eff}$ less than one the reaction may be started by an external neutron but

dies away more or less rapidly. Assemblies in this state are used for experimental purposes, or with $k_{\text{eff}}$ only slightly below unity, as neutron amplifiers.

- If $k_{\text{eff}}$ is exactly one, a condition described as *criticality*, the reaction continues at a steady rate. This is the operating state of a power reactor.

Because of the short time between successive generations of fission (about 0.1 millisecond in this instance), maintaining a reactor at a steady power level requires that the multiplication factor should be kept extraordinarily close to one. But for a small proportion of neutrons released after a slight delay following fission, it would not be practicable at all as control mechanisms could not respond fast enough.

### 5.3.2 Prerequisites for a reactor

Several conditions must be met before the required reaction can occur:

(a) Mass of fissile material. Neutrons near the outside surface are likely to escape into the surroundings without propagating the chain reaction. If a system is physically too small, this neutron leakage will be so large that a self-sustaining chain reaction is not possible. Enough fissile material, the nuclear fuel, must therefore be assembled to sustain a chain reaction. The *critical mass* which is just sufficient under ideal conditions depends on the type and proportion of fissile nuclei (fuel is often artificially enriched in U-235) and on other materials present, while that which is actually necessary depends on the configuration.

(b) Geometry. Since a sphere has the smallest surface area for a given volume, the necessary mass will be smallest for a spherical system. This is why most reactor cores approximate to the shape nearest in practice to a sphere, a cylinder with similar height and diameter. The presence of a neutron reflector, which might consist of a blanket of water or a thickness of graphite, further reduces the critical mass by scattering back into the core neutrons which would otherwise escape.

(c) Neutron spectrum. Neutrons released by fission carry large amounts of kinetic energy, up to 10 million eV each. Unless the fissile nuclei are very concentrated, their interactions with neutrons at these energies are insufficient to sustain a chain reaction. Light atoms, such as those of hydrogen, deuterium or carbon, are very effective at slowing down or *moderating* neutrons by elastic collisions, without absorbing too many of them. Given enough of such material, most neutrons will be slowed almost to thermal equilibrium with it, at energies of an electron volt or less. The probability of neutron-induced fission is then very much increased, and criticality is possible even with low concentrations of fissile atoms. On the other hand, the ability of unmoderated neutrons to induce fission in all uranium and plutonium isotopes is exploited in fast reactors, which however need a correspondingly increased neutron flux.

The physical layout of the fissile and moderating nuclei also matters. In most systems the nuclear fuel and the moderator are separate. Not only is this the best way to ensure efficient transfer of heat from the fuel, among other practical considerations, but it is also more efficient in terms of fuel enrichment to separate the fuel and moderator than to have them intimately mixed (as for example in a system containing an aqueous solution of a uranium or plutonium salt).

The presence of neutron-absorbing materials is also important. Some are included deliberately, as control rods to prevent the reaction from becoming too fierce. Others are structural materials needed to hold the nuclear core together. The nuclear fuel and moderator also contribute to absorption. The designer of a reactor usually tries to minimise such unwanted neutron absorption.

### 5.3.3 Choice of moderator

The material used as a moderator is a very important part of reactor design. In modern power reactors, the most commonly-used moderator is ordinary water. This has the advantage that it can also be used to transfer heat from the reactor to a turbine/generator set, either directly or by way of heat-exchangers.

The other commonly-used materials are graphite (as in the UK's gas reactors) and heavy water ($D_2O$) as in Canada's CANDU reactors. They are superior as moderators to the extent of absorbing fewer neutrons for every one slowed down to the lowest energies, and therefore preferred for use with low-enriched or natural uranium fuels. However, as deuterium and carbon nuclei are respectively twice and twelve times as heavy as a neutron, the energy transfer per collision is smaller than with a proton and a longer path is needed, so that reactors with these moderators tend to be much bulkier than those using ordinary water.

### 5.3.4 Feedback mechanisms

Power reactors must be stable. The essence of stability is negative feedback—an automatic response that tends to cancel any perturbation, in this instance any unwanted change of power level whether overall or in a small region of the core. Slow changes can be corrected by the control system, but its mechanical inertia limits its response, and for events at the atomic level an inherent negative-feedback characteristic is essential.

Temperature effects provide two such mechanisms. Within the fuel, neutrons may be absorbed by U-238 or induce fission in U-235, and the ratio of absorption to fission rises rapidly with temperature. Thus if the rate of fission increases for any reason, the resulting rise in temperature reduces the fission rate and so is self-limiting. Similarly a rise in moderator temperature increases the energy of thermalised neutrons, making them less likely to cause fission and so reducing the power level.

Other feedback effects, such as those due to boiling in a liquid coolant, depend in a more complex fashion upon the particular arrangements. According to whether the coolant serves mainly as moderator or absorber of neutrons, a void may either reduce or increase the power level, and the designer must take this into account.

Still further effects may occur in transient conditions. For instance, the fission product I-135 decays with a 7-hour half-life to Xe-135 which itself decays with a half-life of 9 hours, but while it lasts is a powerful neutron poison (absorber); neither its daughter Cs-135 nor its absorption product Xe-136 absorbs strongly. While a reactor is running at normal power, Xe-135 is kept to a steady low level by a combination of decay and transmutation, but when the reactor is shut down transmutation ceases and the isotope builds up by decay of its parent, reaching a peak in about 10 hours. The reactor operators must plan for this xenon poisoning following a shutdown and for the subsequent burn-out of the xenon once the reactor starts up again, as well as for the changes in xenon concentration which follow any change in reactor power and control rod insertion.

Xenon poisoning is therefore a serious consideration in reactor design, particularly in some specialised applications such as submarine power plant, where the ability to overcome its effects at any time is an essential requirement. It was also an important factor in the accident at Chernobyl, where inappropriate attempts by the operators to compensate for a build-up led to the unstable condition of the reactor.

### 5.3.5 Reactor control

Ensuring that a reactor can be controlled safely is one of the most important preoccupations of designers and operators. There are two aspects. Firstly the design must enable the reactor to be maintained at a steady power level. Secondly, there must be provision for shutting it down, whether routinely (for maintenance or refuelling, for example) or in an emergency.

Section 5.3.1 touched upon the precision with which the multiplication factor needs to be controlled if a reactor is to run at a steady power. The fuel-temperature feedback mechanism can be relied upon for this purpose. The effect of moderator temperature may either reinforce or (for instance if it causes an absorbing coolant to boil) oppose that in the fuel, but in a properly designed reactor the fuel temperature mechanism always wins. Ultimately, the power produced by a reactor is determined by the rate at which heat is removed by the steam turbine. Thus an increase in steam flow tends to cool the reactor. This in turn increases the multiplication factor which increases the power output of the core. The power produced stabilises when it matches the rate at which the turbine extracts heat.

Ensuring safe operation of a reactor is greatly simplified by the so-called delayed neutrons. Most of the neutrons released in a fission event are emitted more or less instantaneously, but some (typically a little below 1%) are emitted after a delay of a few seconds to a few minutes. The designer normally ensures that the reactor can never be critical without the contribution of the delayed neutrons; the delay slows the response of the reactor to small to moderate perturbations of the multiplication factor, to a point that the control mechanism can match.

Shut-down of a reactor is achieved by introducing neutron-absorbing materials such as boron. Normally they are in the form of control rods which move into the core under automatic or operator control, but there are many diverse alternatives, as will be seen in Section 5.4.

### 5.3.6 Decay heat

The need to guarantee safe shut-down of the fission chain reaction is an obvious aspect of reactor safety. The biggest threat to safety in a power reactor, however, is only indirectly related to the chain reaction. During operation, about 7% of the heat output typically comes from delayed neutron fissions or from the radioactive decay of fission products and transuranic elements. When the chain reaction is halted, this source of heat remains, decaying on time-scales dictated by the half-lives of the nuclides concerned. In a modern power reactor with over 3000 MW thermal output, the decay heat can be as high as 200 MW for the first few minutes. The reactor designer has to ensure that there are diverse means of extracting this heat to ensure that the fuel does not become overheated. It was a failure to maintain an adequate flow of cooling water after an automatic shut-down that led to core damage in the Three Mile Island reactor in 1979.

### 5.3.7 Fission products and transuranics

Besides heat, a fission reactor has important by-products. First there are the fission-product elements, with about half the atomic mass of the fissile materials, such as iodine, caesium, xenon and strontium to name just a few of the more important. Although some are stable, the majority are radioactive with half-lives ranging from fractions of a second to many thousands of years. The fission products dominate the radioactivity of discharged nuclear fuel and are potentially the most serious hazard associated with operating a reactor. The greatest hazard is associated with isotopes of medium half-life such as iodine-131 (8 days) and caesium-137 (30 years), which are not only intensely radioactive, but also biologically active.

Secondly the neutron flux generates new elements beyond uranium in the Periodic Table. The most important of these is plutonium, formed on neutron capture by U-238, followed by the beta decay of the product nucleus U-239 to form first neptunium-239 and then plutonium-239:

$$\text{U-238} \xrightarrow{n} \text{U-239} \xrightarrow[23.5 \text{ m}]{\beta} \text{Np-239} \xrightarrow[23.6 \text{ d}]{\beta} \text{Pu-239}$$

Thus a material readily fissioned by thermal neutrons is generated from one that is not so itself, but is thus *fertile*. Pu-239 can also capture a neutron to form Pu-240 and heavier plutonium isotopes. These can in turn form new elements, not otherwise present on Earth, such as americium and curium. Plutonium and heavier elements generally have long half-lives and will eventually dominate the radioactivity of discharged nuclear fuel.

**Fig. 5.1** Windscale pile stack

Finally, the neutron flux can transform nuclei in the structure of the reactor itself into radioactive isotopes, notably cobalt-60, generated primarily from cobalt-59 present in any steel components. Such activation products make an important contribution to the radioactive content of a reactor, enough to influence the choice of structural materials.

## 5.4 Reactor types

The number of reactor designs which have been conceived or actually built since the first experimental pile in Chicago is extraordinary, and there is space here only for the most important. The emphasis is on power reactors, but those occurring naturally or designed for other applications are briefly mentioned.

### 5.4.1 Natural reactors

The basic ingredients of a reactor, namely a large enough concentration of fissile material and a moderator, can occur naturally in a uranium ore body percolated by ground water. Fission chain reactions might then occur naturally, given a sufficient proportion of U-235, but at 0.7%, its current abundance on Earth is insufficient for criticality in any system moderated by ordinary water. Such a hypothetical natural reactor is therefore impossible on Earth today. Since the half-lives of U-235 and U-238 are respectively 0.7 and 4.5 billion years, however, the possibility could have been realised in the past when the relative abundance of U-235 was higher.

In 1972 routine measurements of U-235 abundances in ores from a mine at Oklo in Gabon (West Equatorial Africa) led to the chance discovery of the remains of just such natural reactors that evidently ran some 2 billion years ago.[3] Fission chain reactions took place in many different locations of the ore body where water percolated through, lasting for tens of thousands of years until too much of the fuel was consumed. Since the water was heated and driven away whenever the reaction was too intense, the average power produced by each of these reactors was probably quite small. The Oklo mines represent their fossilised remains.

### 5.4.2 Experimental reactors and reactors for radioisotope production

The first man-made reactor was put together in a squash court in Chicago in December 1942 under the direction of Enrico Fermi,[4] demonstrating a controlled fission chain reaction for the first time. Since the original Fermi pile, many experimental reactors have been built and operated around the world, to designs too numerous to describe comprehensively here. They are of four main types serving different purposes:

(a) Small reactors to test the predictions of computer codes. A typical set-up consists of an array of fuel rods in a moderator, to test the conditions for criticality and measure the neutron distribution in the array. As no appreciable power is produced, and all the components are at ambient temperature, such a reactor is referred to as a zero-power critical facility.

(b) Reactors designed to test new designs of fuel element. Some irradiate test fuel rods to a desired burn-up. Others test their ability to withstand rapid power increases, such as might occur for example in an accident.

(c) Intense neutron sources for many diverse applications, such as testing materials under neutron bombardment, measuring absorption cross-sections and shielding properties.

(d) Small reactors for educational and training purposes, operated by many universities world-wide.

Special-purpose reactors are normally used to produce nuclear materials for medical and other applications, such as radioactive imaging tracers and radiation sources. An example is cobalt-60 which is made by neutron irradiation of natural cobalt (Co-59).

### 5.4.3 Reactors for weapons material production

Some of the earliest reactors were built specifically to produce plutonium and tritium for nuclear weapons. In the UK, this was done on a large scale in the two Windscale piles, until their closure following the fire in 1957. These were air-cooled and graphite-moderated, fuelled with natural uranium metal rods. The heat produced was simply vented to the atmosphere, via filters, from the two tall stacks which are still prominent today. The fuel rods were arranged in horizontal channels, through which air was forced by blowers.

Weapon designers need plutonium with very low fractions of the higher isotopes, which are prone to spontaneous fission and can cause premature detonation with reduced energy yields. Neutron irradiation of U-238 first produces Pu-239 by the fertile capture mechanism described in Section 3.7, but if it is continued, further captures build up significant amounts of the higher isotopes. The fuel is therefore discharged and reprocessed to recover the plutonium after a very short dwell time in the reactor. The throughput of fuel in a reactor being used for military plutonium production is therefore higher by an order of magnitude or more than is normal for commercial power reactors, and this is an important feature which international nuclear safeguards inspectors look for in checking for clandestine plutonium production.

Nuclear weapons may also use tritium (H-3), which in combination with deuterium enhances the yield by nuclear fusion. Tritium is produced by bombarding lithium targets with neutrons from a reactor and since it is radioactive (with a half-life of 12 years), it must be replenished continually if stocks are to be maintained. In spite of the cutbacks in weapons inventories, there remains a role for tritium-producing reactors while the nuclear arsenals are retained.

### 5.4.4 Power reactors—thermal

The majority of reactors operating today fall into the thermal class, so called because they use moderators to reduce the kinetic energies of neutrons to about those due to thermal motion.

There are currently over 400 commercial power reactors in operation worldwide. Most of these belong to one of five main families of design. Within each family the power outputs tend to range from a few hundred megawatts (electrical) for the older variants to up to 1400 megawatts for the largest of recent designs. Table 5.1 presents some of the main core characteristics of typical modern thermal reactors in a concise form.

The reasons for the various designs are largely historical. The PWR, BWR and VVER systems evolved from submarine reactor types, where the main consideration is compact size. The two natural uranium systems, Magnox and CANDU,

**Table 5.1** Typical thermal reactor parameters

| Parameter | PWR | BWR | Magnox | AGR | Candu | RBMK |
|---|---|---|---|---|---|---|
| Thermal output (MW) | 3400 | 3600 | 1875 | 1500 | 2650 | 3200 |
| Electrical output (MW) | 1150 | 1200 | 600 | 660 | 870 | 1000 |
| Fuel type | $UO_2$ | $UO_2$ | U metal | $UO_2$ | $UO_2$ | $UO_2$ |
| Coolant | $H_2O$ | $H_2O$ | $CO_2$ | $CO_2$ | $D_2O$ | $H_2O$ |
| Moderator | $H_2O$ | $H_2O$ | Graphite | Graphite | $D_2O$ | Graphite |
| Core height (m) | 4 | 4 | 9 | 8 | 4 | 7 |
| Core diameter (m) | 4 | 4 | 17 | 9 | 7 | 12 |
| Weight of fuel (tU) | 90 | 140 | 600 | 120 | 80 | 205 |
| Discharge burn-up (MWd/kg) | 35–45 | 30–40 | 5 | 24–27 | 8 | 20 |
| ($10^{12}$ joules/kg) | 3.0–3.9 | 2.6–3.6 | 0.43 | 2.1–2.3 | 0.69 | 1.7 |
| No of fuel assemblies/elements | 193 | 732 | 49000 | 2500 | 5760 | 3400 |
| No of fuel rods per assembly | 264 | 96 | 1 | 36 | 37 | 18 |
| Fuel clad material | Zircaloy | Zircaloy | Magnox | Stainless steel | Zircaloy | Zircaloy-niobium |
| Fuel enrichment (w/o U-235) | 3.0–4.5 | 2.0–4.5 | Natural | 2.5–3.6 | Natural | 2.5 |
| System pressure (bar) | 155 | 72 | 28 | 42 | 90 | 78 |
| Inlet temperature (°C) | 290 | 270 | 250 | 320 | 250 | 265 |
| Outlet temperature (°C) | 325 | 285 | 400 | 650 | 290 | 290 |
| Secondary system pressure (bar) | 70 | N/A | 47 | 170 | 50 | N/A |

were devised in the UK and Canada respectively when neither country had access to isotope separation technology on a commercial scale. In the UK, the AGR was a natural progression from Magnox, while the Soviet RBMK was developed from reactors used for producing military plutonium.

## Pressurised water reactors

As about 60% of the world's commercial power reactors are PWRs, they are given special attention here. Many of their fundamental features are however common to other reactor types, which therefore do not need the same level of discussion. The description which follows applies to the Western PWR designs: the former Soviet Union also developed a PWR called the VVER, with a nuclear core in principle very similar to the Western designs, though there are some detailed differences, most notably a triangular lattice for the fuel rods and hexagonal fuel assemblies.

As shown schematically in Fig. 5.2, a PWR reactor consists of a compact core in a pressure vessel capable of containing ordinary water at high pressure. Water serves as both moderator and heat-transfer medium, under a pressure high enough to prevent boiling at the operating temperatures of about 300 °C. The pressure vessel is connected to between two and four steam generators to form the primary coolant circuit. The steam generators are towers containing many kilometres of

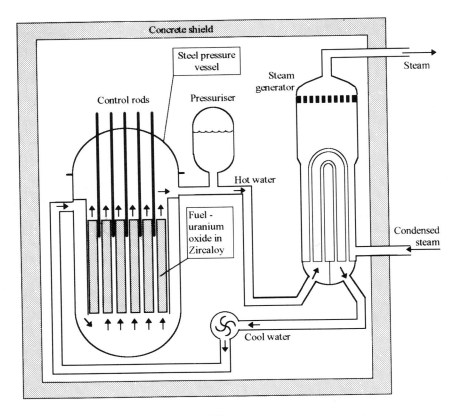

**Fig. 5.2** Pressurised water reactor (PWR)

narrow pipes through which water from the nuclear core is pumped, transferring heat to a secondary coolant circuit in which water at a lower pressure boils to drive a steam turbine/generator set.

In modern PWRs the core typically consists of 157 to 241 fuel assemblies, each a self-standing structure with between 200 and 300 individual fuel rods. A typical modern fuel assembly design contains 24 guide tubes inside which the neutron-absorbing control rods can slide in and out of the core. A central tube can accommodate neutron detectors to measure the flux distribution. The open lattice of fuel rods allows cooling water to flow freely between them, directed into the assembly through a bottom end fitting which is penetrated by an array of holes.

Each fuel rod comprises a stack of isotopically-enriched uranium dioxide fuel pellets in a sealed tube of Zircaloy (slightly alloyed zirconium). The oxide is a ceramic melting in the region of 2800 °C. The fuel pellets themselves form the first barrier to the escape of fission products, mostly trapped in small pores, so the high melting point is an essential prerequisite. The cladding forms the second barrier and Zircaloy is chosen because it combines corrosion resistance with low neutron absorption. A small space near the top of the fuel rod accommodates the 1% or so of gaseous fission products which escape from the ceramic pellets.

PWRs are refuelled at intervals of between one and two years, when a quarter to a third of the fuel assemblies are replaced with fresh ones. This involves taking the reactor off-power for a few weeks, removing the top of the pressure vessel, flooding the upper part of the vessel with water to provide shielding from the radiation, and unloading the fuel assemblies into an adjacent pond.

Normally fuel is replaced when its fissile content has decreased to a level where criticality is only just achievable at full power. The enrichment and number of fresh assemblies is chosen so that the reactor can provide the required power up to the next refuelling. This implies that immediately after refuelling there is excess fissile material that must be prevented from becoming super-critical. In PWRs this important function is performed by a neutron absorber (in the form of boric acid) dissolved in the primary system water. Its concentration is usually adjusted so that the core is just critical at full power with the control rods almost fully withdrawn. This ensures that the full worth of the control rods is available to shut the reactor down in an emergency. Boric acid is also used to supplement the control rods in shutting down the reactor, both in normal operation and in emergencies.

Sometimes neutron-absorbing materials such as boron or gadolinium are incorporated in the fuel assemblies themselves or in free-standing rods which fit inside the guide thimbles. These so-called burnable poisons hold down the multiplication factor of the core by capturing neutrons which eventually destroy the neutron absorber. For example, the isotope boron-10 is a very potent absorber, but the helium-4 and lithium-7 to which it is converted have practically no affinity for neutrons.

Outside the fuel assemblies are additional control rods containing neutron-absorbing materials such as boron, hafnium, or combinations such as silver–indium–cadmium. They are used principally to shut the reactor down in normal operation or in the event of malfunctions, but also in fine-tuning the reactor during operation.

Also present are neutron sources. Although not necessary to start the chain reaction, since spontaneous fission provides enough neutrons for the purpose even in unirradiated fuel, they are essential to monitor the approach to criticality and ensure that it is achieved safely. A sub-critical reactor amplifies an injected neutron flux to an extent which increases as the reactor approaches criticality. By monitoring the neutron count-rate on special detectors either inside or outside the core, the operators can anticipate the point of criticality precisely. The neutron sources must be intense enough to provide a flux clearly detectable above the background noise when the reactor is still well short of criticality.

## Boiling water reactors

BWRs, as illustrated in Fig. 5.3, have much in common with PWRs, being water-moderated, water-cooled and roughly the same size. The main difference is that at

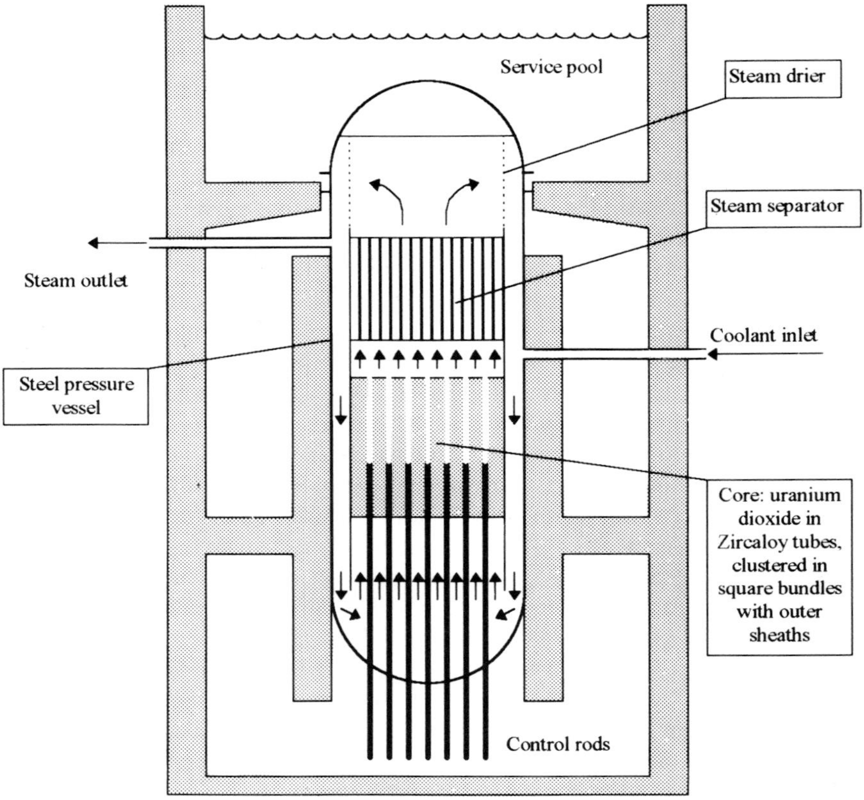

**Fig. 5.3** Boiling water reactor (BWR)

the lower operating pressure, the water passing through the core can boil. The steam passes through drier plates above the top of the core to remove water droplets, and then directly to the steam turbines. This eliminates the need for steam generators, but means that the turbines are effectively part of the reactor circuit. Because the coolant in any reactor is always contaminated with radionuclides to some extent, the turbine must be shielded and radiological protection provided during maintenance. This roughly cancels the capital and operational savings savings due to omitting steam generators, making the overall costs comparable with those of a PWR.

The design differs in detail from that of a PWR. First of all, the steam drier above the core leaves no room for control rods and their drive mechanisms, so the rods enter from below. This arrangement in any case makes them more effective, operating in the region of highest neutron flux; water passes through the core from below, the steam fraction increases as it takes heat from the fuel, and the flux is largest where the moderator is most dense.

The control rods themselves are in the form of large cross-shaped blades which fit into the gaps between assemblies.

Like PWRs, BWRs are refuelled off-load at intervals of between 12 and 18 months. Because the moderator changes phase from water to steam, it is impractical to use boric acid to counter the excess fissile content of the reactor following a refuelling. Instead, BWR core designers rely very heavily on burnable absorbers, usually gadolinia ($Gd_2O_3$) incorporated into the fuel pellets, and on control rods to make the fine adjustments.

The design of the fuel assembly is at first sight quite different from that in a PWR, though there are underlying similarities which reflect their common ancestry. The fuel rods are enclosed in a Zircaloy shroud which serves to guide water through the assembly and prevent excessive cross-flow that might otherwise be caused by steam formation. Most of the heat output of an assembly is transferred to the coolant flowing inside it, so that near the top of the core it is mostly steam, while outside, the heat transfer is small enough for the coolant to remain liquid. The shroud also helps to ensure efficient use of the fuel by confining the steam near the top of the core, where moderation depends largely on water in the wide gap between assemblies.

Modern BWR assemblies also contain an internal water channel in which the coolant stays liquid. It can take the form of either a Zircaloy box which displaces four of the fuel rods, or a cross which effectively divides the assembly into four sub-assemblies. Moderation is most effective near such a box and at the edges of an assembly, or particularly the corners; to prevent rods in these positions from running too hot, they must be less highly enriched than elsewhere, and several different levels may be necessary.

**Gas reactors**

The UK is the only country to have developed a significant programme of commercial gas-cooled reactors. These are graphite moderated reactors cooled by pressurised carbon dioxide. The first generation uses natural uranium metal fuel in a magnesium alloy can which absorbs very few neutrons, as is essential to criticality with natural uranium. The alloy was specifically developed to resist corrosion, hence the name Magnox, for magnesium alloy, no oxidation. Three variants of Magnox alloy are used in different parts of the fuel element, with aluminium and zirconium as the principal alloying components. The second generation, the Advanced Gas Reactors (AGRs), uses uranium oxide fuel in stainless steel cans, with enrichments in the region 2.5 to 3.5%. Uranium oxide has a considerably higher melting point than the metal, allowing higher fuel ratings, higher operating temperatures and improved thermal efficiency.

The most distinctive characteristic of gas-graphite reactors is the bulk of the core, due to the distance needed for slowing down neutrons to thermal energies in graphite. Figures 5.4 and 5.5 show the overall layouts of Magnox and AGR stations. The fuel sits in individual channels through the moderator. In some stations, this arrangement allows the reactors to be refuelled on power. Large blowers

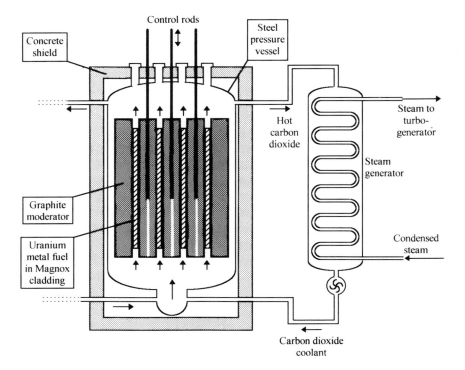

**Fig. 5.4** Magnox reactor

drive the carbon dioxide coolant through the core from bottom to top and into the heat exchangers to raise steam. Passage is mainly through the fuel channels, with some cooling the moderator.

The large size of the core makes the capital cost relatively high, but does have some advantages in safety. Any fault heating up the core tends to proceed very slowly because of the low power density combined with the large heat capacity. The single-phase coolant also helps to ensure that these are intrinsically very safe reactors.

Both Magnox and AGR reactors rely on control rods for fine tuning during operation and for shutdown. The rods slide into holes in the moderator from above, and are designed to fall into the core under gravity should the drive mechanisms lose power. The AGR designs incorporate a system for emergency shut-down by injecting nitrogen, a strong neutron absorber which provides a diverse shutdown system.

A typical Magnox fuel element consists of a uranium metal bar clad with Magnox, in which are formed centralising vanes and cooling fins. Between six and ten elements are stacked end to end in a channel, with the help of top and bottom

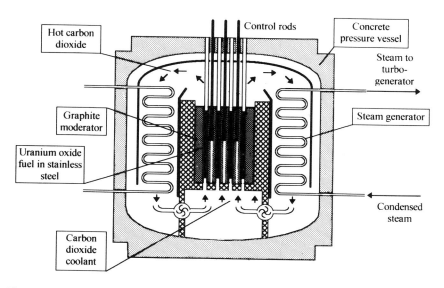

**Fig. 5.5**  Advanced gas-cooled reactor (AGR)

fittings also used in handling the fuel. A complete stack is normally refuelled at a time. A typical element remains in the core for between three and five years before discharge.

An AGR fuel assembly is a much more complicated design with 36 stainless-steel-clad pins located in structural support braces. A graphite sleeve surrounds the assembly, providing extra structural strength and a channel for the flow of the coolant gas. Each pin contains cylindrical uranium oxide fuel pellets similar to those of PWRs and BWRs except that they are larger in diameter and have a central hole. This helps to maintain the integrity of the fuel by accommodating distortion and fission-product gases. Eight assemblies are stacked vertically in a fuel channel to form a stringer, loaded and unloaded as a single unit.

## The high-temperature gas-cooled reactor (HTGR)

Current power reactors are restricted to a working temperature no higher than 700 °C, the limit for the steel used in much of the structure, and the steam generated must perforce be cooler. A higher temperature would permit greater thermodynamic efficiency, according to the formula

$$E = (T_{max} - T_{min})/T_{max}$$

where $E$ is the theoretical maximum efficiency, $T_{max}$ is the input temperature and $T_{min}$ the outlet temperature of the working substance, both on the absolute (Kelvin) scale. Moreover, there would then be wider direct applications for the heat besides electricity generation—steel-making was considered in an extreme case—so avoiding the losses and expense inherent in the electrical intermediates.

Such temperatures bring substantial difficulties. Water would be super-critical and could not be kept in a dense state without recourse to excessive pressures, so a different moderator is needed, with graphite as the obvious choice. Either steam or carbon dioxide would react with graphite, so helium is the most promising coolant. Conventional cladding metals would have to be be replaced, for instance by a ceramic coating, and a rigid structure might be impracticable; fuel could instead form a *pebble bed* through which the coolant would pass and from which the *pebbles* would be circulated to a replenishment system where the most highly-irradiated would be removed and replaced by fresh.

High-temperature reactors with these features were the subject of considerable interest in the 1970s, particularly in Germany with a view to providing process heat and incidentally using thorium fuel which is unattractive in other reactors. A commercial HTGR operated from 1979 to 1989 at Fort St. Vrain (USA) but with considerable problems. Interest has now taken root in China with the start of a 10 MW(th) test unit near Beijing.[5] The cylindrical core chamber will be 1.8 metres across and about 2 metres high. In it, about 27 000 fuel balls of 60 mm diameter will each contain 5 grams of uranium as dioxide initially enriched to 17%. The core will be surrounded by a moderating graphite reflector, with control rods operating in the circumferential wall. In the first trial phase, the gas outlet will be limited to a temperature of 700 °C and feed a steam generator essentially similar to that of a PWR. Later, outlet gas at 950 °C is intended to feed a combined cycle of gas turbine and steam generator.

**Pressurised heavy water reactors**

The Canadian CANDU system is the only pressurised heavy-water reactor (PHWR) developed for commercial power production. The steam-generating heavy-water reactor (SGHWR) was developed to the prototype stage in the UK, but never commercially deployed.

Figure 5.6 shows the core arangement of a CANDU reactor. The moderator is heavy water ($D_2O$) in a tank penetrated by horizontal pressure tubes which form channels for the fuel cooled by heavy water. As in a PWR, steam generators raise steam for the turbine.

As discussed in Section 5.3.3, heavy water is the best moderator available and makes the CANDU system the most efficient in terms of neutron usage. Natural uranium fuel can therefore be used even in the form of oxide (which has a higher neutron absorption than metal). The disposition of fuel in separate channels allows on-load refuelling. This is done from two sides, alternate channels being unloaded and loaded in opposite directions.

A CANDU fuel assembly consists of a simple bundle of typically 37 fuel rods in a support structure, with 12 bundles lying end to end in a channel (see Fig. 5.6).

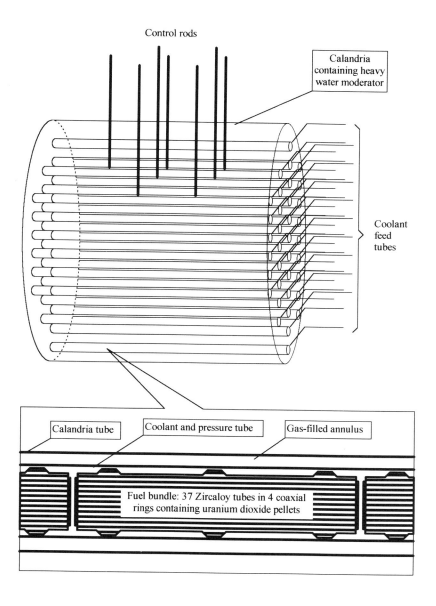

**Fig. 5.6** CANDU reactor core

Conventional absorber rods are used for shutdown, while tubes containing ordinary water are used for flux shaping. A secondary shutdown system, in which gadolinium nitrate solution is pumped into the heavy water moderator, is also provided. The large bulk volume of the moderator provides a natural heat sink for removal of decay heat in the case of reactor accidents, and combined with the docile core characteristics, makes CANDU inherently very safe.

**Water-cooled graphite moderated reactors (RBMK)**
RBMK reactors were developed in the former Soviet Union. Some 27 were built in total. They are graphite moderated, water cooled reactors of some 7 m height by 12 m diameter. The overall layout of the core is similar to that of a gas-graphite reactor such as the AGR, but the fuel channels are separate pressure tubes containing ordinary water under high pressure, nevertheless allowed to boil and directly feed the turbine. Enriched uranium oxide fuel in the form of 3.5 m long bundles of 18 rods makes up each assembly. Two assemblies are stacked on top of one another in a fuel channel.

The principle lends itself well to a modular design, tailored to requirements by extending the size of the core and the number of pressure tubes as necessary. It does however, have a number of features which would not allow it to be licensed in the Western world. With moderation almost entirely due to the fixed graphite, any voids in the neutron-absorbing coolant increase reactivity and raise the level of heating while impairing the ability to extract it; the resulting temperature rise increases the tendency to boil, and so on.

The Chernobyl accident was a direct result of this positive feedback mechanism, and that described in Section 5.3.4. Once the coolant started to boil, power increased out of control, peaking at several hundred times the normal full rating. The temperature rose beyond the melting point of the fuel, the coolant was vaporised and reacted with the zirconium cladding to produce hydrogen, and the resulting explosion destroyed the core, dispersing much of the inventory of radionuclides into the environment. RBMK reactors have since been modified to prevent a similar occurrence, the main change being to use a more highly enriched fuel. This competes more effectively with the cooling water in absorbing thermal neutrons. The void reactivity coefficient depends on the relative balance between fuel and water absorption, and the higher enrichment helps to ensure negative feedback.

### 5.4.5 Power reactors—fast

Fast reactors are so called from the speed of the neutrons propagating the fission chain, not from the rate of reaction. Cores are designed to incorporate the minimum amounts of moderating materials, so that the neutron kinetic energies remain high.

A parameter called eta increases with neutron energy. This is the average number of neutrons released per neutron absorbed in the fuel. Spare neutrons, beyond those needed to propagate the fission chain, can convert fertile to fissile material

**Table 5.2** Fission properties of U-235, Pu-239 and Pu-241

| Isotope | No. of neutrons released per fission | | No. of neutrons released per neutron absorbed in fuel | |
|---|---|---|---|---|
| | Thermal | Fast | Thermal | Fast |
| U-235 | 2.4 | 2.5 | 2.0 | 2.0 |
| Pu-239 | 2.9 | 3.1 | 1.8 | 2.4 |
| Pu-241 | 2.9 | 3.0 | 2.2 | 2.6 |

as described in Section 5.3.7. If there are enough neutrons to replace every fissile atom consumed by at least one other, then the system becomes a breeder. By converting U-238 to plutonium, a breeder can in theory generate over 50 times more energy per kilogram of uranium ore than can a thermal system alone, and is virtually self-sufficient in fuel.

The absolute minimum eta value for breeding is evidently 2.0—one neutron each to continue the chain and to transmute a fertile nucleus. For fissions in U-235, Pu-239 or Pu-241 in a thermal neutron energy distribution, eta is only around 2.0, so that there is no margin to compensate for absorption in the moderator and the structure of the core; a thermal reactor based on the uranium fuel cycle cannot therefore operate as a breeder. But at higher energies the eta values of the plutonium isotopes rise to 2.4 and above, making a breeding cycle possible with plutonium fuel.

Table 5.2 lists the fission properties of the principal fissile isotopes of uranium and plutonium in thermal and fast neutron spectra.

A further advantage of fast neutrons is that they can induce fission in all uranium and plutonium isotopes, while thermal fission is limited to those with odd mass numbers. This becomes important in dealing with plutonium from highly-irradiated thermal fuels.

Until the late 1970s, future scenarios for fast reactors envisaged a world in which uranium supplies would become increasingly scarce and prices high. Fast reactors would then be needed to make maximum use of limited resources. This scenario has failed to materialise and the inherent difficulties of competing economically with thermal reactors have prevented the development of fast reactors to commercial status. The nature of these difficulties is demonstrated by the liquid metal fast reactor.

The proposed European Fast Reactor (EFR) is an example of a modern design for a liquid metal fast reactor. Illustrated in Fig. 5.7, this has a large pool of liquid sodium contained in an unpressurised tank. The central part of the core, which produces the bulk of the power, measures just 4m in diameter by 1m in height, in spite of the thermal output of 3600 megawatts. A liquid metal is well suited to removing heat from this very small volume, as it has high conductivity and boiling point. The absence of very light nuclei also minimises neutron moderation. The mass of sodium in the pool has a large heat capacity and in an emergency can keep the core adequately cooled by natural convection. The low pressure of the

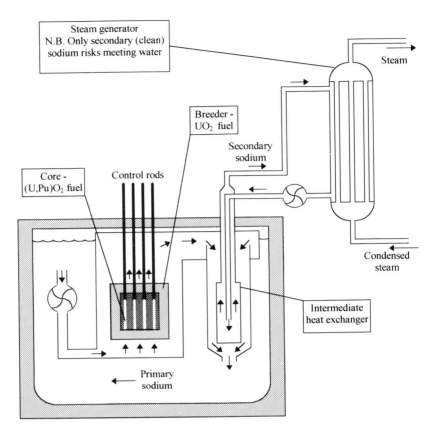

**Fig. 5.7** Fast breeder reactor (FBR)

sodium is a major safety feature, but its chemical affinity for water presents some special problems.

The design provides for a secondary coolant loop to isolate the mildly radioactive sodium flowing through the core from the steam circuit. Sodium is again the coolant, so the liquid metal comes into close proximity with water only in the final heat exchangers, and the possibility of a chemical reaction between sodium and water is confined to a non-radioactive area. Experience shows that such reactions are not unduly violent and the products tend to plug a leak.

The intermediate heat exchange circuit is largely responsible for the capital cost being higher than that of PWRs and BWRs, and is currently the biggest impediment to the introduction of fast reactors.

In the presence of irradiated fuel, decay heat keeps the sodium pool molten. Otherwise electric heaters serve this purpose.

The EFR has two fuelled regions. The central core contains the bulk of the fissile material and generates most of the power. It is surrounded radially and axially by fertile blankets in which escaping neutrons increase the breeding of new plutonium. The blanket utilises uranium *tails* from the enrichment process, typically depleted to about 0.3% in U-235. The fuel in the central core is a mixture of plutonium and depleted uranium oxides clad in stainless steel—the 20 to 30% plutonium provides the main fissile content. The fuel pins are arranged on a hexagonal lattice inside a hexagonal wrapper.

### 5.4.6 Maritime reactors

Since the 1950s nuclear reactors have been used for powering submarines, enabling them to cruise underwater for months at a time without needing to surface. Among surface vessels, nuclear power plants are confined to a few aircraft carriers and icebreakers.

Submarine reactors are usually PWRs fuelled with enriched uranium or plutonium and fundamentally the same as on land, differing substantially only in the small size of the core. For this reason and to reduce or eliminate the need for fuel changes, enrichments are high.

### 5.4.7 Reactors in space

Although most satellites and space probes are powered by solar panels, there are applications where these are not ideal, for instance on missions to the outer planets where the low flux of energy from the Sun reduces their usefulness. Nuclear thermoelectric generators have been used instead. They rely on the conversion to electricity of heat produced from the decay of Pu-238 and not at all on nuclear fission. The heat output is limited, cannot be controlled and declines with the half-life of Pu-238 (88 years). A fission reactor does not suffer from any of these limitations and indeed, the former Soviet Union has considerable experience of space reactors.

Interestingly, a newly-fuelled fission reactor is considerably less of a potential hazard in a launch accident than a nuclear thermo-electric generator. The latter contains a large amount of radioactivity at the time of launch, whereas a fission reactor contains hardly any—it starts operating only in deep space, and until then there are no fission products.

## 5.5 Future perspective

### 5.5.1 Plutonium-fuelled thermal reactors

The principal fuel for thermal reactors in operation today is uranium, but any reactor containing U-238 also generates plutonium which contributes very significantly to the overall energy output, about 30% from a PWR fuel assembly during its lifetime. Fast reactors are specifically designed to use plutonium fuels, but there is increasing interest in using them in thermal reactors, particularly PWRs. By the end of the century about 50 PWRs world-wide are expected to be licensed to operate with up to 30% core loading of plutonium fuel in the form of mixed oxide (MOX) assemblies.

MOX contains up to 8% by weight of plutonium oxide ($PuO_2$) in a carrier of depleted or natural uranium dioxide, reducing requirements for uranium ore and enrichment. At the same time it utilises plutonium recovered from thermal reactor fuel, a serious consideration given the delay in deploying fast reactors. The nuclear characteristics of MOX differ somewhat from those of normal $UO_2$ fuel, but up to 30 or 40% core loading of MOX assemblies can be accommodated in existing PWRs without significant modifications to the overall core characteristics.

Since plutonium is radiologically active, its fabrication into MOX assemblies requires adequate shielding and containment. The consequently enhanced fabrication costs are largely offset by savings on uranium procurement and enrichment.

### 5.5.2 Passive-safety reactor designs

In recent years, reactor vendors have been developing variants of PWRs and BWRs with increased emphasis on passive safety features. Modern designs incorporate very elaborate systems to maintain cooling of the core and containment of the radioactivity following an accident, commonly depending on active devices such as pumps and valves which require external power sources to drive them. Such systems are expensive while complicating safety and licensing assessments. The new designs emphasise passive features to ensure at least equal safety more economically. For example, all penetrations in the pressure vessel are positioned higher above the core. Any break in the pressure circuit is most likely to be in a penetration such as a coolant duct rather than in the pressure vessel itself, and placing them higher increases the volume of water retained for core cooling. Such simple considerations can significantly reduce the number and complexity of safety systems needed to achieve modern safety standards.

One of the more radical passive-safety designs, the Safe Integral Reactor (SIR™), is a PWR with a conventional core but all the primary circuit components (steam generators included) contained within the large pressure vessel, in which the number of penetrations is thereby minimised. Eliminating high-pressure pipework greatly reduces the probability of a leak in the primary circuit. The large mass of water in the pressure vessel also helps to provide passive cooling of the core in the event of a malfunction.

### 5.5.3 Thorium-based reactors

An alternative to the conventional uranium/plutonium fuel cycle is based on thorium, which is much more abundant than uranium, and potentially a very significant energy resource. Th-232, virtually the sole natural isotope, is not itself fissile but like U-238 is fertile. On capturing a neutron it decays by way of protoactinium-233 to U-233 which is fissile:

$$\text{Th-232} \xrightarrow{n} \text{Th-233} \xrightarrow[22.3 \text{ m}]{\beta} \text{Pa-233} \xrightarrow[27.0 \text{ d}]{\beta} \text{U-233}$$

In a thermal neutron flux, U-233 fissions with a higher eta value (Section 5.4.5) than U-235, Pu-239 or Pu-241, yielding on average more than 2.2 neutrons after allowing for absorption in the fuel. There is thus a margin to cover extraneous absorption, and a thermal reactor based on the thorium-uranium cycle can in principle be self-sufficient in fissile material or nearly so. This is one major attractive feature. The other is that the cycle generates only trace amounts of plutonium, and so is arguably less of a proliferation risk than a conventional uranium-plutonium system (although U-233 would be an excellent material for weapons).

Re-fabrication, as in other recycling schemes, is complicated by needing radiological shielding and containment to protect against decay products, particularly thallium-208. The system has proved generally unpromising and has been developed beyond the early stages of feasibility studies only in India, which has little uranium, large thorium reserves and a great concern for strategic independence. It has developed thorium fuel cycles for its CANDU reactors, but demonstrated them only on a very limited scale.

One area where interest in the thorium system has to some extent revived is in the disposal of minor actinides formed in the uranium-plutonium cycle. They might be destroyed by fission in fast reactors, were these commercially deployed; alternatively, neutrons for the purpose might be generated by the impact of a high-energy proton beam on a suitable target. Currently-attainable beam currents are not directly adequate for the purpose, but by amplifying the neutron flux in an assembly of fissile material just short of criticality, useful destruction rates might be achieved. To use uranium or plutonium as the amplifier would generate more of the actinides that are meant to be destroyed; accordingly, the thorium-uranium system has been proposed instead.

## 5.6 References

1 O. Hahn and F. Strassemann. Concerning the observation of alkaline metals resulting from the irradiation of uranium by neutrons (in German). *Die Naturwissenschaften*, **27**, p. 11 (1939)—the name of L. Meitner was omitted for political reasons (R. Sime. *Chemistry in Britain*, 30, pp. 482–4 (1994).

2 L.Meitner and O.R.Frisch. Disintegration of uranium by neutrons: a new type of nuclear reaction. *Nature* **143**, p 239 (1939).

3 A. Weinberg. Assessing the Oklo phenomenon. *Nature*, **266**, p. 206 (17 March 1977).

4 *Enrico Fermi—The man, his theories.* Souvenir Press (1965).

5 Xu Yuanhui and Sun Yuliang. Nuclear heats up in China. *Nuclear Engineering International*, **40 (496)**, pp. 45–6 (November 1995).

# 6 Transport and storage of irradiated fuel

D. E. Haslett
*BNFL, THORP Division, Sellafield, Cumbria*

## 6.1 Introduction

The storage and transport of irradiated nuclear fuel, although not strictly process stages in the cycle, are nevertheless essential links between those stages. As finally discharged from the reactor, fuel represents to the operating utility a liability requiring disposal, whether directly or through a reprocessing plant that can recover materials for re-use. Meanwhile it must be kept under conditions protecting public and work-force from its radioactivity, while preserving the integrity of its structure. The world stock of stored irradiated fuel in 1993 is estimated to have been 95 000 metric tons of heavy metal (MTHM), that is the amount of uranium plus plutonium.[1]

Nuclear power stations were formerly designed with integral storage sufficient only for their short-term requirements plus spare (contingency) capacity in case of reactor emergencies. In the longer term, more extensive facilities are needed. Both may be either wet (ponds) or dry, and although ponds have always been predominant, dry storage schemes do exist and in some countries are becoming the favoured option. The transport system needed to carry fuel between immediate and longer-term stores must be suitably interfaced at both ends.

The industry distinguishes between supervised storage, a temporary measure however prolonged, and disposal considered permanent. Of the two, this chapter is concerned only with the former; permanent disposal is covered in Chapter 10.

## 6.2 At the reactor

### 6.2.1 Requirements

Storage facilities at the reactor site—AR storage according to IAEA terminology—need to handle the intense radioactivity of freshly-discharged fuel, shielding it to protect the operators and dispersing the heat generated by radioactive decay (which might otherwise damage the fuel structure). Operationally, they must provide for receiving fuel from the reactor, locating it identifiably for its sojourn, and eventually exporting it for carriage to its next destination. Records of location must be kept accurately to ensure that fuel is not prematurely selected for onward transit.

Shielding is provided by the store structure, and in ponds by a water filling. The necessary thickness of concrete, typically 1.5 to 2.5 metres, is generally sufficient to withstand hazards to structural integrity such as earth tremors, tornado-borne missiles, or the impact of crashing aircraft, but this has to be confirmed by calculations specific to the site. Confirmation is particularly necessary in any proposal to increase the permitted capacity of an existing store, although the effect of such increases is generally slight and within the design margins.

Although they were originally designed to store fuel for minimal periods before transport to another site, uncertainties and delays in providing the further reprocessing or disposal facilities have prompted extensions to capacity at many reactor sites. Countries already possessing reprocessing sites can use the buffer storage facilities there; others are providing centralised stores.

### 6.2.2 Wet storage

Water at least 2.5 metres deep is both an effective shield and a good heat-transfer medium; moreover it permits fairly easy manipulation and (depending on its purity) clear visibility. Reactor stores are thus predominantly water-filled, and as they allow the fuel to *cool* ready for easier transport, by decay of the shorter-lived radionuclides, they are commonly known as cooling ponds.

Ponds are reinforced concrete structures incorporating sealing membranes and in some instances stainless steel linings to ensure leak-tightness. If any leak should nevertheless occur, means are provided to detect and contain it before it becomes serious. The water must be of high quality and is usually demineralised; a suitable pH must be maintained, corrosion inhibitors may be added (e.g. a trace of NaOH for Magnox fuel, clad in a corrodible alloy), and a biocide was formerly used in external ponds to prevent the growth of algae that would impair visibility. To restrict the deposition of airborne contaminants, such as salt spray in coastal sites or bird droppings anywhere, the whole structure is now enclosed. Effective ventilation is therefore needed to prevent excessive humidity due to the large expanse of water at above-ambient temperature.

Dispersing the heat load is an important consideration, both to restrict corrosion of the fuel and to limit stresses on the fabric of the pond itself. With fuel of relatively low radioactivity, natural convection may suffice; more commonly heat-exchangers are incorporated, whether in the original design or back-fitted to accommodate larger stocks or more highly-irradiated fuel.

Equipment includes cranes and hoists, long grappling tools, access platforms above the pond, storage racks, radiological instrumentation, cameras and lighting. All systems must be integrated with operational procedures or interlocks (for instance, to keep fuel below a sufficient depth of water to maintain biological shielding during movements); document control is designed to ensure that safety is demonstrably maintained and that accurate records are kept. A key requirement is to avoid criticality without wasting space. The geometry of the storage racks, supplemented where necessary by neutron-absorbing materials of construction, ensures safety in this respect, and handling operations are carefully controlled with normally only one movement allowed at a time.

Besides mere storage, the pond may accommodate special operations. Leaking elements detected during reactor operation require examination to determine the implications for storage and transport and for future fuel design. Some operators extend the utilisation of fuel by combining the least-irradiated elements from several sub-assemblies and returning the rebuilt structure to the reactor.

Export facilities need to handle the massive steel flasks used for transport, weighing perhaps 100 tonnes or more and so requiring heavy-duty cranes to move them around the loading area. The fuel is transferred from the storage rack to the internal flask basket designed to accommodate it without risk of criticality. Sometimes, as with the UK Magnox and AGR fuels, the storage container or skip is transferred complete with contents to the empty flask. In any case it is then removed from the pond, washed and checked for contamination before despatch.

Radiation from a cooling pond may be due not only to the fuel contained within it, but particularly in older facilities, also to radioactive contaminants dissolved or suspended in the shielding water. These are chiefly Co-60 from steel cladding or fittings and Cs-137 from the fuel itself where it has been exposed by a failure of the cladding. The water must be continuously purified by filtration and ion exchange, or purged with a clean supply. Ion exchangers are usually of the conventional organic type. Besides radioactive contaminants, the process removes other species that might enhance the corrosion of the fuel itself and of immersed equipment.

### 6.2.3 Dry storage

While ponds are well proven, water is a corrosive medium and can attack some fuel cladding (or if that should fail, the fuel itself) at an appreciable rate. That is particularly true of the magnesium alloy-clad metal fuel of the first-generation British power reactors, fuel which therefore needs to be reprocessed quickly. Moreover the hydrostatic pressure places considerable stresses on the pond structure, and leakage is always a possibility. Both problems are avoided or mitigated by dry storage, which however requires more elaborate arrangements to shield, cool and handle the fuel, with a dry or inert cooling medium, and is less flexible in operation.

In the UK, Nuclear Electric operates dry storage facilities for the Magnox station at Wylfa in Wales, while Scottish Nuclear for a time considered the possibility of similar facilities for its two AGR stations as an alternative to early reprocessing, and is still interested in developing the technology. Wylfa in fact has two stores of different design; in the first, elements in groups of 12 are held within sealed tubes containing carbon dioxide at slightly more than atmospheric pressure (3 psig, about a tenth of a car tyre pressure); the tubes are cooled by natural circulation of air. In the second, elements are held in open-topped cans, themselves held in open skips (metal boxes), cooled by air which in turn is pumped through water-cooled heat exchangers. Fuel freshly discharged from the reactor is kept in the first section for up to 150 days, after which it may be transferred to the second.

Other countries have or propose dry storage, but generally for longer-term purposes.

Handling arrangements in dry stores are generally similar to those in ponds, except that the use of a solid rather than a fluid shielding medium requires more elaborate manipulating systems for fuel containers.

## 6.3 Transport

### 6.3.1 General requirements

Carrying fuel from the reactor site for centralised storage or reprocessing must satisfy stringent safety precautions, particularly as the transport infrastructure, being designed to link centres of population, may make passage through them unavoidable. Nevertheless it must not be unduly expensive. Carriage may be by road, rail or sea (often in combination), within or across national boundaries.

The fuel must demonstrably be securely contained and shielded to ensure integrity under both normal and accident conditions, while dissipating the residual decay heat efficiently enough to prevent excessive internal temperatures.[2] Tests are performed on any design of flask to show that it will withstand:

- in its most vulnerable orientation, a drop from 9 metres on to an unyielding surface, a drop from 1 metre on to a vertical bar of 15 cm diameter;
- an enveloping fire at 800 °C for 30 minutes;
- immersion simulating a head of 200 metres for one hour,
- all while maintaining containment and shielding.

Typical systems consist of heavily-shielded containers known as flasks or casks, generally carried by purpose-built road vehicles, rail wagons or ships. Some flasks are now being designed to serve the dual purpose of transport and interim storage, and will be considered later.

Because flasks must pass through areas open to the public, external radiation must be strictly limited, no more than 2 mSv/hr at the surface of the flask, or 0.1 mSv/hr at a distance of 2 metres from it. The less penetrating alpha and beta radiations are easily stopped by the flask structure; shielding requirements are determined by gamma and neutron emission, the latter particularly from highly-

**Fig. 6.1** Unloading fuel flasks at Barrow dock

irradiated or plutonium-enriched fuels containing significant amounts of higher actinides. Protection against gamma radiation is provided by the flask body, typically of steel some 30 cm thick (or a thinner shell with a 15–20 cm lead lining). Neutrons are more effectively absorbed by lighter elements, such as a water filling or an external layer of boron-loaded silicone rubber.

### 6.3.2 Flask types

The many types of flask in use may, like the stores, be divided into wet or dry categories according to whether the internal heat-transfer medium is water or a gas (inert or air). They must be easy to operate, decontaminate and maintain while meeting the physical constraints of the reactor station and destination store.

### 6.3.3 Wet transport

In the UK, irradiated fuel in any significant quantity is always transported in wet

*Transport and storage of irradiated fuel* 107

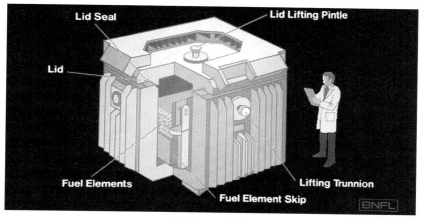

(a) Magnox/AGR

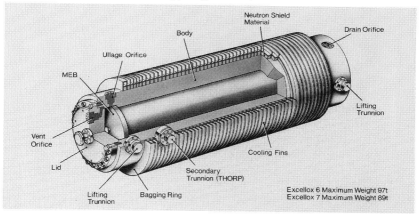

(b) LWR

**Fig. 6.2** Transport flasks, Magnox (AGR) and LWR

flasks. Those for Magnox or AGR fuel are roughly cuboidal, and for LWR fuels cylindrical (Fig. 6.2).

The cuboid type was developed by the former Central Electricity Generating Board and is now used by the Nuclear Electric and Scottish Nuclear companies; also by BFNL for fuel from the Japanese Tokai reactor. Fuel elements or assemblies are held in rectangular skips within the flask cavity, which is filled with water except for an ullage space to allow expansion. The earliest form comprises a thick-walled steel box with a bolted lid sealed by two concentric O-rings. External fins help to dissipate the heat, while valves in the lid and body permit venting, filling with water and establishment of the ullage space. For the more highly-irradiated

AGR fuel, a lead liner was added to improve the shielding. This fuel is however now transported in flasks of the A2 design, with very thick monolithic walls (no liner), an inset lid and integral shock-absorbers; these modifications reduce the payload from twenty to fifteen elements.

A system for LWR fuels, operated by the two commercial transport firms Pacific Nuclear Transport Limited (PNTL) and Nuclear Transport Limited (NTL), is essentially similar except that the assemblies are much longer at 4–5 metres and are carried in long, cylindrical (*Excellox*) flasks, of various detailed designs according to the type of fuel.[3] The main shell is of mild steel, finned circumferentially, with an open-ended cylindrical lead liner in the early versions. At first fuel was carried in an open frame that ensured safe geometry; more recently a sealed multi-element bottle (MEB) has been used to simplify unloading and contain loose radioactive material (such as might leak from damaged elements or be shed as activated corrosion products from the cladding) and so prevent it from contaminating the internals of the flask or larger volumes of water in the receiving pond. Compartments of the MEB are separated by neutron-absorbing materials such as Boral or boronated stainless steel to allow closer packing without risk of criticality.

Depending on the intended service, external neutron shielding may be applied in the form of hydrogen-rich resins or other organic material, particularly where an internal ullage space reduces the shielding provided by water. By acting as thermal insulation and reducing the exposed area of cooling fins, such external shielding reduces the permissible heat loading, but not to any troublesome extent. Mechanically vulnerable fittings such as valves are protected against impact damage by crushable shock-absorbers at each end of the flask, formed from balsa or other wood encased in steel. A side impact would be absorbed by the cooling fins.

GNS in Germany uses the Castor S1 flask, similar to the Excellox 6 design but made in monolithic cast iron without a lead liner. Neutron shielding is by rods of resin in the flask walls.

### 6.3.4 Dry transport

Dry transport flasks have an advantage in carrying capacity over wet designs. Air or more often an inert gas such as nitrogen or helium serves as heat-transfer medium. The absence of water is perceived, realistically or not, to avoid a possible hazard under accident conditions; on the other hand, its neutron-absorbing function must be performed by some other material, and the higher temperature that the fuel can reach presents its own problems.

Flasks of this kind are used in the USA, the former USSR, Germany and France. Typical designs are the TN12 and TN17 used by Pacific Nuclear Transport Limited (PNTL) for transport from Japan to the Cogema reprocessing plant in France, the number denoting the number of assemblies that each is intended to hold—12 PWR or 17 BWR respectively, although the TN12 can alternatively take 32 BWR assemblies. Both are basically similar, but the TN17 is smaller. A cylindrical body made of forged carbon steel has a stainless steel lining. Various orifices permit draining, venting and purging. The lid is bolted to the body and sealed with two concentric O-rings. The base and lid are protected by shock-absorbers. Radial fins of nickel-chromium-plated copper dissipate heat, and around their base is a thick layer of neutron-absorbing resin.

Flasks are loaded under water in the vertical attitude, then drained, dried internally and filled with nitrogen at sub-atmospheric pressure. The fuel is located in an open-topped basket with compartments separated by neutron absorbers. For transport, the flask is of course carried horizontally. Unloading requires either a dry shielded cell or a pond; underwater operation requires the flask to be first filled with water, a delicate operation requiring careful control to prevent excessive thermal shock to the hot fuel or violent chemical reactions with the cladding.

### 6.3.5 Safety and licensing

The safety issues surrounding the transport of irradiated fuel are crucial to national and international acceptance. In fact the record is excellent, with no instance of significant radioactive release in many millions of transport miles, thanks to the high standard required of both flask design and operating procedures.

Most countries have adopted the regulations laid down by the International Atomic Energy Agency (IAEA),[2] and if engaged in reprocessing operations are obliged to observe them; other countries use these regulations as the basis for their own. They apply directly to international transport.

The regulations cover the design, construction, operation and maintenance of transport flasks, specifying standards of leak-tightness, ability to withstand major accidents, external radiation limits etc. Justification by assessment of flask design must be supported by model or full-scale tests, or by validated calculations, on leak rates and resistance to impact or fire (Section 6.3.1). One of the most spectacular of these tests was to crash a 100-ton locomotive at 100 mph into a cuboid flask, which maintained its integrity satisfactorily. The impact proved less severe that the regulatory 9-metre drop test.

Licensing flasks for fuel transport depends on demonstrating to the competent authority that the appropriate national or international regulations on design, construction, operation and maintenance have been met in full.

### 6.3.6 Maintenance

Transport flasks must be kept in first-rate condition as designed. Most routine maintenance operations, such as checking the condition of seals and lifting features, are performed after unloading at the receipt facility. *Basic maintenance* with more detailed checks is carried out in dedicated facilities typically at intervals of three years, with major maintenance and complete refurbishing of LWR flasks every six years. The basic maintenance schedule includes detailed gauging checks on screw threads, tests on the lifting trunnions, and full leak tests. In the major maintenance, flasks are completely stripped down, repainted, and re-assembled; definitive tests on pressure resistance, leak-tightness and trunnion soundness ensure that the flask is restored to mint condition.

### 6.3.7 Future trends

With the tendency towards increased irradiation of fuel and the use of mixed oxide (MOX), new types of flask are now in service. Those denoted as Excellox 6 and 7 have been designed jointly by BNFL and NTL for fuel assemblies respectively 5.0 and 4.5 metres long.[4] Building on the design and operating experience of both companies, they take advantage of improved forging and fabrication techniques by adopting a monolithic body of forged carbon steel, thick enough to provide all the necessary gamma shielding and so dispensing with the lead liner. This makes the flask easier to decontaminate and maintain.

These flasks are complemented by a new type of MEB, with boronated stainless steel between the fuel compartments, fitting neatly into the flask cavity. As the neutron-shielding layer of water is thus eliminated, the function is taken over by an external layer of boronated silicone rubber between the cooling fins.

### 6.3.8 Transport vehicles

Transport by road is usually limited to the distance between reactor and railhead or port. Massive low-loaders are used with flasks firmly fastened down, and speeds are restricted.

Rail is the normal mode of transport by land. Flasks are designed to fit limits on weight and loading gauge for the British network; as these are the most restrictive in the world, for historic reasons, the flasks can therefore be carried on any other standard or broad-gauge system. The wagons are designed for the purpose, open low-loaders used by BNFL and a model with a sliding canopy by NTL European for longer journeys.

Ships used for marine transport are again designed for the purpose and for exclusive use. They have specially strengthened hulls, watertight subdivisions, and extensive duplication in power and navigation systems.

## 6.4 Interim storage

Whether the discharged fuel is to be reprocessed, directly disposed or held pending a decision, interim storage (Away From Reactor or AFR storage in IAEA terminology) is a necessary requirement. With over four hundred commercial power reactors in operation and limited reprocessing facilities, capacities must be substantial; for instance, the THORP receipt and storage facility at Sellafield can accommodate 3000 MTHM of LWR fuel, with a further 2300 MTHM elsewhere on

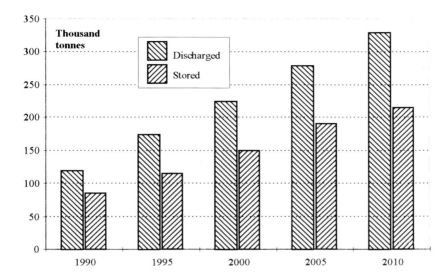

**Fig. 6.3** Irradiated fuel, reprocessed or in storage: expected arisings

the site, besides capacity for Magnox and AGR fuels. Amounts involved worldwide are estimated to be more than doubled (to 215 000 MTHM) by the year 2010, the earliest date by which permanent repositories are expected to open (Fig. 6.3). Many reactor operators have already maximised their existing local storage capacity and without additional local or centralised provision would be unable to continue functioning.

Reprocessing plants in turn need interim storage to allow further cooling and so reduce problems of radioactivity in the chemical processes; also as a buffer to permit fuel arisings to be grouped into economic process campaigns, or to be held while plant is unavailable, e.g. during maintenance, refurbishment or replacement of equipment.

As with storage at reactors, ponds have been the norm, remain predominant and are well proven, but dry stores have their attractions and are increasingly favoured in some countries. For 1992 the total AFR storage capacity was 49500 MTHM, of which 94% was wet, with 53800 MTHM in 1993. Either type must have receipt facilities to accommodate the transport flasks carrying fuel from reactor sites, and in turn to permit onward transfer to reprocessing plant or permanent repository.

### 6.4.1 Wet stores

Fuels from water-cooled reactors have been stored in ponds for over twenty years in some instances, and despite increased burn-up should remain in good condition for twice as long; such periods are proposed for the Swedish CLAB underground ponds.

Modern facilities take advantage of the more compact modes of storage developed in many countries, for instance after consolidation and with neutron absorbers such as boronated stainless steel between units to permit closer packing without risk of criticality.

Fuel can be unloaded from the transport flasks in a pond deep enough to accommodate the height of the vertical flask plus the length of the fuel assemblies while still maintaining the necessary depth of water above them, or in a dry inlet cell with suitable shielding. Both methods are used by BNFL at Sellafield for LWR and for Magnox or AGR fuels respectively. Details of movement to storage racks depend on how the fuel is shipped, as individual assemblies in an open frame or packed into MEBs.

All combinations of wet or dry receipt with wet or dry unloading are adopted in one or other of the British and French installations. Equipment is essentially much the same as described for reactor stores, though with increasing reliance on mechanical handling and closed-circuit television to supplement or replace manual operations and direct observation.

Fuel unloaded from the transport flask is transferred in its rack to the storage pond where it may be held for tens of years. Underwater trolleys or rack-handling machines may be used, and handling machines place the racks in predetermined positions on a grid location system.

The principles of construction and monitoring are similar to those of reactor stores. Drainage systems direct any leakage to sumps for collection and monitoring. Capacity is determined by floor area and the scope for multiple stacking (itself determined by the length of the fuel and depth of the pond—thus AGR and Magnox fuels are stacked, but LWR fuel is not), provisions against criticality, and the depth of water. Within these physical limitations, further restraints may be imposed by the heat loading of fuel more highly irradiated than originally envisaged, unless the cooling system is up-rated to match. Ponds may be modular with provision for future extension.

Again as in reactor stores, maintaining the purity of the water is a prime consideration. Receipt and handling operations tend to spread radioactive contamination, chiefly in the form of corrosion products dislodged from the cladding. It is restricted in some facilities such as THORP by containerisation in order to protect operators from the consequent radiation doses, reduce the volumes of water needing treatment, and so limit the arisings of waste ion-exchange material. A further advantage is to reduce the number of handling operations required. The containers must however be loaded and water in the sub-pond used for the purpose is exposed to the naked fuel elements; it therefore requires purification or replacement at an appropriate rate to keep doses to operators within acceptable limits. THORP has a once-through purge system, with treatment of the water before discharge, while

Cogema recirculates water through ion exchangers and filters. To economise on water usage, the purge from a relatively clean facility may serve as the feed to an older, more heavily-contaminated pond.

### 6.4.2 Dry stores

In the past dry storage, although used for fuel of all kinds, has been limited to small quantities. With the current requirements for extended interim capacity, many countries are considering such methods for interim bulk storage or final disposal, in preference to wet methods on the economic grounds (real or perceived) of:

- low maintenance and high reliability;
- low arisings of secondary wastes;
- easier decommissioning;
- reduced corrosion over long periods;
- low radiation doses to operators and public;
- readier compatibility with passive cooling concepts.

The principle may be applied to vaults, dry wells, concrete or metal casks, all of which have been demonstrated and used to varying degrees.

### 6.4.3 Vaults

The concept of vaults has been used in the UK, France and the USA. The vault itself is a reinforced concrete structure, shielding and housing the spent fuel in open or sealed canisters. Although essentially simple, it needs separate receipt, unloading and perhaps containerising facilities. Closed containers, if used, are of metal, with leak-tight welded lids and a filling of inert gas. Decay heat is removed by natural convection, conduction or if necessary by forced circulation of air, although this detracts from simplicity and safety assurance while increasing maintenance requirements.

Two forms of such a system, in use at the Wylfa power station in Wales since the early 1970s and 1980s respectively, were described in Section 6.2.3. In France, dry vaults of various designs are used for heavy water and fast breeder reactor fuels, while a design for LWR fuel also exists. The HWR store is designed for a 50-year life, with cooling by natural convection in helium; FBR fuel is first cooled in cans under water, then transferred to dry vertical channels encased in concrete and cooled by a forced air draught. The LWR design also supposes a preliminary 5-year cooling period, followed by containerisation of the fuel and natural convection cooling. All these facilities provide for eventual retrieval of the fuel.

### 6.4.4 Dry wells

This concept has been demonstrated in the USA and Japan, encompassing both sub-surface and deep wells. A lined cavity can accommodate one or more assemblies, shielded by the surrounding earth and a plug above, and normally cooled by conduction to the surroundings.

The Japanese system was a demonstration facility accommodating about 15 tonnes of uranium metal fuel from a research reactor, held in sealed helium-filled canisters themselves placed in wells formed in a massive concrete structure. In this instance cooling was by air circulated between the canister and wall.

Demonstrations in the USA included both shallow and deep wells for PWR fuel, but led to no commercial-scale applications and the system is no longer favoured.

### 6.4.5 Concrete casks

This system has been extensively demonstrated in Canada and the USA and is being introduced on a commercial scale. Canadian CANDU fuel is held in a sealed canister with a welded lid and filled with helium. The concrete cask, providing shielding and secondary containment, is kept on hard standing in the open air with natural cooling. Following a demonstration programme, this is now a routine practice at the Whiteshell Nuclear Research Establishment; it is also used at the Gentilly and Douglas Point stations for both BWR and pressurised heavy-water reactor fuels, with helium or air fillings.

A similar arrangement is being considered as one form of modular retrievable storage (MRS) in the USA. The Nutech Horizontal Modular Storage System (NUHOMS) has LWR fuel assemblies in a basket in a stainless steel container, a horizontal concrete storage silo and a transfer cask. The canister is loaded at the storage pond, lidded, drained, vacuum-dried, filled with helium and sealed. By way of the transfer cask it is moved in a horizontal attitude to the interim storage facility and loaded into the silo by a ram incorporated in the cask; a shield door is then closed and sealed. Canisters holding 7 or 24 PWR assemblies have been used in demonstrations, and several US reactor operators have selected the system for interim storage at their own sites.

### 6.4.6 Metal casks

Since the requirements of shielding, containment, heat dissipation and high integrity under accident conditions are much the same for interim storage as for transport, it is a natural development to use the same or similar dry casks for both purposes. Sometimes it can eliminate a complex handling operation, otherwise restrictions at the reactor site may require a smaller design for transport. The concept is finding favour in several countries, notably the USA and Germany.

In Germany a dual-purpose transport and storage design has been based on the monolithic square-shaped Castor transport flask, and is in use at both reactor sites and the Gorleben central facility, where a storage hall can accommodate 420 casks or 1500 MTHM of fuel. This requires one or two years' preliminary cooling in the reactor pond, where it is loaded into a basket and so into the cask. This is then dried internally and filled with helium to minimise corrosion. The lid is double-sealed. The largest current form of cask holds 22 PWR assemblies, although larger versions are being developed along with designs for higher burn-up and MOX fuels.

At Gorleben the receipt area is divided from the storage hall by a shield door. The shielding provided by the massive casks requires supplementing only by a relatively light-weight concrete building. Pressure in the casks is monitored to demonstrate continued integrity of the package. The facility is licensed for LWR fuel for up to 40 years, and similar central stores are being built elsewhere in Germany.

In the USA, dry metal casks are regarded as the most advanced of the interim storage systems and have been in use since the mid 1980s. They can each accommodate 21–33 PWR or 45–70 BWR assemblies. Four different types have been successfully tested, and a reactor site storage facility licensed to hold up to 81 Castor V21 casks in the open air is already operating.

Developments are centred around increasing capacity by reducing the spacing between assemblies. This depends on taking credit for the reduced reactivity, instead of assuming the fuel to be at its original enrichment level, and so in turn on gaining acceptance that the burn-up can be reliably determined. Equipment for the purpose of fuel assay is in use or being developed for both reactor stations and reprocessing plants, for instance THORP at Sellafield where the fuel parameters are recorded before entry.

Dry storage in metal casks is a technical proven option and commercially available, but expensive for significant quantities of fuel owing to the capital cost of the flasks. Its future thus depends on economic comparison with the alternatives for the amount of fuel involved..

## 6.5 References

1  (a) C. Bristol *et al. Experience of the design, operation and technology of spent fuel storage facilities.* To be published. (b) C. Bristol, *IAEA International Symposium on Spent Fuel Storage—Safety, Engineering and Environmental Aspects.* Vienna (1994).

2  *Regulations for the safe transport of radioactive material.* IAEA Safety Series No. 6 (1985, amended 1990) plus supporting documents Nos. 37 and 80.

3  R. Gowing. Transport of irradiated LWR fuel elements in water-filled flasks. *Nuclear Energy* (1990) p. 359.

4  R. Gowing. Designing new flasks for high burnup spent fuel. *Spent Fuel Management and Transport 1994*, p. 10. (Supplement to *Nuclear Engineering International*, December 1994.)

# 7 Reprocessing irradiated fuel

I. S. Denniss and A. P. Jeapes
*BNFL, UK Group, Sellafield, Cumbria*

## 7.1 Introduction

All thermal power reactors operate by converting the energy released by fission of actinides such as U-235 and Pu-239 into electrical energy. Fertile isotopes such as U-238 (the bulk of the fuel) are partly transmuted into additional fissile material some of which also undergoes fission. On balance the initial fissile content of the fuel is reduced during irradiation, and accumulated fission and transmutation products change its physical and neutronic properties. Some products such as neodymium absorb neutrons and *poison* the fuel whilst others alter its structure, and the gases (e.g. xenon and krypton) may pressurise it. Eventually further irradiation of the fuel becomes uneconomic and it is removed from the reactor.

It now contains residual fissile material as well as the vast bulk of the original fertile isotopes, with a relatively small amount of fission products and higher actinides that are either useless or not currently worth recovering. After irradiation to say 40 GWd(th)/t these waste elements would amount to about 3% of the initial heavy metal content. Reprocessing can recover the residual uranium and plutonium for re-use and condition the wastes into suitable forms for disposal. To this end the most recent reprocessing flowsheets are designed to maximise the recovery of uranium and plutonium and to avoid adding any indestructible species to the waste streams.

Irradiated uranium was first processed in the 1940s to recover plutonium for the military programme which required a very pure product. Current plants recover material for recycle in power reactors, the uranium is generally enriched before recycle, and this also requires a pure product. The degree of purification achieved by a process is denoted by decontamination factors (DFs), defined as the ratio of a stated impurity to desired component in the feed divided by the equivalent ratio in the product; values of $10^6$–$10^8$ are typically needed and demand a very specific and efficient separation process. Few if any techniques are selective enough to give such separations in a single stage, so multistage processes are needed and must be operable under remote control in a shielded plant since the fuel is highly radioactive. Precipitation techniques were initially applied to some military fuel,[1]

**Table 7.1** Planned or operating civil reprocessing plants

| Location (and designation) | Reactor, fuel type | Capacity | Start-up date |
|---|---|---|---|
| UK, Sellafield (Magnox) | Thermal, metal | 5 t/day | 1964 |
| UK, Sellafield (THORP) | Thermal, oxide | 5 t/day | 1964 |
| UK, Dounreay | Fast, oxide | 0.025–0.05 t/day | 1960, 1980* |
| France, Marcoule (UP1) | Thermal, metal | 1–2 t/day | 1960 |
| France, La Hague (UP3) | Thermal, oxide | 4 t/day | 1990 |
| France, La Hague (UP2–800) | Thermal, oxide | 4 t/day | 1994 |
| Japan, Tokai Mura | Thermal, oxide | 0.7 t/day | 1977 |
| Japan, Rokkashamura | Thermal, oxide | 4 t/day | 1988? |
| Russia, Mayak (RT1) (besides submarine and HEU lines) | Thermal, oxide | 400 t/day | 1976 |

*With intermediate reconstruction

but were soon displaced by solvent extraction which is better suited to continuous, large scale, remote operation.

This exploits the unusual propensity of uranium and plutonium to form neutral nitrate complexes that can be extracted by nucleophilic (electron donor) solvents. The first was diethyl ether, soon replaced by less flammable species such as methyl *iso*-butyl ketone (hexone) or dibutyl carbitol (Butex).[1] These in turn were replaced by the more efficient extractant tri-*n*-butyl phosphate (TBP) dissolved in a largely inert hydrocarbon diluent, a combination which has now been used for over 40 years in a variety of flowsheets known generically as the Purex process.[2]

At present commercial scale civilian reprocessing plants are planned or operating in Britain, France, Japan and Russia (Table 7.1). The second French plant at La Hague, UP2, has been refurbished and expanded as UP2-800.

Most fuel comes from thermal reactors. All types are clad in protective metal to prevent attack by reactor coolant or leakage of fission products. For oxide the cladding is stainless steel (AGR) or the zirconium alloy Zircaloy (LWR); for metal it is a magnesium alloy known as Magnox—a term also applied to the fuel as a whole and by extension to the reactors that use it—or Sicral in France. The CANDU type, natural uranium oxide in Zircaloy, is not normally reprocessed; others are natural uranium as metal or enriched uranium as oxide. The different characteristics of the basic types influence the flowsheets and techniques used to reprocess them.

Purex plants are divided into two major areas, the head end where the fuel is dissolved in nitric acid and solvent extraction where uranium and plutonium are separated and purified. Two other important aspects of the plant described elsewhere are the ventilation system and the equipment and procedures necessary to comply with international safeguards criteria.

## 7.2 Head-end processes

Head-end processes usually comprise:
- removing or penetrating the fuel cladding;

## 118  Reprocessing irradiated fuel

**Fig. 7.1**  Construction work, THORP

- dissolving the fuel in nitric acid;
- treating the dissolver off-gases;
- monitoring and disposing of the cladding;
- clarifying the dissolver solution from insoluble fission products;
- conditioning the solution to facilitate solvent extraction.

The first two stages depend strongly upon the nature of the fuel, in particular whether it is essentially uranium metal or oxide. Metal, clad in softer magnesium alloy, is in the form of single massive rods. Oxide, with its lower thermal conductivity and higher operating temperature, is packed as small ceramic pellets into narrow cylindrical tubes of stainless steel or Zircaloy, about 10 mm in diameter and from 1 to 5 m long, combined into assemblies of 36 to 289 *pins* with suitable spacing grids or wire wrapping. Typical designs are illustrated in Fig. 3.1.

**Fig. 7.2** Fuel receipt pond

## 7.2.1 Decladding, dissolution and cladding disposal

The functions of decladding, dissolution and cladding disposal are treated together since the first two may in practice be a single operation and the third is closely associated.

## Magnox

Cladding is removed from uranium metal in one of two ways. In the French UP1 plant, it is selectively dissolved in dilute nitric acid and subsequently vitrified with the high-level waste. In the British plant at Sellafield, Magnox is stripped off the uranium rod as fragments of alloy by ramming the fuel element through a set of slitter wheels. The fragments were for many years stored under water, where occasionally rapid corrosion aroused concern about the resulting hydrogen, but now are encapsulated in cement within steel drums. The uranium rods are then dissolved continuously in nitric acid. Because of the low fissile content, a large pot-shaped dissolver can be used without risk of criticality.

## Oxide

The complex construction of oxide fuel elements, the relatively fragile nature of the long, thin, ceramic-packed pins and the hardness of the cladding prevent mechanical stripping. Selective dissolution has been practised in the Eurochemic plant at Mol (Belgium), but no common dissolvent has been found for both Zircaloy and stainless steels. The growth of external oxide films during irradiation passivates the cladding and can lead to excessively long dissolution times.

The more customary and successful approach is to cut the fuel pins into short pieces so that the contents can be leached out with nitric acid. Pins may be separated and fed to a small shear, as in the pilot WAK facility in Germany. Alternatively, as with LWR fuels in the UK, France and previously the USA, higher throughputs can be achieved by bulk shearing of the entire fuel assembly (in the French UP2–800 and UP3 plants, the heavy end appendages are first cropped and disposed separately). British AGR fuels are first dismantled[3] and the pins from three assemblies packed into a stainless steel can; four of these are thereafter treated like a single LWR assembly. Parts of the shear exposed to the fuel are subject to severe stresses and become heavily contaminated, so that they need remote maintenance to replace items such as bearings and blades.[4]

The cut pieces of fuel pins, with fragments of spacing grids and any other remaining fittings, are allowed to fall into nitric acid in the dissolver. After leaching, the *hulls* (chopped cladding) and other solid debris are rinsed and monitored for undissolved fuel content. Dissolvers, whether batch or continuous, are designed to ensure that the fuel is thoroughly leached from the cladding; losses are mainly due to gas locks preventing acid from reaching into closed ends, and limited to some 0.1%. Amounts are checked in THORP (UK) by neutron interrogation. Together with other cladding fragments, hulls are then tipped into drums and encapsulated in cement for disposal.

The dissolver cycle is complicated by the need to provide for removing the empty cladding hulls etc. In THORP,[5] the chopped fuel is dropped into a perforated basket which allows access by acid but retains the cladding, apart from small fragments that may escape and have to be removed subsequently from the base

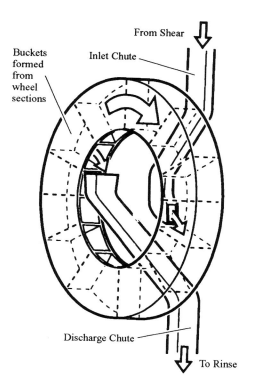

**Fig. 7.3** French continuous dissolver (schematic)

of the dissolver. Batch dissolution is completed by a period of several hours at near-reflux temperature after all the fuel load has been added. The rate of dissolution is not easily predicted, and to prevent any possible criticality due to large if transient inventories of undissolved fissile material, the nitric acid charge can be dosed with gadolinium nitrate as a soluble neutron poison.

Inserting and removing baskets, neutron interrogation and tipping the hulls into waste containers require much remote mechanical handling, and a continuous dissolution process would offer considerable advantages. The French UP3 and UP2–800 plants incorporate a continuous dissolver in the form of a wheel,[6] with the rim divided by radial partitions into twelve sectors closed at the periphery but open inwards (Fig. 7.3). The wheel rotates in a vertical plane with the lower part immersed in hot nitric acid in a slab tank, dimensioned to avoid risks of criticality. Chopped fuel is loaded by a chute into a compartment entering the acid, the rate of revolution allows essentially complete dissolution before it emerges, and as the compartment rises to the top of the wheel the hulls fall into a discharge chute for rinsing and disposal.

## 7.2.2 Off-gas treatment

Dissolution, whether of uranium metal or dioxide, involves oxidation to the hexavalent state, and oxides of nitrogen ($NO_x$) are generated by reactions such as:

$$3UO_2 + 8HNO_3 \rightarrow 3UO_2(NO_3)_2 + 2NO + 4H_2O \quad (7.1)$$

$$UO_2 + 4 HNO_3 \rightarrow UO_2(NO_3)_2 + 2NO_2 + 2H_2O \quad (7.2)$$

Volatile fission products are also released—principally tritium, carbon-14 monoxide and dioxide, krypton-85, iodine-129 and to some extent ruthenium-106 which can form a volatile tetroxide. The off-gases therefore need to be treated to ensure that their impact on the environment is acceptable.

In the UK Magnox plant, gaseous effluents are led through a reflux condenser and two packed columns countercurrently to the incoming acid feed, together with injected oxygen to convert nitric oxide to dioxide which is efficiently absorbed as regenerated nitric acid. This avoids the wastage and pollution or disposal problem that would arise if the nitrogen oxides were vented or scrubbed out to be discarded.

A large flow of air is needed in an oxide plant to ensure that dust formed in the shearing process is drawn into the dissolver rather than released into the shear cave. This dilutes the dissolver off-gases and increases the flow-rate, complicating the subsequent treatment.

In THORP,[5] the off-gases are first scrubbed with dilute nitric acid which removes the remaining fuel dust and most of the nitrogen oxides; the liquor is added to the dissolver product after sparging with air to strip out absorbed iodine. A second scrub with caustic soda removes all but traces of the nitrogen oxides from the off-gas, but more importantly iodine-129 (mostly elemental) and carbon-14 dioxide. Finally, filtration removes any remaining dust and ruthenium-106 that may have been reduced to particulate form.

Tritium, occurring predominantly as tritiated water, is largely removed from the off-gas by the aqueous scrubs. Krypton-85 is evolved in large amounts in terms of radioactivity, along with a greater mass of inactive xenon, and owing to its inert nature is not absorbed by aqueous scrubs but has a low radiological impact on discharge. Processes to remove at least 90% of it do exist, although the gas cannot readily be immobilised and the reduction in the slight risk that it poses is not considered commensurate with the expense (see Chapter 13); nevertheless the position is frequently reviewed.

## 7.2.3 Clarification of dissolver liquor

The wide range of fission products formed in the reactor includes elements of the platinum and adjacent groups which remain wholly or partly metallic. At the low irradiations and operating temperatures of metal fuel, they remain dispersed and subsequently pass into solution without evident difficulty. In the much more highly irradiated oxide,[7] some of them (notably ruthenium, rhodium, technetium, palladium and molybdenum) tend to form an immiscible metallic phase as particles of insoluble alloy; moreover, any fuel grains particularly rich in plutonium may fail to dissolve completely, and a mixed zirconium-molybdenum oxide can subsequently precipitate from the dissolver solution. While particles are generally small (less than five microns) and amount only to some 0.5% of the fuel mass, the substantial fraction of total ruthenium-106 in the insoluble fission products (IFPs) constitutes a source of high radiation and heat release.

A small proportion of the fuel cladding is also reduced to fine particles by the shear. The various solids, suspended in the dissolver, could settle in regions of low fluid velocity and accumulate at later stages of the process, where the heat and radiation might be troublesome or the plutonium content constitute a criticality hazard. To prevent this, the dissolver product liquor is clarified in both Britain[8] and France by a semi-continuous centrifuge from which the solids are flushed out by high-pressure acid sprays every 24 hours or so and encapsulated for disposal.

### 7.2.4 Conditioning of dissolver liquor

The dissolver product liquor needs conditioning before solvent extraction to ensure that:

- the bulk of fission-product iodine is desorbed into the off-gas system (once in the solvent it is liable to contaminate too many streams with intractable species);
- plutonium is predominantly in the tetravalent state, rather than the less extractable hexavalent form to which it tends at high temperatures in nitric acid;
- acidity and uranium concentration are correct.

The first two aims may best be achieved, as in THORP, while the liquor is still in the dissolver. During the period of near reflux, the dissolver is sparged with air which leaves less than 2% of the initial iodine content in solution. Afterwards the liquor is cooled to 60 °C and sparged with nitrogen dioxide which rapidly reduces Pu(VI) to Pu(IV).

The composition is adjusted after clarification by addition of 12M nitric acid and, if necessary, recycled uranyl nitrate. Accountancy starts here, since this is the first point at which the amount of fissile material entering the process can be measured rather than estimated from fuel history.

## 7.3 Solvent extraction

The solvent extraction aspects of the Purex process have been studied intensively for 40 years, and there have been several recent and comprehensive reviews of the literature[1,2] with extensive lists of references. This section therefore covers only the general principles involved and brief descriptions of the major reprocessing flowsheets.

### 7.3.1 General principles of process and equipment

The separation of a mixture by means of solvent extraction depends on the selective transfer of components between two immiscible liquids. In the Purex process the major components transferred are uranium and plutonium, and the two phases are aqueous nitric acid and TBP/diluent solutions. The transfer occurs on mixing the two liquids which then have to be separated. Products are required to be very pure and recovery virtually complete, so the processes must be readily reversible and very efficient.

The overall efficiency of any solvent extraction process is optimised by carrying it out in a multi-stage countercurrent contactor. Those used in reprocessing plants have been reviewed and described briefly by Batey,[1] and are also described in standard chemical engineering textbooks.[9] The fundamental requirement is to mix the two phases well enough for efficient mass transfer between them but not so finely that the dispersed phase fails eventually to settle out or is carried in the wrong direction (*flooding*). There are several secondary design criteria:

- to accommodate the required throughput;
- to achieve the required flowsheet performance (e.g. acceptable losses of U and Pu to waste streams);
- to avoid any risk of a nuclear criticality;
- to meet any process limitations (e.g. to limit residence times to restrict degradation of the solvent).

There are two general classes of contactor; *stage-wise* and *differential*. Both of these can be divided into many different types but current reprocessing plants use only two; mixer-settlers (stagewise) or pulsed columns (differential).[10] A mixer settler consists of a series of discrete stages in which the phases are mixed to reach equilibrium, then separated and passed in opposite directions to the adjacent units. Because they operate effectively as discrete equilibration stages, mixer-settlers are easy to design, scale-up, control and operate, but the inter-stage separation occupies a large volume. Criticality control is therefore difficult and residence times are long, increasing solvent degradation, and so they are generally used in plants processing fuel with a low plutonium content (U metal fuel as in the Sellafield Magnox reprocessing plant) or for the separated uranium stream.

A simplified example (in practice there could be four times as many units) is shown in Fig. 7.4, which represents a compound extract and scrub contactor (Section 7.3.5). An aqueous feed containing two components, '**a**' (extractable) and '**b**' (much less extractable), is fed into the centre of a 5-stage cascade of mixers ($M_n$) and settlers ($S_n$). Fresh solvent is fed into stage 5 and passes through the

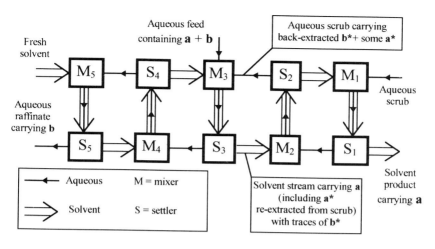

**Fig. 7.4** Mixer-settler (simplified) in extract and scrub mode

cascade in the opposite direction to the aqueous (counter current), so that as the aqueous solution moves through stages 3, 4 and 5 it is extracted sequentially with solvent containing progressively less 'a'. This maximises the amount of 'a' which can be extracted by a given volume of solvent. The solvent leaving stage leaving stage 3 contains all of component 'a' plus traces of 'b' (b*), which are then removed in stages 1 and 2 by contact with fresh 'scrub' acid. The small amount of 'a' (a*) which is also removed by the scrub is returned to stages 3–5 where it is re-extracted.

A pulsed column is shown in Fig. 7.5. The two phases pass through it in opposite directions, one dispersed in the other and separated only at the top or bottom. In the body of the column the two phases never reach equilibrium and so continuous mass transfer occurs, equivalent overall to several theoretical stages. The factors controlling the performance of pulsed columns interact strongly and in a complex fashion,[11] so they are more difficult to design and scale up than mixer-settlers, but they have a smaller ratio of volume to surface area and so criticality control is easier. Columns up to 6 inch (15 cm) diameter are ever-safe, and above that neutron poisons can be incorporated into the design. The columns are packed with perforated plates or an alternating series of rings and discs, and an alternating hydraulic pulse is applied to maintain dispersion and enhance mass transfer.

### 7.3.2 Extraction by TBP

TBP is a neutral extractant which requires a negative counter-ion to allow extraction of metallic cations. Certain valence states of the actinides complex with

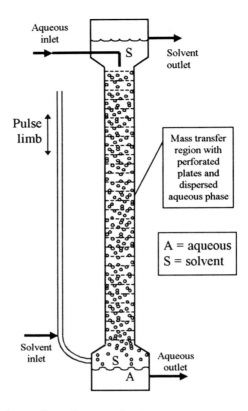

**Fig. 7.5**  Pulsed column with continuous solvent

many anions (e.g. sulphate and chloride) but nitrate complexes have proved to be uniquely suitable for the extraction of uranium and plutonium:

$$UO_2^{2+} + 2NO_3^- + 2TBP \rightarrow UO_2(NO_3)_2(TBP)_2 \quad (7.3)$$

$$Pu^{4+} + 4NO_3^- + 2TBP \rightarrow Pu(NO_3)_4(TBP)_2 \quad (7.4)$$

The extraction can be reversed by controlling the nitrate or nitric acid concentration in the aqueous phase, and the extractive power of TBP is reduced by dissolution in a diluent, normally a paraffinic hydrocarbon. The diluent is also needed to decrease the density and viscosity of TBP and so aid phase separation in the solvent extraction contactors. The effect of nitric acid concentration on the distribution ratio (D—the ratio of total concentration in the solvent to that in the aqueous phase) of hexavalent uranium is illustrated in Fig. 7.6(a). As acidity increases, so at first does D owing to the increasing nitrate level (equation 7.3). Above about

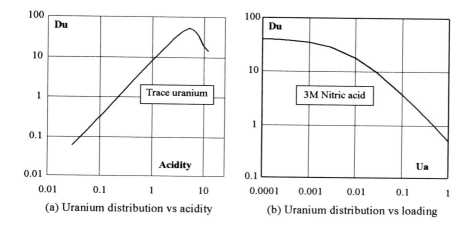

**Fig. 7.6** Extraction of uranium by 30% TBP at 20 °C: dependence on (a) acidity and (b) loading of solvent (molar units).

6M, however, acid competes for TBP according to equation 7.5, and the distribution ratio falls.

$$\text{TBP} + \text{HNO}_3 \rightarrow \text{TBP.HNO}_3 \tag{7.5}$$

This competition is eliminated if nitric acid is replaced by inextractable nitrate salts (e.g. sodium nitrate) so that higher D values are obtained for the metals; however, the salts then contaminate the effluent streams from the plant and complicate waste treatment, whereas nitric acid can be recovered from the effluents by distillation and is recycled to minimise the waste arisings.

The extraction involves stoicheiometric compounds and so the solvent has a limited and calculable capacity to extract U and Pu before reaching saturation. As the loading of the solvent increases, so reducing the amount of TBP which is free to extract other metals ($[\text{TBP}]_{\text{free}}$), all D values decrease. This effect is illustrated in Fig. 7.6(b) and is used to 'squeeze out' impurities by maintaining a high loading of uranium in the solvent.

The behaviour of any species in a contactor depends on the Extraction Factor (E) which is the product of the distribution coefficient (D) and ratio of solvent to aqueous flow rates (E = D.S/A). Broadly, if E is greater than one the species extracts, if it is less than one the species does not extract or is backwashed.

Under certain conditions of high metal loading the solvent can separate into two immiscible phases, virtually pure diluent and an extractant-rich phase containing a high metal concentration. This is termed 'third phase' formation and occurs most readily in the presence of tetravalent actinides.[12] The phenomenon

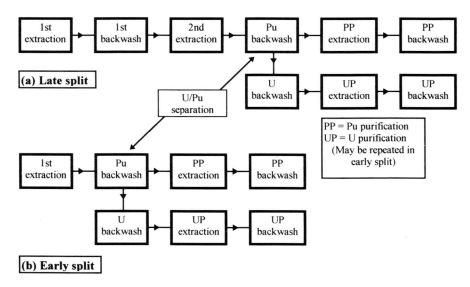

**Fig. 7.7** Late (a) and early (b) split separation

is well understood and flowsheets are designed to prevent its occurrence since it would seriously impede or altogether prevent the proper operation of contacting equipment.

TBP is chemically stable but undergoes some thermal and radiolytic degradation in the process to form dibutyl phosphate (HDBP) and mono-butyl phosphate ($H_2MBP$).[13] HDBP forms strong extractable complexes with plutonium and zirconium in particular, whilst $H_2MBP$ tends to form precipitates. Both by-products complicate the flowsheet chemistry,[2b] although the effects are controlled by washing the solvent with alkali to remove the acidic degradation products.

The TBP content of the solvent controls the D values and the amount required to process a given amount of fuel, but an increase raises the density and viscosity. 30% is the normal maximum as used in THORP (plutonium areas) and UP3. 20% is used in the Magnox plant and the uranium purification section of THORP. The lower the proportion the higher the flow-rate required for a given throughput, but there have been suggestions that higher fractional solvent loadings and better decontamination from fission products may be achieved at lower percentages of TBP, with improved hydraulic qualities and consequently higher volume throughputs compensating for the reduced absolute loading.[14]

### 7.3.3 Behaviour of individual elements

Current reprocessing plants separate U and Pu from each other and from all other components of the fuel, both tasks aided by the specificity of TBP. The fuel is

**Table 7.2** Extractabilities of various elements in fuel

| Strong extraction ($D \gg 0.5$) | Weak extraction ($D \ll 0.5$) |
|---|---|
| $U(VI)^+ > U(IV)$ | |
| $Np(VI) > Np(IV)$ | $Np(V)$ |
| $Pu(VI) < Pu(IV)$ | $Pu(III)$ |
| $TcO_4^-/U(VI)$, $/Pu(IV)$ or $/Zr$ | All other species |

dissolved under oxidising conditions and most elements are present in solution in a single, high oxidation state; few are extracted strongly under normal Purex conditions.

Uranium emerges solely as the extractable hexavalent uranyl ion which has been described above. The less extractable tetravalent uranous ion can be used elsewhere in the flowsheet (§ 7.3.5). Plutonium can exist as the extractable Pu(VI) and Pu(IV) and inextractable Pu(III). Dissolution produces $PuO_2^{2+}$ which is less extractable than $UO_2^{2+}$, and to ensure virtually complete recovery, the feed is first conditioned with nitrous acid to generate the more extractable Pu(IV). The inextractability of Pu(III) is commonly used to separate U and Pu (Section 7.3.5).

Neptunium can exist as the extractable Np(VI) and Np(IV), and the inextractable Np(V). These are interconvertible and $NpO_2^+$ disproportionates at high acidity.[2c] Dissolution produces a mixture of the three states with an appreciable amount of Np(V), but in conventional plants this is oxidised in the extraction contactor and so most of the neptunium is extracted.

The transplutonium actinides are almost exclusively trivalent and virtually inextractable.

Most of the fission products are scarcely extractable,[15] although ruthenium forms a mixture of interconverting complexes with a wide range of extractabilities,[2b,16] while technetium and zirconium can co-extract as complexes. Technetium exists as the pertechnetate $TcO_4^-$ ion which by itself is extracted weakly, but its complexes with U(VI), Pu(IV) and zirconium are much more extractable.[2b,17] The extractability of zirconium is also enhanced by the presence of technetium. The extractabilities of the various elements are summarised in Table 7.2.

### 7.3.4 Design criteria for plant and equipment

The plant is designed to achieve the desired throughput (usually 4 or 5 tonnes heavy metal per day) although the solvent extraction area normally has a wider operating range (e.g. 3–5 tonnes/day). Buffer storage is provided to decouple different areas of the plant, such as the head end which may be a discontinuous process from solvent extraction which is always continuous.

Extensive instrumentation is provided to give several types of information:

- accountancy—to safeguard the nuclear material present;
- process control—to show that flowsheet conditions are achieved;

130  *Reprocessing irradiated fuel*

- safety—to protect against and warn of potential maloperations.

The process is designed to be inherently safe and where maloperations could occur, however improbably, extensive safety measures are built into the plant to prevent or detect the initiating event, detect its effects and shut down (*trip*) the plant automatically if necessary. The formal procedures to ensure safety are described in Chapter 13.

### 7.3.5 Separation techniques

All reprocessing plants have the same sequence of process steps: co-extraction to separate U and Pu from the maximum amount of impurities, separation of U and Pu and finally purification of the two products. Nevertheless the detailed design varies from plant to plant. Early plants such as that for Magnox fuel at Sellafield and the Japanese Tokai Mura use a *late split* flowsheet where the uranium and plutonium are separated in the second solvent extraction cycle, but more recent plants such as THORP separate the two in the first cycle (*early split*—see Fig. 7.7).

The reasons for the difference are related to the process chemistry. The Magnox plant uses ferrous sulphamate as reductant to separate plutonium from uranium. Although effective, it leaves residues in the waste stream preventing it from being concentrated much for storage and hampering disposal; to permit it to be discharged as low-level waste after treatment, it is therefore preceded by a full cycle and a second extraction to separate radioactive impurities.

In THORP and UP3, in order to reduce the amount of salt-laden wastes, the reductant is U(IV) which follows the uranium product stream after serving its purpose, leaving no inextractable residues. The particular incentive for the late split thus disappears, and the advantages of an early split (reducing both the volume of waste and the amount of equipment which has to be designed to avoid criticality in the large volume of mixed uranium and plutonium) become paramount. The total amount of equipment may be reduced if no further cycles of product purification are needed.

The processes occurring in each contactor, or group of contactors, are described below.

**Co-extraction in the first contactor**

The feeds to the extraction contactor are designed to achieve at least 75% loading of TBP by uranium to maximise the rejection of other species by *squeezing* them out. The loaded solvent is then contacted with a low volume flow of fresh acid in a scrub contactor (*strip* in American terminology) to remove further contaminants. Decontamination factors (DFs) of at least 1,000 are achieved for all species apart from neptunium and technetium. The feed to commercial reprocessing plants contains a mixture of all three valence states of neptunium and the inextractable

$NpO_2^+$ is oxidised in the contactor so that the bulk follows the uranium and plutonium.[2c]

Technetium co-extracts completely with the quantity of zirconium present in modern oxide fuels, and is then carried through the scrub column as a uranium complex.[17] It can interfere with the chemical reduction step used to separate U and Pu (below) and in the UP3 plant at Cap La Hague a further scrub process removes technetium from the solvent stream.[18,19]

**Separation of uranium and plutonium**

The separation depends on a selective reduction in the extraction factor of plutonium well below unity, so that it is transferred into the aqueous phase whilst uranium remains in the solvent. Most flowsheets achieve this by reducing plutonium to the almost inextractable Pu(III), although it is also possible to use reagents (e.g. $H_2SO_4$) which complex preferentially with Pu(IV) to reduce its distribution ratio; a comprehensive review is given by Miles.[2a]

Pu(III) is unstable in nitric acid solution through oxidation catalysed by nitrous acid, but can be stabilised by compounds such as hydrazine or sulphamic acid which destroy the catalyst:

$$2HNO_2 + N_2H_4 \rightarrow N_2 + N_2O + 3H_2O \tag{7.6}$$

$$HNO_2 + NH_2SO_3H \rightarrow H_2SO_4 + N_2 + H_2O \tag{7.7}$$

The process therefore needs a combination of reductant and stabiliser. Early flowsheets such as in BNFL's Magnox plant use ferrous sulphamate which is very efficient and easy to prepare, but generates ferric and sulphate ions and so complicates management of the wastes from the plutonium purification cycle. Modern plants such as THORP and UP3 use uranous (U(IV)) nitrate stabilised by hydrazine as a *salt-free* reagent:[2a,20]

$$U(IV) + 2Pu(IV) \rightarrow U(VI) + 2Pu(III) \tag{7.8}$$

The U(IV) is generated by hydrogenation or electrolytic processes and eventually follows the main uranium stream.

In catalysing the oxidation of hydrazine by nitric acid, technetium interferes with the use of U(IV)/hydrazine.[21] It is therefore removed by a special contactor in UP3 (above); in THORP it is made tolerable by operating the separation contactor in solvent-continuous mode to reduce the time available for reaction in the aqueous phase.[22] The reaction reduces technetium to inextractable Tc(IV) which follows the Pu stream in both plants.

Neptunium is reduced to extractable Np(IV) by both Fe(II) and U(IV), and therefore follows the uranium stream. Other impurities partition between the product streams.

### Backwashing
Uranium and plutonium are backwashed into one stream in the first cycle of a late split process, otherwise separately. The two have different responses to acidity. Uranium is backwashed at very low acidity (less than 0.1M) and at high temperature since its extraction is exothermic; plutonium(IV) is hydrolysed and precipitated at low acidity, high temperature and high concentrations, so is generally backwashed at higher acidity (typically 0.2M or more) and often in the presence of a reductant to generate Pu(III) which does not hydrolyse so readily.

### Uranium purification (UP) cycle
Current plants are required to produce a product suitable for re-enrichment, i.e. virtually free from radioactive contaminants capable of forming volatile fluorides. The major objective is to remove alpha emitters (Pu and Np) but the fission products Ru-106 and Tc-99 are also significant.

Neptunium is removed by a combination of selective oxidation and reduction to produce inextractable Np(V); the reductant also removes plutonium as Pu(III).[23] Np(IV) is oxidised by heating, which also improves the DF for ruthenium by promoting the equilibration between species. The behaviour of technetium is controlled by its co-extraction with uranium; however, since the feed to this cycle contains a relatively low concentration of uranium (60–80 g/l), the solvent to aqueous ratio in the extract contactor is around one, too low for efficient extraction of Tc.

### Plutonium purification (PP) cycle
This cycle is designed for decontamination from fission products and to concentrate the plutonium. As the minor component of all thermal reactor fuel its daily throughput is small (e.g. up to 50 kg in THORP), allowing the solvent extraction equipment to be made ever-safe against criticality.

The plutonium in the feed to this area is inextractable, either complexed or trivalent. It is therefore conditioned by oxidation if necessary (typically with $NO_x$ gases), with additional nitric acid to increase extractability, then extracted into solvent, scrubbed, and backwashed with a salt-free reductant such as hydroxylamine to minimise losses to the solvent wash cycle.

### Diluent washing
The aqueous streams leaving solvent extraction contactors contain small amounts of dissolved TBP and entrained TBP/diluent. These can complicate subsequent waste treatment and particularly the handling of streams containing high concentrations of plutonium (e.g. plutonium product). Solvent can be removed by steam-stripping from streams which are to be evaporated, but to avoid the risk of irreversible hydrolysis, plutonium streams are always contacted with a small volume of pure diluent. The diluent wash contactors are designed for efficient contact between two streams at very different flowrates (*S/A* in feeds typically 1:20).

**Table 7.3** Comparison of typical metal and oxide commercial reactor fuels

|  | Metal | Oxide (PWR) |
|---|---|---|
| Initial enrichment in U-235, % | 0.7 | 4 |
| Burn-up (irradiation), GWd/t | 5 | 40 |
| Composition in kg/t irradiated fuel: |  |  |
| Uranium | 994 | 955 |
| Plutonium | 3 | 10 |
| Neptunium | 0.023 | 0.460 |
| Total fission products | 3 | 35 |

**Solvent wash cycles**

These are designed to remove the acidic degradation products of TBP (HDBP and $H_2MBP$) by means of alkaline reagents, typically sodium carbonate and hydroxide. Since the solvent also contains traces of uranium and plutonium which could be precipitated by caustic alkali and so cause a criticality hazard, while the carbonate complexes are soluble, the solvent is always washed with carbonate before hydroxide. To give long residence times allowing slow reactions to occur and efficient contacting at high flow ratios of solvent to aqueous, special designs of mixer settler which recycle the aqueous phase are used.

### 7.3.6 Description of specific flowsheets

This section highlights the differences between various operating plants (Table 7.1).

**Sellafield Magnox**

This plant was commissioned in 1964 and has processed over 35 000 tonnes of metal fuel from the Magnox reactors. It has a late split flowsheet (Fig. 7.5) and uses the *salt* reagent ferrous sulphamate to separate uranium and plutonium. The plant uses mixer-settlers which are acceptable for the low concentration of plutonium in the fuel and has four separate solvent extraction cycles.

**Sellafield THORP**

This plant began active operation in 1994 and processes oxide fuel irradiated up to 40 GWd/t (reference fuel). An early split flowsheet using pulsed columns for the main plutonium-bearing streams has been adopted to suit the characteristics of the fuel (Table 7.3).

It also uses the salt-free reagents U(IV)/ hydrazine to separate U and Pu and hydroxylamine to backwash Pu and control neptunium. The major innovation compared with other reprocessing plants is the development of flowsheets which achieve the product specifications with single cycles of product purification; the plant therefore has only three solvent extraction cycles (one common and one each for the U and Pu products).[23]

### La Hague UP3 and UP2–800

The UP3 plant at Cap de la Hague, France, was commissioned in 1990 and the UP2–800 plant in 1994 at the same site, with similar early split flowsheets and using salt-free reagents. Both use pulsed columns for the plutonium-bearing stream, apart from a mixer-settler battery to separate U and Pu in UP3. There are two cycles of purification for each product and so five solvent extraction cycles overall. The plutonium purification cycles concentrate the product by recycling a large fraction, so as to allow *finishing* without any evaporation. Many of the aqueous raffinate streams are subjected to diluent washes (Section 7.3.5) and the used diluent is added to the main solvent inventory. A fraction of this is then distilled under vacuum to recover pure diluent and TBP, adding to the clean-up of the solvent achieved by the conventional solvent wash processes.[19]

### 7.3.7 Future developments in reprocessing

The main concerns for the future are to reduce the cost and environmental impact of reprocessing, and two approaches are being pursued. The first aims to recover uranium and plutonium and prepare all other radioactive components for disposal as waste, as at present but more cheaply and efficiently; the current generation of reprocessing plants such as THORP and UP3 have been examined to identify possible areas for improvement. The second involves the partitioning of the components of the waste stream to allow more selective methods of waste disposal, including transmutation of the most toxic components (partitioning and transmutation, P&T as it is sometimes abbreviated).[24] This has aroused considerable interest, particularly in Japan and France, though not entirely on technical or economic grounds; Britain remains sceptical (Chapter 14).

#### Head end

The major aspects of current head end processes which could be optimised are the size of the equipment and facilities, the amount of mechanical handling required and the amount of waste generated (whether in terms of volume, tonnage or radioactivity).

Current shear machines are large, and need maintenance and replacement of the cutting blade. This generates active waste, requires large active maintenance cells and contributes largely to occupational radiation doses. The leached hulls also form a bulky waste. Future development might:

- reduce the size of shear machinery and the associated cell;
- reduce or eliminate the need for maintenance;
- reduce the volume of solid waste produced;
- develop processes which combine exposing the fuel with reducing the volume of cladding wastes.

The size of the shearing machinery can be greatly reduced if the fuel is first dismantled to single pins, since the force needed to shear them is then much smaller, and complications due to grid structures etc. will be avoided. The size of any associated maintenance facility will also be much reduced, as will the proportions of fuel dust and cladding fines.

Alternative methods of dissolving the fuel from single pins without shearing are also being investigated to reduce or eliminate active maintenance. These include piercing the cladding mechanically, electrochemically or thermally (e.g. by lasers) to allow penetration by acid. Japanese workers have also proposed a rolling operation which expands the cladding and shatters the fuel which can then be tipped out into a dissolvent solution.[25]

A disadvantage of such methods is the need for preliminary dismantling. Other investigations therefore address non-mechanical methods of processing whole assemblies, such as lasers, plasma torches or inductive heating to penetrate the cladding.[26] Alternatively the cladding might be embrittled by hydrogen, nitrogen or iodine and then crushed.[27]

Leached hulls from a conventional head end occupy some 0.38 m$^3$/t uranium, about ten times the actual metal volume. Options for melting or mechanically compacting them are therefore being pursued.[28]

### Solvent extraction

Current plants employ three or five cycles of pulsed columns, requiring expensively large, tall, shielded buildings and generating many effluent streams. Future developments may therefore be expected to aim at reducing the number of cycles and the size, particularly the height, of individual contactors (process intensification).

Process intensification implies the adoption of equipment with increased throughput per unit volume. Smaller, lower plant should be possible with units such as centrifugal contactors, i.e. mixer-settlers with accelerated phase separation.[1] Attempts at process simplification include the German Impurex system which aims to achieve adequate purification in a single cycle by operating at very high saturation.[29] Less radical approaches include optimisation of scrub temperatures and further development of salt-free separation techniques.[30]

## 7.4 Conclusion

Reprocessing is a technically successful operation but not the only option for managing discharged fuel nor always the one preferred. Costs, although a relatively small part of the total for the generating cycle, are still substantial, while the recovery of fissile material and the attendant generation of wastes are commonly perceived as politically sensitive, unnecessary for the present and posing allegedly unsolved problems for the future. Power-generating utilities, themselves subject to short-term market forces, certainly see attractions in at least postponing any decision to have their fuel reprocessed so long as storage for a generation or two appears a cheaper and less controversial option.

One principal aim of any reprocessing operation is therefore to cut costs; another to become more acceptable by further reducing waste arisings. Such considerations are likely to dominate this sector of the industry for the foreseeable future.

## 7.5 References

1 W. Batey. Nuclear fuel reprocessing using solvent extraction. In *Science and practice of liquid-liquid extraction*. Oxford Engineering Science Series 27, Vol. 2 (ed. J. D. Thornton) pp. 102–93, (1992).

2 W. W. Schulz, L. Burger and J. D. Navratil (eds) *Science and technology of tributyl phosphate*, Vol. III. CRC Press (1990).

a J. H. Miles, Chapter 1, Part II.

b D. J. Pruett, Chapter 2.

c V. A. Drake, Chapter 3.

3 R. H. Allardice. Nuclear fuel processing in the UK. in *German Atomic Forum*, Nuremburg (16 May 1990).

4 R. W. Asquith, P. I. Hudson and M. Astill. The oxide fuel shearing system for THORP. *International Conference on Nuclear Fuel Reprocessing and Waste Management RECOD 87*, Paris pp. 533–40 (1987).

5 D. Milne and P. G. Grant. The development and design of the plant for spent oxide fuel dissolution and treatment at Sellafield. In *Symposium on Energy Production Processes*, I. Chem. E, London (14 April 1988).

6 C. Bernard, J. P. Moulin, P. Lederman, P. Pradel and M. Viala. Advanced Purex process for the new French reprocessing plants. *Global '93*, Seattle, pp. 57–62 (1993).

7 H. Kleykamp. Post-irradiation examinations and compositions of the residues from nitric acid dissolution experiments of high burn-up LWR fuel. *J. Nuclear Materials*, 171, pp. 181–8 (1990).

8 P. I. Hudson. Developing technology to reprocess oxide fuel. Nucl. Engineering International Special Publication *British Reprocessing* (October 1990).

9 R. E. Treybal. *Mass-transfer operations*, 3rd edn. McGraw-Hill (1980).

10 C. Godfrey, C. Hanson and M. J. Slater. The design of solvent extraction equipment. *Chemistry and Industry* 1977, pp. 713–8.

11 G. Sage and F. W. Woodfield. Pulse column variables . Chemical Engineering Progress, 1954, pp. 396–402.

12 P. D. Wilson and J. K. Smith. Third-phase formation by plutonium(IV) in 30% TBP with various diluents. In *Extraction '87, I. Chem. E. Symposium Series no. 103* pp. 67–74 (1987).

13 L. L. Burger and E. D. McLauchlan. *Industrial and Engineering Chemistry* 50, p. 153 (1958).

14 A. L. Mills and E. Lillyman. Review of metallic fast-reactor fuel reprocessing at Dounreay. In *Proceedings of the International Solvent Extraction Conference (ISEC 74)*, SCI (1974) p. 1499.

15 J. V. Holder. A review of solvent extraction process chemistry of fission products. *Radiochimica Acta*, 25 pp. 171–80 (1978).

16 A. A. Siczek and M. J. Steindler. The chemistry of ruthenium and zirconium in the Purex solvent extraction process. *Atomic Energy Review*, 16 p. 575 (1978),

17 J. Garraway and P. D. Wilson. Coextraction of pertechnetate and zirconium by tri-*n*-butyl phosphate. *Journal of the Less-Common Metals*, 106 pp. 183–92 (1985).

18 B. Boullis, J-P. Gué and C. Bernard. Process for the separation of technetium present in an organic solvent with zirconium and at least one other metal such as uranium or plutonium, utilisable notably for the processing of nuclear fuels. *European Patent* 0270453 (1988).

19 W. Fournier *et al.* Purex process improvements for the UP3 spent fuel reprocessing plant at La Hague, France. *Solvent Extraction 1990*, Elsevier pp. 747–752 (1992).

20 I. S. Denniss and C. Phillips. The development of a three-cycle flowsheet to reprocess oxide nuclear fuel. In *Solvent Extraction '90* (ed. T. Sekine) Part A, Elsevier (1992), pp. 549–54.

21 J. Garraway and P. D. Wilson. The technetium-catalysed oxidation of hydrazine by nitric acid. *Journal of the Less-Common Metals*, 97 pp. 191–203 (1984).

22 I. S. Denniss and C. Phillips. The development of a flowsheet to separate uranium and plutonium present in irradiated oxide fuel. In *Extraction '90*, Intitution of Chemical Engineers, Symposium Series 119 pp. 187–97 (1990).

23 I. S. Denniss and C. Phillips. The development of a three-cycle flowsheet to reprocess oxide nuclear fuel. in *Solvent Extraction '90* (ed. T. Sekine) Part A, Elsevier. p. 549 (1992).

24 L. H. Baetslé. Burning of actinides: a complementary waste management option? *IAEA Bulletin*, 3/1992, p. 32.

25 Y. Takashima *et al.* New concept on the head end of Purex process towards the next century. In *International Conference on Nuclear Fuel Reprocessing and Waste Management, RECOD '87*, Paris, pp. 569–72. (1987).

26 M. Nakatauka. Potential application of Zircaloy chemical embrittlement to volume reduction of spent-fuel cladding. *Nuclear Technology* 103, pp. 426–33 (1993).

27 B. Roger. Procédé pour découper un tube métallique par induction at application à la séparation des combustibles irradiés de leur gaine métallique. French Patent FR2 667 533A1 (1992).

28 P. Lederman, P. Miquel and B. Boullis. Optimisation of active liquid and solid waste management at La Hague plant. In *4th International Conference, Nuclear Fuel Reprocessing and Waste Management RECOD '94*, Vol. II, London (1994).

29 H. Schmieder *et al.* IMPUREX: a concept for an advanced Purex process. Radiochimica Acta, 48 p. 181 (1989).

30 S. Tachimoro and W. Kubo. Low-temperature partitioning of plutonium with high U(VI) loading for fuel reprocessing. *Journal of Nuclear Science and Technology*, 26 pp. 852–60 (1989).

# 8 Recycling uranium and plutonium

J. P. Patterson and P. Parkes
*BNFL, UK Group, Sellafield, Cumbria*

## 8.1 Introduction

The uranium and plutonium recovered by reprocessing irradiated fuel from a modern power reactor represent some 97% of the original uranium content, with no major use but as a source of energy. Fabricating reactor fuel from them is a way of husbanding resources that are currently plentiful but cannot be assumed to remain so; it has a long and established history which emphasises the maturity of both the technology and the concept.

Naturally occurring uranium contains 0.71% by weight of the isotope U-235 which is fissionable by thermal neutrons, as described in Chapters 1 and 5. For many current designs of reactor with relatively small cores, a greater proportion is needed to maintain the fission chain against losses of neutrons by non-fissioning absorption or leakage. During fuel manufacture the U-235 level is therefore deliberately enhanced to 2–4%.

This proportion inevitably falls during irradiation to a residual level often in the range 0.4–1.1%, indeed often in the upper part of the range. It is then still higher than natural, so to that extent there is an advantage (somewhat diminished by the adverse effects of new isotopes and trace fission products) in using recycled rather than fresh uranium for new fuel.

These adverse effects are substantial but not overwhelming. Uranium-232, typically in the range 0.03–3.5 ppb, decays with a 70-year half-life to yield strongly gamma-emitting daughter products such as thallium-208, from which operators must be protected in the fuel manufacturing plants. Uranium-234 and trace transuranic impurities similarly give rise to undesirable levels of activity. Uranium-236 is not radiologically troublesome, but even in small quantities (typically 0.2–0.6% by weight) is a very effective absorber of neutrons and acts as a poison or negative enrichment which must be compensated. Generally speaking, with current fuel burn-ups, the amount of enrichment needed is still somewhat less for reprocessed than for natural uranium, and since the enrichment process is an expensive step in the fabrication of fuel, there is considerable scope for cost savings by recycling.

Plutonium generated in the reactor from U-238 also consists of several isotopes, primarily Pu-239 but also the others from Pu-238 to Pu-242, formed by further nuclear processes to an extent that increases with irradiation. Plutonium-238 has a high specific heat output as a result of its radioactive decay and can be used as a power source in locations remote from external supplies, for instance spacecraft; in storing or recycling large quantities, this property is a nuisance requiring specific measures to prevent over-heating.

Pu-239 and Pu-241 (roughly 60–70% of the total in a typical present-day LWR) are fissionable by thermal neutrons, and towards the end of the fuel's life contribute some 30% of the total heat output, so that in a sense the use of plutonium as a fuel has been recognised from the earliest days of nuclear power. Plutonium-238 and Pu-240 are fertile, easily capturing neutrons to yield fissile material. The small quantities of Pu-242 present act as a neutron poison in much the same way as U-236.

In the presence of more energetic neutrons, such as may be found in a fast breeder reactor, all the isotopes of plutonium are fissionable. A much greater energy yield can then be realised and indeed plutonium fuels are particularly attractive in such fast reactors. Even in thermal fuel, plutonium can serve instead of enrichment in U-235 to provide the main fissile content in mixed-oxide (MOX) fuel, at the same time utilising the otherwise almost useless enrichment tails of U-238 as a fertile matrix.[1]

## 8.2 Rationale for recycling

Technical feasibility is not sufficient by itself to justify recycling uranium and plutonium; there must be some advantage in doing so. Reasons may in fact be a combination of economic, strategic and political.

### 8.2.1 Cost

Even allowing for the poisoning effect of the inevitable U-236, less enrichment work is generally required to achieve a given reactivity from recycled than from new uranium. A significant cost saving is therefore possible, although offset by the extra costs in conversion, enrichment and fuel fabrication due to the provision for shielding, remote operation and decommissioning.

Any remaining advantage will clearly depend very much on the unit costs for the relevant processes, which in turn vary around the world, with order size and contractual commitments etc. However, assuming free-issue reprocessed uranium at a typical LWR composition of 0.85% U-235 and 0.3% U-236, converted into reactor fuel for roughly 40 GWd/t irradiation, then savings are possible when new uranium ore costs more than about \$11/lb (as $U_3O_8$). Whilst the spot market may drop below this figure, the majority of ore is obtained on secure long-term contracts at prices significantly higher, and fuel made from reprocessed uranium becomes attractive.

Recycled plutonium and the depleted tails from uranium enrichment may also be considered free-issue products of completed processes, capable of being made up into fuel without the costs of fresh uranium, isotopic enrichment and the associated conversion steps. The savings are all the greater with fuels designed for higher burn-ups, with a correspondingly high fissile content. The capital and operating costs of manufacturing plants are however currently about four to five times higher than for uranium fuels, offsetting the potential benefits to some extent. They can still offer a saving of up to about 30%, depending on the assumed price of new uranium ore, although as present-day LWRs can accommodate only about 30% plutonium fuel, the total saving is limited to about 10% over the whole core. Benefits can be reinforced if the fuel manufacturing plant is adjacent to the reprocessing facility to avoid the cost of inter-site plutonium transport, public concern about it and coordination problems between the production and use of plutonium.

### 8.2.2 Resource conservation

Manufacturing fuel from reprocessed material, thus increasing the electricity generation from a given amount of raw uranium, reduces the requirement for new ore, which is a finite natural resource. The uranium and plutonium from all reprocessed oxide fuel could each save about 4% of projected world requirements up to the year 2000, if both were used in thermal reactors.[2] The total of 8% could be increased to about 15% if all fuel were reprocessed for recycling. Even this worthwhile figure is dwarfed by the potential saving if recovered plutonium is used in a fast breeder reactor, where transmutation or direct fission of all uranium and plutonium isotopes permits a 60-fold increase in the energy yield from the original uranium.

A further factor that is important to some countries is the reduced dependance on imported uranium ore due to recycling. This emphasises the position of uranium and plutonium as strategic energy resources.

### 8.2.3 Improved proliferation resistance

*Proliferation* is the unlawful diversion of fissile material for military purposes. Plutonium in particular is potentially attractive to a terrorist organisation and considerable precautions are taken to ensure its safe storage in accordance with national and international regulations and agreements. However, if the plutonium is manufactured into fuel, then any would-be terrorist would not only have to divert some heavy (500 kg each), large (4 m long) fuel assemblies, but would have to dismantle them and chemically separate the plutonium from the uranium. This is an unlikely scenario. Furthermore, what safer place to store plutonium than in a reactor, as fuel?

### 8.2.4 Use of ex-military material

The reduction in nuclear arsenals provides an opportunity to burn the plutonium

**Table 8.1** Waste volume arisings

| Waste Arising: | | Waste Volume, m³/GW(e)y | |
|---|---|---|---|
| | | Recycle | Direct Disposal |
| Uranium mining & milling | | 17 500 | 20 000 |
| Reactor operations | ILW | 50 | 50 |
| | LLW | 200 | 200 |
| Irradiated Fuel | HLW | 4.5 | 40 |
| Management | ILW | 34 | 6 |
| | LLW | 134 | ? |

HLW = High Level (heat generating) Wastes
ILW = Intermediate Level Wastes
LLW = Low Level Wastes

and highly-enriched uranium from redundant warheads as reactor fuel. About 230 tonnes of plutonium is estimated to be available from contemporary warheads in the US and the CIS (former USSR), plus that already in store from warheads progressively retired since the mid-1960s. This would be sufficient for over 4000 tonnes of LWR fuel, worth about £1M ($1.5M) per tonne at present.

The transfer to civil use under strict safeguards would strengthen the non-proliferation regime and demonstrate that the superpowers were genuinely disarming and not merely reorganising their nuclear arsenals. For the utilities, not only are there cost savings in using free-issue fissile plutonium, but also a sense of moral satisfaction in contributing to a safer world.

### 8.2.5 Reduced environmental impact[3]

Compared with direct fuel disposal, recycling uranium and plutonium also brings significant environmental benefits, which may be illustrated as follows for a modern PWR of 1 GW(e) capacity, discharging fuel at a burn-up of 42 GWd/t. Plutonium is assumed to be recycled twice, and the second-generation MOX fuel directly disposed after discharge.

#### Reduced waste volume arisings

Table 8.1 compares the volumes of waste in different categories arising with and without recycling. With recycling, mining wastes are 12.5% lower, reflecting the reduced demand for uranium. Elsewhere the long-lived, heat-generating high-level waste is reduced nine-fold, which in view of its nature, may be judged to make up amply for the increase in intermediate and low level waste.

#### Reduced radioactive content

Direct disposal will add approximately 250 kg of plutonium per year to the fuel awaiting burial. The assumed recycling regime at equilibrium would halve this amount. Since the radioactivity in the medium to long term is dominated by plutonium isotopes and daughters, this too would be reduced, by over 30% after 10,000 years.

**Reduced toxicity**

Whilst the simplest method of comparing direct disposal and recycling scenarios is in activity content, it gives insufficient weight to the longer-lived isotopes of relatively low activity, such as the actinides. A more realistic method involves the concept of radiological toxic potential.[3] In simple terms, this can be thought of as 'the volume of water needed to dilute a radionuclide so that it would be safe to drink.' Integration of the toxic potential over time demonstrates that after the short-lived isotopes (e.g. Cm-244) have decayed, the medium and long-term toxicity is again dominated by plutonium and its daughters, notably Am-241 and Np-237. By 100,000 years, the integrated toxic potential for a recycle scenario is some 35% lower than for direct disposal. This reflects the consumption of plutonium and conseqent formation of fission products with a shorter half-life.

An exercise has been started at the University of Surrey (UK) to assess the overall impact of recycling on various aspects of the environment including resource depletion. The methods are still being developed, data are not yet adequate for a rigorous treatment, and so far only uranium has been considered whereas a greater benefit is expected from plutonium particularly in fast reactors, but in due course the process should yield a more completely explicit comparison with the once-through concept than has hitherto been possible.

## 8.3 Brief history of recycling

Large quantities of civil reactor fuel have been reprocessed around the world, almost 40 000 tonnes of irradiated uranium metal in the UK alone. Of this, over 15 000 tonnes has been isotopically enriched and made into new fuel, including over 1600 tonnes of oxide fuel for the UK's AGR programme—about 60% of all AGR fuel production. Reprocessing of oxide fuels from commercial power reactors has been taking place since the mid-1970s, and new plants presently building up to full capacity are rapidly increasing the rate. Some of the uranium product has been re-enriched and fabricated into fuel for reactors in Germany, France, Belgium and the UK. As yet the total is only 60 tonnes, sufficient nevertheless to demonstrate amply the validity of the principle. Further use is planned in France, Japan and the UK.

Although plutonium is always formed within the reactor, the first commercial fuel in which it was deliberately included at manufacture was loaded into a Belgian reactor in 1963. Demonstration in the prototype British AGR followed in the same year, and in Germany and the United States also in the 1960s. Other countries, notably France, Switzerland, Japan and the CIS, have more recently used plutonium fuels in thermal reactors. The indigenous Japanese Advanced Thermal Reactor, which can run on a 100% MOX core, has

**Table 8.2** Production of thermal reactor plutonium fuels

| Organisation | Country | Capacity (t/y) | Status | Cumulative production (tonnes to end 1993) |
| --- | --- | --- | --- | --- |
| Belgonucleaire | Belgium | 35 | operational | [230] |
| Belgonucleaire | Belgium | 40 | planned | – |
| BNFL/UKAEA | UK | – | shut down | 3 |
| BNFL/UKAEA | UK | 8 | operational | 0 |
| BNFL/UKAEA | UK | 120 | design | – |
| COGEMA/CEA | France | 15 | operational | [30] |
| COGEMA | France | 120 | construction | – |
| PNC | Japan | 10 | operational | 0.5 |
| Consortium | Japan | 100 | design | – |
| Siemens | Germany | 25 | shut down | 153 |

been successfully operated in prototype since 1979, using over 100 tonnes of fuel.

In total, over 400 tonnes of plutonium-based fuel has been fabricated for LWRs alone. Current usage is limited by the manufacturing capacity; new plants are therefore being planned, designed or constructed to meet the demand by utilities as they take title to increasing quantities of plutonium from the new oxide fuel reprocessing plants at La Hague (France), Sellafield (UK), Rokkashomura (Japan) and Krasnoyarsk (Russia). Table 8.2 lists the status of current and future facilties to produce plutonium fuel for thermal reactors.

Plutonium-enriched fuels have also been used very successfully in fast breeder reactors since the mid-1960s. Over 150 tonnes of fuel has been made in a similar manner to LWR fuel, but typically with four or five times the plutonium content. Table 8.3 shows the fuel produced by the current suppliers.

## 8.4 Recycling uranium

Process stages to convert reprocessed uranium into new oxide fuel are mostly similar to those for virgin material (Chapters 3 and 4) and will not be repeated except where necessary. The prime requirement is a reactive oxide suitable for fluorination (as a preliminary to re-enrichment) or a free-flowing stock for pelleting.

**Table 8.3** Production of fast reactor plutonium fuels

| Organisation | Country | Status | Cumulative production (tonnes to end 1993) |
| --- | --- | --- | --- |
| Belgonucleaire | Belgium | operational | 6 |
| BNFL/UKAEA | UK | shut down | 18 |
| CEA/COGEMA | France | operational | [115] |
| PNC | Japan | operational | [5] |
| Siemens | Germany | shut down | 6 |
| ? | Russia | ? | ? |

### 8.4.1 Uranium finishing

Uranium emerges from the purification cycles of a reprocessing plant as a solution of uranyl nitrate in dilute nitric acid. It may be converted to oxide either by thermal denitration or by precipitation followed by thermal decomposition.

**Thermal denitration**
The most straightforward way of converting uranyl nitrate from a Purex process to oxide is by direct thermal degradation, i.e.

$$UO_2(NO_3)_2 \cdot 6H_2O \rightarrow UO_3 + 2NO_2 + \tfrac{1}{2}O_2 + 6H_2O \qquad (8.1)$$

Denitration under a reducing atomosphere, such as pure hydrogen or argon/4% hydrogen, yields the dioxide directly,

$$UO_2(NO_3)_2 \cdot 6H_2O + 4H_2 \rightarrow UO_2 + 2NO + 10H_2O \qquad (8.2)$$

but in practice reduction always forms a second stage in the process.

There are eight known phases of $UO_3$ and of these the alpha, beta and amorphous phases reduce to dioxide of *ceramic grade* which is suitable for fluorination or fuel fabrication. Unfortunately, the gamma phase produced by direct denitration does not yield an oxide with good powder properties, so further proceessing is required before fluorination (or fabrication) as discussed in Chapter 3.

In early production processes, large batches of uranyl nitrate were directly denitrated in stirred pots. The concentrated solutions would solidify on drying, but the partially hydrated salts melted before decomposition, yielding a glassy product adhering to surfaces:

$$\text{solution} \xrightarrow{\text{evaporation}} \text{solid hydrate} \xrightarrow{\text{melting}} \text{molten hydrate} \xrightarrow{\text{denitration}} \text{solid oxide}$$

To compound the problem, two molecules of hydrating water are strongly bound directly to the uranyl ion, and during the decomposition form hydroxy bridges leading to sticky polymeric intermediates. The difficulty was partly overcome by a continuous process using a rotary tube calciner incorporating an internal Archimedes screw with countercurrent gas flow which kept the product mobile and in better contact with the cover gas. Similar equipment is still used today for thermal decomposition of plutonium oxalate to oxide.

Cover gases which complexed the metal cation and displaced water, such as $NO_2$ and $CO_2$, were partially effective in improving powder properties. Workers in the USA discovered, almost by accident, that when strongly acidic solutions were neutralised before denitration to limit acid attack on the equipment, uranyl nitrate and ammonium nitrate formed a complex which did not go through a low-temperature melting stage. Its decomposition is again partially successful in achieving a better powder product, but milling is still necessary to produce acceptable fuel and the process is not used in current reprocessing plants.

In the UK, uranyl nitrate was mixed with a substrate, such as carbon wool, on to which the nitrate solution would dry to give better contact with the cover gas before the carbon was oxidised, to give a more free-flowing product. Although this process has been used at laboratory scale to recover enriched uranium (and for finishing plutonium, as this type of matrix evaporation restricts air-borne contamination), it is too expensive for commercial use.

Fluidised bed processes are commonly used elsewhere for high-throughput, low-cost conversions. The technology was developed to convert concentrated uranyl nitrate solution to oxide, and is still in use today for purified uranyl nitrate whether virgin or recycled, although for the THORP product the bed must be narrow enough to avoid any risk of criticality. The high heating rates and constant motion improve the powder properties, but further processing is still required before reduction to achieve adequate fluorination. In practice, the fluidised *dryway* process incorporates 0.1% sulphate in the product to aid formation of the $UO_3$ beta dihydrate. When this is reduced to $UO_2$ the resulting change in the crystal lattice breaks down the powder by a process known as chemical milling, and the subsequent product is then suitable for hydro-fluorination. It still needs further processing before the product is suitable for fuel fabrication.

The Japanese have now extended the fluidised bed process to incorporate microwave heating, which overcomes problems of the high heat input to the fluidised bed and avoids attack by nitric acid on the heaters. This version of the process is installed in the Tokai reprocessing plant and is planned for the commercial scale plant at Rokkashomura.

Incorporating the latest technology of the time, the Russians have updated denitration by passing nitrate solution through a plasma, as in the large-scale production of titania for pigments. The process has been used at demonstration scale to co-finish uranium and plutonium for experimental fast reactor fuel rods.

The French company Comurhex has likewise patented a type of freeze-drying process, as used in instant coffee production, which dehydrates uranyl nitrate under vacuum before denitration and so avoids the problems of melt formation discussed earlier. The use of vacuum is in fact only an extension of the reduced-pressure evaporation stage used to concentrate the nitrate stream to 1000 g/l before the fluidised bed.

## Precipitation

The reaction of uranyl nitrate with ammonia to precipitate ammonium diuranate (ADU) was the first process to produce an oxide suitable for fuel fabrication without further powder processing. More power reactor fuel has been produced by this route than by all the others combined, and all current fuel specifications are written around its well-characterised properties.

The fact that the oxide can be used directly for fuel fabrication is no advantage if it is first to be fluorinated, but the process is also suitable for converting fluoride to oxide with low levels of residual fluorine. Thus the ADU process can be used for any conversion of nitrate or fluoride to oxide. The French still use it to convert reprocessed uranyl nitrate, while in the UK it is retained to recover enriched uranium oxide and fluoride residues despite being superseded by a dry route for fuel production (see Chapter 3).

Precipitation:

$$2UO_2(NO_3)_2 + 6NH_3 + 3H_2O \rightarrow (NH_4)_2U_2O_7\downarrow + 4NH_4NO_3 \qquad (8.3)$$

Thermal decomposition reactions:

$$(NH_4)_2U_2O_7 \rightarrow 2UO_3 + H_2O + 2NH_3 \qquad (8.4)$$

$$2NH_4NO_3 \rightarrow 2N_2 + 4H_2O + O_2 \qquad (8.5)$$

The conditions of precipitation are a compromise, as pH values below 5 form coarse sediments that settle rapidly and are easily filtered, but fine sediments yielding high-density pellets occur at pH above 6. In practice, most of the uranium is precipitated at low pH followed by a second stage of filtration at high pH.

GEC in the USA extended ADU precipitation by passing the slurry directly to a fluidised bed and so avoided the problems of separating fine precipitates suitable for fuel production. The fluidised bed process was demonstrated at 100 kg/d, but never exploited commercially as it was intended for co-finishing with plutonium.

The ADU process was improved by adding carbonate, as either solid or carbon dioxide, which results in a free-flowing oxide after thermal decomposition. The powder is suitable for pressing into fuel pellets without any further processing. While this is an advantage in the conversion of enriched hexafluoride, it is irrelevant in preparing recycled uranium for fluorination, and carbonation is not used for this purpose. However, the ammonium carbonates of uranium and plutonium are iso-structural, so that they can be precipitated as a solid solution, which makes the route ideal for MOX production.

### 8.4.2 Fluorination

Fluorination (also known as *conversion*)—the process of converting the solid uranium trioxide via uranium tetrafluoride to uranium hexafluoride by direct reaction successively with hydrogen fluoride and fluorine gases in fluidised bed reactors—has already been covered for virgin material in Chapter 3. Only those aspects peculiar to recycled material are considered here.

Reprocessed uranium has an artificial radioactivity due to traces of fission products and transuranic species, but chiefly to U-232 which of course is not chemically distinguishable from natural isotopes. From Magnox (metal) fuels, irradiated to less than 5 GWd/t, the amount is too small to prevent conversion in the same facilities as for new uranium, although there is a build-up of gamma-activity, typically increasing five-fold on ageing from 6 months to 10 years in the store. Uranium recovered after a higher burn-up (e.g. from LWR fuels at 30–50 GWd/t) has a larger proportion of U-232, hence a more significant activity. Whilst this is more serious in later stages of fuel fabrication, some small precautions are necessary during conversion to ensure that dose rates to operators remain low. Such measures include increased shielding and special disposal routes for the wash-out liquors from transport cylinders. Both BNFL in the UK and COGEMA in France have commercial-scale facilities dedicated to the conversion of reprocessed uranium.

### 8.4.3 Enrichment

The enrichment process was described in Chapter 4 and again only the effects of recycling are presented here.

The first operation at the enrichment plant is to vaporise the uranium hexafluoride. The traces of gamma-emitting daughter products that have been growing in since reprocessing are generally not volatile at the temperatures used, but remain concentrated in the transport cylinder and must be removed by special procedures. It is usual to enhance the shielding around the vaporising stations, but very few other plant modifications are required and many centrifuge plants have handled commercial quantities of reprocessed uranium.

It is unclear whether any of the world's diffusion plants routinely enrich reprocessed uranium. The capability is claimed, but not necessarily practised.

### 8.4.4 Fuel design

The structural designs, geometry and components for reprocessed uranium fuel are identical with those for new uranium. The only difference lies in the nuclear design where the enrichment must be enhanced to take account of the poisoning effect of U-236. The amount can be calculated by well-validated design codes, but as a rule of thumb is roughly an extra 0.3% U-235 for each 1% U-236.

### 8.4.5 Fuel manufacture

Although the process stages are the same as for new stock, as described in Chapter 3, design and operation must take account of ingrown gamma-emitting daughter products, principally of U-232. If the dose rate to plant operators is to be minimised, then so must be the time elapsed between enrichment and loading into the reactor. Modern production-scheduling tools and close co-ordination between enricher and fuel fabricator can help to achieve this aim. Additionally, modern plants designed to handle reprocessed uranium tend to have increased shielding, greater use of remote or automated processes, enhanced residue and effluent treatment systems etc.

### 8.4.6 Transport

Fuel is transported to the reactor site with no additional features or precautions.

### 8.4.7 Reactor operations

Reprocessed and new uranium fuels are designed to perform identically. Actual operating experience bears this out, though details of reactor performance are seldom reported openly.

### 8.4.8 Reprocessing

A second reprocessing is chemically possible, with the same general operations as described in Chapter 7, but under a constraint. Because of the high specific heat output of Pu-238, most modern oxide fuel reprocessing plants have a limit on the amount present, usually expressed as an average within any one fuel transport flask. To ensure a sufficient *time-to-boil* for all tanks and vessels should cooling be interrupted, the limit is typically 4% of all plutonium.

Now, reprocessed uranium includes U-236 which is enriched together with U-235. During irradiation in a second cycle, some is converted via Np-237 to Pu-238:

$$\text{U-236} \xrightarrow{n} \text{U-237} \xrightarrow[6.75\ d]{\beta} \text{Np-237} \xrightarrow{n} \text{Np-238} \xrightarrow[2.12\ d]{\beta} \text{Pu-238}$$

For example, after two typical 40 GWd/t PWR cycles, the Pu-238 level on discharge from the reactor is 7.7%, increasing with burn-up. A transport flask of such assemblies would therefore be unacceptable by today's standards. In practice however the reactor core will contain mixtures of new and reprocessed uranium, so mixing will occur naturally in reprocessing. It can be shown that in a closed PWR cycle at equilibrium, reprocessed uranium will account for only about 9% of total fuel, so that the dilution requirements are almost guaranteed.

However, there are other limitations on multiple recycling. Uranium-232 and U-234 are also enriched by current processes depending directly on mass number, and would cause unacceptable doses in present-day fuel fabrication plants. Most utilities loading reprocessed uranium will probably use it for only one cycle, then store it as the oxide pending selective enrichment techniques such as AVLIS (see Chapter 4) which if brought into economic operation are expected to be capable of removing U-232, U-234 and U-236.

## 8.5 Recycling plutonium

Plutonium was initially produced for military use, but as this requirement was met, attention in the civil nuclear power programme in many countries turned to fast breeder reactors, which utilised mixed plutonium and uranium ceramic fuel and offered the maximum energy returns from uranium ore (Chapter 5). Carbide and nitride fuels were initially considered as offering the highest breeding gains, but this became less urgent as delays in the nuclear programme postponed the expected shortage of plutonium for several decades; these options were then dropped because reprocessing nitrides and carbides, if practicable at all, would be severely troubled by carbon-14 content. Thus oxide fuel was required. Eventually, plutonium came to be used in thermal reactors.

### 8.5.1 Plutonium finishing

The different options for finishing are described in a roughly chronological order to show both why and how they have developed from the initial weapons programmes, through fast reactor development, to today's commercial utilisation of plutonium in MOX fuel for use in civil power reactors.

Any finishing process for uranium can also be used for plutonium. In some, e.g. the ammonium uranyl plutonium carbonate process, the plutonium intermediates are iso-structural with those of uranium and the two can be finished together if desired. In others, like ammonium precipitation or de-nitration, the intermediates do not have the same crystal structure and the resulting oxide is composed of discrete but intimately-mixed uranium and plutonium oxide particles.

**Early thermal de-nitration (TDN) processes**

As with uranium, the most straightforward way from nitrate to oxide is by direct thermal degradation:

$$Pu(NO_3)_4 \cdot 6H_2O \xrightarrow{heat} PuO_2 + 4NO_2 + O_2 + 6H_2O \qquad (8.6)$$

The product adheres to equipment surfaces, probably as a result of the solution-to-solid phase change. This process was modified to have the nitrate absorbed on a carbon wool matrix before decomposition, with greatly reduced product sticking and lower activity carried in the gas stream. It became explosive when the form of carbon was changed from wool to colloidal graphite.

Fluidised-bed processes as used for uranium received some development in the USA for co-conversion with plutonium. However, the problems of scrubbing plutonium from the amounts of off-gas required for fluidisation are daunting on a large plant, the powder produced is again difficult to process for fuel fabrication, usually resulting in low-density or weak pellets, and the process never progressed beyond small-scale trials despite substantial development.

**Precipitation**

A further drawback with direct decomposition processes is that the product contains any involatile impurities in the feed. The initial driving force for plutonium finishing was military, and the performance of weapons-grade material depends on isotopic and chemical purity. Denitration thus gave way to precipitation techniques which offered further purification as well as better handling properties, and produced powder in a finer, more reactive form for subsequent fluorination en route to metal.

Various precipitants have been used.

(a) *Peroxide* gives a good decontamination from metal cations such as are produced during irradiation and decay:[4]

$$2Pu(NO_3)_4 + 3H_2O_2 + (x+1)H_2O \rightarrow Pu_2O_7.xH_2O + 8HNO_3 \quad (8.7)$$

$$Pu_2O_7.xH_2O \xrightarrow{heat} 2PuO_2 + 1.5O_2 + xH_2O \quad (8.8)$$

The powder quality is not suitable for fuel fabrication and this route has been used only as an intermediate purification, e.g. from in-grown americium. The plutonium must all be tetravalent to limit losses. Peroxides have the potential to decompose violently, although the process has been used commercially to purify uranium ore.

(b) The most widely-used method of converting plutonium nitrate solution to dioxide is by precipitation of plutonium (IV) *oxalate*[5] (Fig. 8.1):

$$Pu(NO_3)_4.6H_2O + 2H_2C_2O_4 \rightarrow Pu(C_2O_4)_2.6H_2O + 4HNO_3 \quad (8.9)$$

It achieves significant decontamination factors (DFs) for most impurities and, in fact, modern reprocessing plants have come to rely on them to meet current specifications, particularly for removal of residual uranium (DF of 60). The plutonium stream is concentrated in evaporators and conditioned to Pu(IV) with hydrogen peroxide before precipitation, and an excess of oxalic acid is used, in order to reduce the solubility of the product. The particle size and structure of the oxide are governed by the initial precipitation, and in turn by the temperature and by concentrations of plutonium, nitric acid and oxalic acid. The effluent stream from filtration, known as the oxalate mother liquor (OML), contains residual plutonium and is recycled, but the excess oxalic acid must first be destroyed, usually by refluxing with concentrated nitric acid.

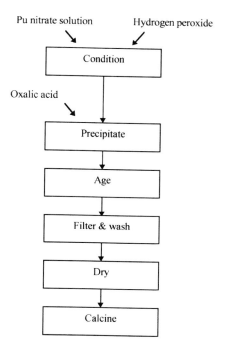

**Fig. 8.1** Oxalate precipitation

The oxalate is decomposed to oxide,

$$Pu(C_2O_4)_2 \cdot 6H_2O \xrightarrow{heat} PuO_2 + 2CO + 2CO_2 + 6H_2O \quad (8.10)$$

suitable for fluorination and for ceramic fuel fabrication. Since the process was optimised to give a high surface area for fluorination, particles are small and tend to disperse as dust which, adhering to surfaces, adds to radiation levels. For the same reason the product has an inconveniently high capacity for adsorbed moisture, which undergoes radiolysis and may pressurise containers, so for storage the surface area is usually reduced by calcination.

(c) Fabrication of mixed uranium-plutonium fuel initially relied on available oxalate-derived $PuO_2$ powder, dry blended with urania from the ammonium diuranate (ADU) process. However, a solid solution would be better produced by starting from mixed aqueous solutions rather than by dry-blending oxides and relying on solid-state diffusion during sintering.

Hydrous plutonium oxides can be precipitated from nitrate solutions with ammonia, and owing to their insolubility this process remains important today for waste treatment (e.g. in the actinide recovery plant at Sellafield). Precipitation of uranium with ammonia as ADU is still a commercial finishing step. Unfortunately plutonium and uranium precipitate at different pH values, so the UK adopted a multistage co-precipitation route with ammonia in a pH gradient to produce experimental fast reactor fuel in the early 70s.[6] This still did not give an intimately mixed product, and even after a subsequent blending stage the resulting fuel tended on dissolution to leave skeletons in the form of the whole pellet. *Ammonium co-precipitation* was therefore dropped in this form in favour of dry blending.

(d)   The US government was keen to avoid separated plutonium in the civil fuel cycle and so funded attempts to overcome problems in finishing a combined uranium and plutonium stream. One of these was a combined CO-PREcipitation and CALcination (COPRECAL) route in which a slurry from ammonium co-precipitation was fed directly to a fluidised bed for thermal decomposition, thus avoiding filtering difficulties and a poorly mixed product.[7] Tens of kilograms of a mixed product were made on a laboratory scale and uranium trials at 100 kg/day demonstrated scale up. Despite claims that it produced acceptable fast-reactor fuel pellets, the route has never been implemented commercially, perhaps because it was overtaken by development of the gel sphere co-precipitation routes.

(e)   The next development, in the late 1970s, was *gel-supported co-precipitation*. Drops of a mixed metal nitrate solution, with an organic polymer such as polyacrylamide and a structure-modifying agent such as formamide, were hydrolysed to form a precipitate within the spheres. The hydrolysing base could be introduced externally by allowing the drops to fall through a column of ammonia solution,[8] or internally by including a compound such as hexamethylene tetramine which decomposed while falling through a heated column of an organic solvent.[9] The spheres were aged, washed and dried in further columns before calcination and sintering. Nearly every country with a nuclear development programme developed variations of internal or external gel sphere routes. One of their main advantages was that, being carried out in liquids, they avoided formation of dust and the consequent radiation problems in an otherwise dose-intensive part of the cycle.

Another novelty in these routes was that the densified spheres were intended to be vibro-compacted into pins directly, rather than pressed into pellets. They were sintered up to around 95% of theoretical density and in mixed sizes (e.g. 800 and 80 microns) achieved packing (smear) densities of around 80% in pins. Over 200 kg of mixed oxide spheres were produced at Harwell and fabricated into pins at Windscale for testing but, although the sintered densities and smear densities were equivalent to those with annular fast-reactor pellets, the different distribution of void space led to anomalous corrosion or fuel slumping at the very high ratings in experimental fast reactors. A pilot plant was commissioned with uranium and thorium at Sellafield, but in the light of fuel performance never progressed to plutonium operation. Some fast reactor fuel is however still made by vibro-packing routes, alongside others, in Russia.

(f)   Thus fuel fabrication partially reverted to dry blending. Since the expected

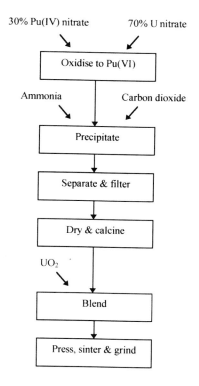

**Fig. 8.2** Carbonate flowsheet

demand for fast-reactor fuel did not materialise, the larger MOX production plants in Belgium, France and Germany adapted their fabrication process to produce thermal reactor fuel by diluting a Masterblend with additional urania.

Germany persevered with co-precipitation and extended the ammonium uranyl carbonate precipitation route to form the AUPuC (*ammonium-uranyl-plutonyl-carbonate*) process (Fig. 8.2)[10] Unlike ammonia, which precipitates ammonium di-uranate and hydrous oxides of plutonium at different pH, carbon dioxide precipitates mixed, isomorphous compounds of uranium and plutonium, provided that the latter has previously been oxidised so that both are hexavalent:

$$MO_2(NO_3)_2 + 3H_2O + 6NH_3 + 3CO_2 \xrightarrow{steam} (NH_4)_4MO_2(CO_3)_3 + 2NH_4NO_3 \quad (8.11)$$

$$(NH_4)_4MO_2(CO_3)_3 + H_2 \xrightarrow{heat} MO_2 + 3CO_2 + 4NH_3 + 3H_2O \quad (8.12)$$

Uranyl nitrate was mixed with 40% plutonium nitrate solution straight from the reprocessing plant for direct production of a Masterblend. This took the form of fairly large soft spheres that could be diluted to the appropriate concentration with free-flowing AUC-derived uranium dioxide and directly pelleted, thus avoiding dusty and dose-intensive milling and conditioning stages. With the closure of the small-scale WAK plant and the reprocessing plant at Wackersdorf, the German supply of plutonium nitrate is no longer available and the process has been abandoned.

(g) Development now seems to have come full circle and, since MOX fuel does not require the same purity as military material, is centred on reducing costs by *applying modern technology to denitration*. The French freeze-drying, Russian plasma, and Japanese microwave-assisted denitration proceses have already been mentioned.

**Storage**

In the early days of civil reprocessing it was assumed that all plutonium would be used in fast reactors within a decade. As the relative economics changed and the implementation of fast reactors moved back, it became evident that long-term storage would be required pending the installation of fabrication capacity for thermal MOX fuel.

Plants designed to package plutonium dioxide from reprocessed oxide fuels operate remotely to minimise operator doses. The oxide from the finishing lines is calcined and metered remotely into storage cans under a blanket of argon to minimise the amount of moisture and other gases that can be absorbed on to the product. The primary containment is normally a stainless steel screw-top can which holds 5–10 kg. It is inserted into an intermediate sleeve and outer can, both of stainless steel with welded closures to provide three layers of containment.

The store consists of large thick-walled concrete cells containing re-entrant tubes accessible only through loading ports fitted with shield plugs in the operating face of the cell. Containers are loaded into the channels by means of a shield block and remote mechanical system. With highly-irradiated plutonium, decay heat may render the cans too hot to touch physically; such heat is removed by a forced draught, although the tall racks generate their own chimney effect. For protection against external hazards such as earth tremors, the walls are of massive reinforced concrete.

The cans within racks are inspected routinely through an endoscope arrangement, or by a remotely-operated trolley carrying a camera and thermocouples which enter the storage tubes.

### 8.5.2 Fuel manufacture

This section will describe primarily the manufacture of plutonium-based fuel for thermal reactors (e.g. LWRs and the UK's AGRs). The various stages closely parallel those for uranium fuel as described in Chapter 3. The design of the plant however is somewhat different to allow for higher levels of radiation and the smaller volumes capable of forming a critical mass.[11] The process route for fast-reactor

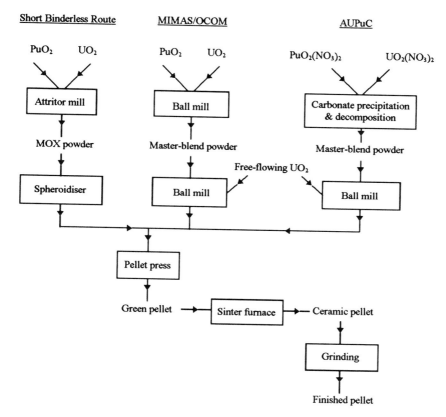

**Fig. 8.3**  Process routes for mixed-oxide powder

fuels is very similar, save that the plutonium contents, ratings and burn-ups are much higher and the design characteristics again adapted accordingly.

Plutonium dioxide is never used alone, but diluted with uranium dioxide to limit reactivity and in thermal reactors to simulate as far as possible the nuclear properties of enriched uranium fuel. Concentrations range from about 1% for the Japanese Advanced Thermal Reactor (ATR), through typically 5–10% in thermal reactors, to some 25–35% for fast reactors.

The uranium may be new stock, depleted tails from the enrichment process, or recycled. New uranium is used for this purpose in Germany where licensing considerations restrict alternatives. Tails, typically containing 0.2–0.3% U-235, are the normal choice elsewhere.

Particles containing more than about 40% plutonium dioxide dissolve very slowly in boiling nitric acid, and on reprocessing any remnant could settle out in the dissolver. In any case, local concentrations of plutonium would form hot spots in the reactor where they might promote fission gas release, weaken the cladding or in an extreme case rupture it during a loss-of-coolant accident (LOCA). A uniform solid solution of uranium and plutonium oxides is therefore desirable. It can be achieved by mixing nitrate solutions and then co-finishing, or by finishing the two streams separately and milling the products together to form an intimate mixture that will form a solid solution on sintering.

Alternatives for co-finishing are covered in the previous section. There are two strategies for dry mixing: mix the oxides directly in the desired proportions, or make a Masterblend of high enrichment (originally as suited to fast reactors) and subsequently blend it down (Fig. 8.3).

The Masterblend route, variously known as the Optimised Co-Milling (OCOM) or Micronisation of Master Blend (MIMAS) process, is currently used in Belgium, France and Germany. A mixture of oxides, typically containing 30–40% $PuO_2$, is ball-milled for up to about 8 hours to achieve the required particle size, homogeneity and specific surface area. This Masterblend is then diluted with more free-flowing uranium dioxide to the required concentrations. Most of the MOX fuel in the world has been made by this process, which despite early set-backs causing poor solubility on reprocessing and necessitating refinements, is capable of producing acceptable fuel.

An alternative dry mixing route operated in the UK is the *short binderless route*, so called as derived from the binderless route to AGR fuel.[12] It is similar except for omitting the intermediate Masterblend. Uranium and plutonium dioxides are milled for about 20 minutes in a high-energy Attritor mill, consisting of a stationary vertical mill pot with a vertical shaft and paddle arrangement to stir the powders and ball charge by means of a powerful motor. The process gives an excellent homogeneity and correct particle size in a fraction of the time taken by a ball-mill. However, because the Integrated Dry Route does not yield a free-flowing uranium dioxide suitable for diluting a micronised Masterblend, the mixture is made up directly to the intended final composition, and after milling, granulated in a patented *spheroidiser* in which the powder is slowly rolled down a scalloped blade with a little zinc stearate as a die lubricant. In about 20 minutes, this gentle tumbling action produces free-flowing agglomerates suitable for pressing. The whole process is now in routine commercial production.

After mixing, fuel manufacture follows a conventional route. Pellets are pressed and sintered in a hydrogenous atmosphere at temperatures around 1650 °C. They are then ground to diameter and closely inspected for dimensions, cracks, chips etc. Acceptable pellets are loaded into fuel tubes, end plugs are welded in place and after inspection, these fuel rods are fabricated into an assembly.

MOX fuels emit alpha, beta, gamma and neutron radiation and the processes have to be contained in gloveboxes or cells to protect the operators. The amounts of the more penetrating gamma (due mainly to Am-241 grown in from Pu-241) and neutron radiation depend on several factors, such as the reactor type, the burn-up of the original fuel and the age of the plutonium since reprocessing. Using ex-military plutonium greatly reduces the radiation levels owing to its isotopic purity.

Gamma radiation, which tends to be dominant at lower burn-ups, is attenuated by steel and lead composite shielding. Neutrons predominate with plutonium from higher burn-up fuel and are attenuated by hydrogenous material such as polythene or impregnated woods (Jabroc). Automated and remote or robotic techniques are increasingly used to reduce human involvement and dose. Current plants tend to limit the age of the plutonium to 2–3 years, assuming contemporary burn-ups. New plants soon to come on stream will increase the permitted storage time and 10-year-old plutonium from 60 GWd/t uranium fuel is an achievable limit, thus maximising the customers' flexibility in recycling plutonium.

### 8.5.3 Fuel design

LWR MOX fuel is mechanically identical with the equivalent enriched uranium fuel. Indeed, they can be distinguished externally only by their unique identification markings, and geometrically MOX fuel assemblies can substitute exactly for uranium fuel. It is therefore unnecessary to revalidate the mechanical design features or the supporting thermal hydraulic design codes.

The neutronic designs of the two fuel types however are different, reflecting the basic nuclear differences between plutonium and U-235 as fissile content. Specifically, plutonium has higher cross sections than uranium for both fission and absorption of thermalised neutrons. The neutron multiplication factor, often referred to as $k_\infty$ and given by

$$k_\infty = \frac{\text{(No. of neutrons per fission)} \times \text{(fission cross section)}}{\text{absorption cross section}}$$

is initially lower for MOX fuel than for the equivalent uranium fuel, but falls more slowly with burn-up. To match the performance of adjacent MOX and uranium fuels, equivalence formulations can be used which take into account the contributions of the individual isotopes.[13]

One important consequence of the higher neutron absorption by plutonium is that it competes more effectively for neutrons with the reactor control and shutdown mechanisms, which are therefore less effective in the presence of MOX fuel. Their number and strength is determined by the design of the reactor, and cannot in practice be greatly modified thereafter beyond up-rating absorbents. To preserve adequate safety margins, the amount of MOX fuel in typical present-day LWRs must therefore be limited to about 30%, either within individual assemblies or by suitable disposition of all-MOX and all-uranium assemblies. The latter option (the *island* design) is commonly adopted since otherwise the whole core, rather than a mere third of it, would be subject to the MOX manufacturing premium.

Because neutrons are used more effectively in a MOX than in a uranium assembly, pins at the edges or particularly the corners of the former, exposed to a neutron flux more appropriate to uranium fuel, would tend to run hotter than the rest if they were of the same enrichment. Such power peaking is undesirable, and minimised by having fuel rods at differing plutonium enrichments within the assembly, with the lowest at the corners, the highest inside, and an intermediate level along the edges.

### 8.5.4 Reprocessing

Although fast breeder reactor fuels have been reprocessed and recycled in the UK, and thermal reactor MOX fuel to some extent in Germany, contemporary thermal reactor fuel reprocessing plants have been mainly designed for uranium fuel. If LWR MOX fuels are to be reprocessed in such plants or their like, the differences in properties must be taken into account.

The head-end, where the fuel is chopped and dissolved, will not be significantly affected, although the presence of more plutonium will require a re-assessment of the criticality safety cases and possibly poisoning of some process streams with gadolinium nitrate. Dissolver cycle times may have to be increased. In the chemical separation area, constraints are likely to be on throughput, radioactivity and heat output of the active liquor, and perhaps on some aspects of process chemistry affected directly or indirectly by plutonium concentration. These constraints can be met by increasing the cooling time (typically from 5 to 7 years), and by co-processing MOX and uranium fuels, thus effectively diluting the MOX fuel. Various dilution schemes may be necessary or possible, but a dilution of at least three-fold with uranium is usually necessary and up to 12-fold may be desirable. Some such dilution is already necessary in the reactor core, and may be required to meet heat limitations in transport before reaching the reprocessing plant.

In plutonium finishing, there may be a throughput limitation. MOX fuel contains more plutonium per kilogram of irradiated fuel, and as the burn-up of the MOX fuel increases, the proportion of fissile plutonium decreases but the quantity of plutonium goes up.

MOX fuels will not give rise to any new effluent streams, though the activity (particularly alpha) and volume of existing waste streams are increased by the presence of more plutonium, americium and curium.

## 8.6 Future trends in recycling

### 8.6.1 Higher burn-ups

Operating economics are significantly improved as burn-up of the fuel increases. Hence there is likely to be a requirement to fabricate recycled fuels from higher burn-up parent fuel and subsequently to drive these recycled fuels to a higher burn-up themselves. A target of 60 GWd/t is seen as achievable for LWR fuels in present-day reactors.

The effects include the need to fabricate fuel with high plutonium contents (criticality and dose considerations), design of such fuel (validation of design codes outside the current range), reprocessing (higher heat output, dose, plutonium content), increased poison content (U-236 in reprocessed uranium) etc.

### 8.6.2 Use of ex-military materials

With the welcome destruction of nuclear warheads as part of the global disarmament process, the enriched uranium and plutonium are newly available for civil use, and indeed their high fissile content makes this an attractive option as long as they can be guaranteed not to re-enter the military cycle. The dilution with militarily deleterious isotopes would itself be an obstacle to such diversion, and there is considerable interest in the idea, particularly in the United States where government agencies may fund the supply of suitable materials.

### 8.6.3 New enrichment processes

Current techniques enrich the undesirable isotopes such as U-236, as well as the desirable isotopes. Techniques such as AVLIS (Atomic Vapour Laser Isotope Separation—Chapter 4) would enrich only the specific isotope of interest. If such processes reach commercial maturity, radiation from U-232 or U-234 and the poisoning effect of U-236 may be eliminated from consideration in the manufacture of reprocessed uranium fuel.

### 8.6.4 Plutonium finishing

The earlier driving forces in reprocessing and finishing have been superseded by the need for clean technology, process intensification, dose reduction, tighter safeguards, and cost reduction.

*Clean technology* aims to eliminate and simplify effluent streams, avoiding the need for complex treatment to meet discharge limits. This trend favours routes like thermal denitration which minimise liquid effluents.

Process intensification—increasing the throughput per unit plant volume—will favour continuous processes (like fluidised beds and plasma conversions) over batch operations.

Dose reduction will favour the use of Masterblend, which affords a degree of self-shielding and minimises dry, dusty process stages with the consequent mechanical maintenance and radiation dose. Processes like gel precipitation are essentially dust-free and suitable for remote operation.

Safeguards considerations will also favour Masterblend to lessen the risk of diversion and so open up the potential market for reprocessing. Relaxed specifications for products, allowing them to retain more fission-product activity, will both reduce the costs of separation and, by requiring remote operation, also guard against diversion. Combined finishing and fabrication, as in gel fuel or melt casting, will remove opportunities for diversion as well as cutting the overall number of process stages.

Apart from requirements for remote handling, all these considerations should also lead to cost reductions by simplifying processes.

## 8.7 References

1. *Plutonium fuel: an assessment*. OECD, Paris (1989).
2. T. Williams. Recycling experience in the UK—past, present and future. *Atom*, February 1991, p. 12.
3. Beaumont *et al*. The environmental benefits of MOX recycle. *9th Korean Atomic Industry Forum meeting, Seoul., KAIS Conference report* pp. 359–77. (April 1994).
4. M. Haas. *Transactions of the American Nuclear Society*, **40**, p. 46 (1982).
5. R. E. Lerch and R. E. Norman. Nuclear fuel conversion and fabrication chemistry. *Radiochimica Acta* 36, pp. 75–8 (1984).
6. S. E. Smith. *Proceedings of the 3rd Conference on Peaceful Uses of Atomic Energ,* **10(1)**, pp. 161–7 (1964).
7. L. B. Kincaid, E. A. Aitken and I. N. Taylor. COPRECAL—co-conversion of U/Pu mixed nitrate to mixed oxide. *Transactions of the American Nuclear Society 33*, Winter Meeting, San Fransisco, p. 470 (November 11–15, 1979).
8. E. S. Lane. Manufacture of microspheres by gel precipitation. *GB Patent* 1313750 (1973).
9. R. B. Mather and P. E. Hart. Nuclear fuel pellets fabricated from gel-derived microspheres. *Journal of Nuclear Materials* 92, pp. 207–16 (1980).
10. V. W. Sneider, F. Hermann, and W. G. Druckenbrodt. The AU/PuC process. A co-precipitation process with good product homogeneity to the full scale of plutonium concentration. *Transactions of the American Nuclear Society*, 31, p. 176 (1979).
11. J. Edwards and J. Stanbridge. A review of mixed oxide fuels for light water reactors. *The Nuclear Engineer*, 31, p. 162.
12. W. Baxter and P. Parkes. The Short Binderless Route for mixed oxide fuel production. *European Nuclear Conference*, Lyon (September 1990).
13. K. Hesketh, G. M. Thomas and C. Robbins. Elimination of the homogenisation step in the manufacture of mixed oxide assemblies for light water reactors by means of a reactiviy equivalence formulation. *Proceedings of the Annual Meeting of Nuclear Technology*, p. 225. German Nuclear Society, Karlsruhe (May 1992).

# 9 Waste treatment

G. V. Hutson
*BNFL, UK Group, Sellafield, Cumbria*

## 9.1 Introduction

Any industry generates waste materials besides the desired product. In the nuclear industry the predominant waste-producing operations are:

- mining and milling of uranium ores;
- reprocessing of fuel discharged from reactors;
- decommissioning of redundant or obsolete facilities.

The first of these is covered in Chapter 2, the last in Chapter 12. This chapter will therefore concentrate on reprocessing wastes, with particular reference to the principal British site at Sellafield as a representative modern undertaking; waste management elsewhere and in other parts of the cycle is conducted on similar principles. Wastes from military programmes outside the UK, sometimes posing specialised problems, are not covered, nor are the much less substantial amounts from the thousand or so minor producers such as hospitals and research laboratories.

Wastes may be either:

- primary, i.e. inherent in the material to be processed;
- secondary, derived from the process itself.

In fuel reprocessing, primary wastes are the fission products, unwanted transuranic elements (*minor actinides*) and the remains of the cladding. Secondary wastes range from reagents and solvent that are no longer economically recoverable, through worn-out equipment and protective clothing, to rubbish of domestic types such as paper towelling that may have become more or less contaminated with radioactive material in the course of use.

The objectives of waste management are to:

- minimise the arisings of secondary waste;
- convert as much as possible of both radioactive and inactive wastes into the

least practicable volume of a solid that is suitable for long-term storage and ultimate disposal;
- keep the environmental impact of the inevitable aerial and liquid discharges as low as is reasonably attainable (ALARA), and in any case within strict limits set by the regulatory bodies.

Reprocessing plants are therefore designed to direct the vast majority of the fission products and minor actinides into an aqueous stream amenable to concentration by evaporation, followed by eventual vitrification. Almost all the remaining potentially harmful materials are routed to appropriate treatment plants, the products of which can be incorporated into a monolithic matrix, with an effluent sufficiently decontaminated for discharge to the environment. The pressure to reduce such radioactive discharges has led over the past decade or so to an increase in abatement technology that itself generates additional solid wastes, besides some actual or potential radiation dose to operators. The benefits thus have to be weighed against the drawbacks.

## 9.2 Waste types

In the UK, solid wastes are conventionally classified by radioactive content as high-, intermediate-, low- or very low-level wastes (HLW, ILW, LLW and VLLW respectively). The formal definitions are:

- HLW—wastes in which the temperature may rise significantly as a result of their radioactivity, so that this factor has to be taken into account in designing storage or disposal facilities.
- ILW—wastes with radioactivity levels exceeding the upper boundaries for low-level wastes, but which do not require heating to be taken into account in the design of storage or disposal facilities.
- LLW—wastes containing radioactive materials other than those acceptable for disposal with ordinary refuse, but not exceeding 4 GBq/t alpha or 12 GBq/t beta/gamma.
- VLLW—wastes which can be safely disposed of with ordinary refuse (dust-bin disposal), each 0.1 $m^3$ of material containing less than 400 kBq beta/gamma activity or single items containing less than 40 kBq beta/gamma activity. Such wastes need not be considered further.

In practice, HLW is essentially the bulk of the fission products etc. after vitrification; ILW is largely the stripped or leached remains of the cladding, or plutonium-contaminated material (PCM) such as filters or process residues carrying significant amounts of plutonium that cannot be removed economically; LLW consists largely of discarded equipment, tools or protective clothing, plus a considerable amount of material merely suspected of being contaminated, such as waste paper from offices in a controlled area.

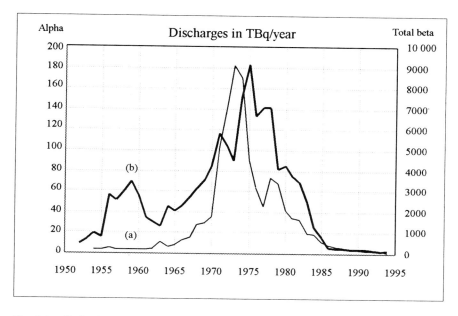

**Fig. 9.1** Reduction in radioactive discharges from Sellafield: (a) alpha, (b) beta-gamma

The definitions are recognised to be less than ideal, in that they take account only of the current radioactivity without regard to half-life. Changes are possible in the foreseeable future. In the USA, classification is based effectively on the source of the material rather than its content. This has the advantage of avoiding any question about the categorisation of a given package, but may lead to unnecessarily expensive disposal routes for relatively innocuous consignments, and again there are moves for revision.

Liquids are similarly divided into high-, medium- and low-active streams. Aerial effluents form a special category. Liquid arisings are conveniently considered first, in the specific case of Sellafield and the progressive reduction in activity discharged from the site.

## 9.3 Liquid wastes

Over the last 20 years, existing operations have been improved and new treatment plants commissioned in response to increasingly stringent requirements to limit discharges. Figure 9.1 shows that the activity discharged to sea from the site has thus been reduced by several orders of magnitudes, despite increasing throughput and irradiation of Magnox fuel and the commencement of oxide reprocessing, which doubles reprocessing capacity.

A schematic illustration of the effluent system is given in Fig. 9.2. Most of the

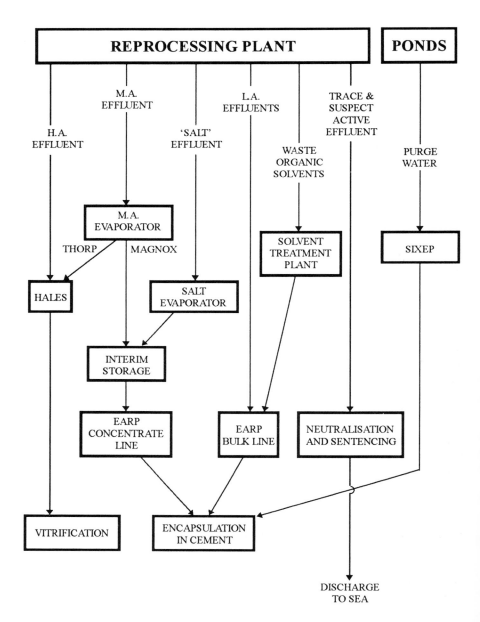

**Fig. 9.2** Liquid effluent treatment scheme, Sellafield site

nitric acid in the feeds to evaporation processes is recovered and recycled to the reprocessing plant.

The *highly active raffinate* from the first stage of fuel reprocessing, containing the vast majority of the fission products and minor actinides, is concentrated up to 100-fold by evaporation. The concentrate, strongly heated by the decay particularly of caesium-137 and strontium-90 (both with half-lives of about 30 years), is initially stored in cooled, high-integrity double-walled stainless steel tanks. Eventually it is converted into monolithic cylinders of glass, each contained in a transportable stainless steel flask. The process is described in more detail later.

*Medium-active* effluents principally arise from the purification of uranium and plutonium. Some can follow the same path as the highly-active liquor; those which are unsuitable, by virtue of either chemistry or volume, are fed to a separate evaporator and the concentrate stored. Improved operating practices in the reprocessing plant have allowed additional streams to follow this route as time passed.

To help reduce the level of contamination in effluents, fuel is delay-stored or *cooled* for at least a certain minimum period before reprocessing. Short-lived radionuclides such as ruthenium-103 (half-life 39 days), zirconium/niobium-95 (64 and 35 days respectively) and particularly iodine-131 (8 days) can thus decay to very low levels. Delay periods have also been used after reprocessing to reduce the activity of somewhat longer-lived radionuclides in effluents destined for discharge. Until about 1980 the medium-active concentrate was discharged to sea after at least three years' decay to reduce amounts of relatively short-lived radioisotopes, such as ruthenium, zirconium and niobium which constituted most of the activity. However with the ever falling levels of activity in the aqueous discharges, longer-lived isotopes, notably actinides, contributed an increasing proportion of the total. The decay-stored liquors were thenceforward retained until they could be treated in an enhanced actinide removal plant (EARP) which commenced operation in 1993. Inevitably the remaining activity in the effluents derived from these historic wastes adds somewhat to those from current arisings, but the levels are small and the total amount limited.

*Salt raffinates* arise from the alkaline washing of process solvent (tri-*n*-butyl phosphate in *odourless* kerosene, TBP/OK) before it is recycled. The sodium content of these raffinates is incompatible with the normal medium-active evaporation process, and they are therefore concentrated in a separate Salt Evaporator which was commissioned in 1985. Like the medium-active concentrate, the salt concentrates were originally intended to be delay stored before discharge to sea, but eventually it was decided to retain these liquors until they could be processed through the new EARP plant.

With the start-up of the salt evaporator, all suitable effluents were now being concentrated in this way. Those remaining, and containing significant levels of activity, were too voluminous or also contained comparatively high levels of iron. To treat these streams, ion-exchange and chemical precipitation processes were developed.

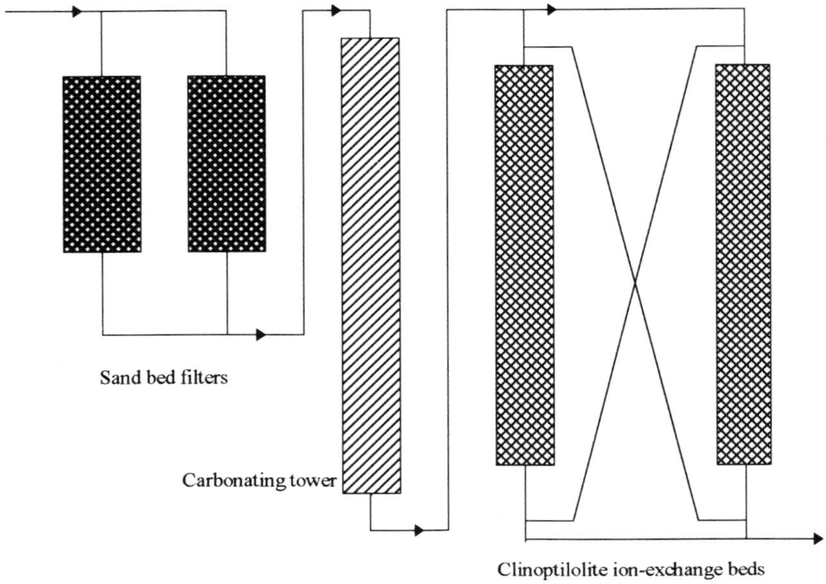

**Fig. 9.3** Schematic flow diagram of SIXEP

### 9.3.1 Removal of caesium and strontium

These elements are the main contaminants in effluents from fuel storage ponds and decanning caves. There are also significant quantities of magnesium hydroxide from corrosion of the fuel cladding. To restrict corrosion, and to avoid competition with strontium for active sites in the subsequent ion-exchange process, the solubility of magnesium is minimised by keeping the pH at approximately 11.

Figure 9.3 is an outline flow diagram for the plant which has an average throughput of about 2000 m$^3$ per day.

Magnesium hydroxide is first removed by deep sand bed filters. The pH is then adjusted in a carbonating tower by passage of carbon dioxide, which cannot reduce the pH too far and does not add any ions that could complicate further treatment, nor produce additional solid waste.

The ion exchanger chosen after an exhaustive selection process was clinoptilolite (a natural zeolite extracted from the Mojave desert in California) because it is

- selective enough to absorb the caesium and strontium ions in preference to inactive sodium and calcium ions which are present at much higher concentrations;
- strong enough mechanically for use in a large column, with a suitable particle size;

- available in adequate quantity at an acceptable price;
- amenable to a convenient disposal route when spent.

The ion exchange beds are operated in series, the second acting as a polishing filter to ensure maximum removal of the radioactive ions. When the leading bed approaches saturation, feed is stopped and the bed is fluidised for discharge to interim storage and eventual encapsulation. The functions of the beds are then interchanged, to maximise activity removal and minimise waste generation.

The effluent from the ion-exchange process is discharged to sea. It cannot be recycled, as repeated basification and neutralisation would cause an unacceptable build-up of salts, detrimental to both fuel storage and the ion-exchange process. Similarly the ion exchange material is loaded only once and then discharged. This practice, rather than regenerating it for repeated use, may appear to generate excessive active solid waste for disposal. However, when considered as part of the total waste management strategy it does not, since:

- clinoptilolite can replace raw fillers as an essential ingredient of the cement used to encapsulate other solid wastes, so that the total waste volume is not increased;
- regeneration would produce radioactive liquors themselves needing treatment before disposal.

Thus the once-through practice reduces costs, secondary waste and radiation dose to operators.

The ion exchange plant typically decontaminates effluents by factors of 2000 and 500 for caesium and strontium respectively, and treats 20 000 bed volumes per charge of clinoptilolite.

## 9.3.2 Removal of residual actinides

The commissioning of the ion-exchange plant and of the salt evaporator in 1985 was the culmination of that stage in the discharge reduction programme. The only streams containing significant activity still routed to sea were unsuited to evaporation or to ion exchange, owing mainly to inactive chemical species present; they amounted to approximately 250 m$^3$/day. There were also the concentrates from the medium-active and salt evaporators, stored in high-integrity stainless steel tanks whilst a treatment process was developed, and amounting to about 1000 m$^3$/year. Liquors of both groups contained significant quantities of iron, nitrates and acid. The major requirement for decontamination was to remove alpha-emitting species, although some reduction in beta/gamma activity was also desirable.

A study of methods available for the treatment of both types of feed showed a flocculation process to be the most suitable.[1] Increasing the pH to 10–11 by adding sodium hydroxide precipitates ferric hydroxide in a flocculent form that carries down the majority of the alpha activity. Adding relatively small volumes of

168  *Waste treatment*

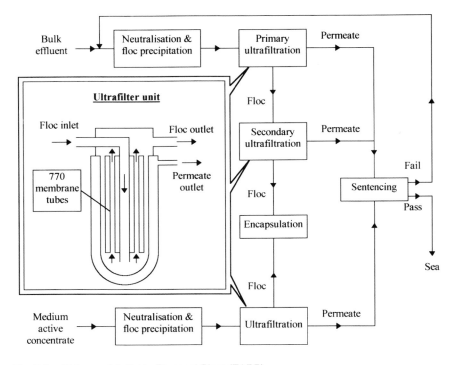

**Fig. 9.4**  Enhanced Actinide Removal Plant (EARP)

specific chemicals, such as nickel ferrocyanide for increased caesium removal, also improves the beta-decontamination, particularly of the concentrates, which are most effectively treated separately in batches. The resulting aqueous phase is virtually inactive.

The floc as first produced is voluminous, containing only a few hundred ppm of ferric hydroxide, and needs to be concentrated for economic long-term storage and disposal. Ulatrafiltration was chosen for the purpose as it provides the most effective mutual separation between solid and surplus water, and its use has been the key to the success of the operation.

A simplified process flow diagram for the plant is shown in Fig. 9.4, with a relatively enlarged illustration of a filter unit inset. The ferric floc containing nearly all the feed activity is circulated through tubular graphite/zirconia membranes with pores typically a few hundredths of a micron in diameter. The scouring action of the cross-flowing liquor thus prevents the fouling that would follow any attempt at straight-through filtration. High concentration factors are achieved in two stages, the first reducing the volume by up to 90% without greatly increasing the viscosity, so that the final thixotropic product containing up to 100 g/l iron can be produced in a smaller, specialised unit. The resulting sludge is suitable for

**Table 9.1** Decontamination factors in the Enhanced Actinide Removal Plant

| Isotope | Bulk Effluents | Concentrates |
|---|---|---|
| Co-60 | 50 | N/A |
| Sr-90 | 10 | 150 |
| Zr/Nb-95 | 20 | 20 |
| Tc-99 | 1 | 1 |
| Ru-106 | 2 | 15 |
| Cs-134/137 | 10 | 50 |
| Ce-144 | 20 | 100 |
| Np-237 | 30 | 30 |
| Pu-alpha | 100 | 500 |
| Pu-241 | 100 | 500 |
| Am-241 | 100 | 500 |

encapsulation in cement, while the filtered solution is checked for radioactivity and discharged to sea.

Decontamination factors as found in development work are given in Table 9.1. During the initial period of plant operation they have been largely substantiated; indeed, for most elements they appear significantly better. Significant decontamination from beta-emmitters is also achieved, with the significant exception of technetium which is most stable in the anionic pertechnetate form.

### 9.3.3 Solvent disposal

Commissioning of the enhanced actinide removal plant left only one untreated effluent stream with significant potential for adverse environmental impact, namely the relative small quantity of waste process solvent (nominally 20% TBP/OK) that is not suitable for recycle, stored on site since 1983. Arisings of 30% TBP/OK will come from THORP. A process to treat these liquors has been developed, the plant is under construction and is due to be commissioned in 1997.

No commercial-scale treatment has yet been successfully applied to waste solvent within the nuclear industry world-wide, so the first stage was to specify the requirements for a suitable process. These are:

- a reasonable chance of successful development on an acceptable time scale;
- demonstration at pilot scale with active species;
- assured safety;
- conventional engineering with a minimum of novel or untried equipment;
- demonstrated compatibility with associated plant;
- ease of control with the fewest possible process steps and unit operations;
- acceptable lifetime costs.

Several possible processes that have been investigated, with their potential advantages and the major reason for rejection, are given in Table 9.2.

**Table 9.2** Solvent destruction processes

| Process | Advantages | Problems |
| --- | --- | --- |
| Direct incineration | Conceptually straightforward | Burnout, plant life, ventilation clean-up |
| Pyrolysis | Avoids corrosive fume | Intractable residues |
| Phosphoric acid split ('Eurowatt' process) | Possible recycling | High-temperature moving parts; undesirable distribution of activity |
| Encapsulation: a) Direct encapsulation | Conceptually straightforward | Low fractional incorporation, weepage |
| b) Absorption and cementation | Conceptually straightforward | High volume of solid waste |
| Ultraviolet irradiation | No reagents needed | Highly inefficient, hence high energy costs |
| Gamma radiation | Conceptually elegant process, radiation available, continuous degradation process | Very high energy required, practical demonstration lacking, potentially serious safety engineering problems |
| Microbial degradation | Potential low temperature process, modern technology | Large waste volumes, very large plant required |
| Dealkylation (Friedel–Craft) | Chemically elegant | Poor performance, difficulties in handling by-products |
| Distillation | Simple process, reduces interim storage requirement | Partial process, OK distillate not suitable for recycle |
| Emulsification and sea discharge | Simple and cheap | Extremely large effluent volumes, discharges all activity and organics |
| Silver II electrochemical | Potentially versatile | Energy costs, needs excessive development with no guarantee of success |

These processes were all judged to be unsuitable without significant development work. Those selected for serious study are now described in some detail.

**Complete hydrolysis process**
The first stage of this process is alkaline hydrolysis, under reflux with strong aqueous sodium hydroxide, to sodium dibutyl phosphate:

$$(C_4H_9O)_3PO + NaOH \rightarrow (C_4H_9O)_2P(O)ONa + C_4H_9OH$$

The process yields three phases:
- lowest, aqueous sodium hydroxide containing the vast majority of radioactivity and other metal contaminants;

- middle, sodium dibutyl phosphate (NaDBP), containing virtually all the remaining contaminants;
- uppermost, odourless kerosene and butanol, virtually free of phosphatic species and other contaminants.

The aqueous sodium hydroxide is a suitable feed for the Enhanced Actinide Removal Plant, and the kerosene might be fed to a relatively straightforward incinerator. However, although the sodium dibutyl phosphate phase contained only about 10% of the initial radioactivity, this was too much for sea discharge. The material could be converted to inorganic phosphate by acid hydrolysis,

$$(C_4H_9O)_2PO(O)ONa + HNO_3 \rightarrow NaH_2PO_4 + \text{organic fragments}$$

but the reaction proved strongly exothermic and difficult to control. Reaction might be incomplete without addition of sulphuric acid, which is corrosive and potentially incompatible with downstream aqueous effluent treatment plant.

### NUKEM pyrolysis process

The heart of this process is a high-temperature reaction of TBP with calcium hydroxide in a stirred pebble ball reactor, followed by gaseous incineration of the volatilised organics:

$$2(C_4H_9O)_3PO + 2Ca(OH)_2 \rightarrow Ca_2P_2O_7 + 6C_4H_8 + 5H_2O$$

The attractions of this process were the prospect of forming essentially a single major product in a solid form containing all the radioactivity, and the existence of a well-engineered pilot plant in Germany. The remaining engineering difficulties however proved formidable, the product was inconsistent, presenting problems in encapsulation, and there were concerns about meeting safety standards in the incinerator.

### TBP oxidation process

Oxidation by hydrogen peroxide was originally developed in Japan. Its initial attraction was in giving in a single stage an aqueous product potentially compatible with downstream plant:

$$(C_4H_9O)_3PO + 36H_2O_2 \rightarrow H_3PO_4 + 48H_2O + 12CO_2$$

Little or none of the kerosene was affected, but it emerged relatively clean and could be burnt easily, while the aqueous product contained virtually all the radioactivity. However, the oxidation of TBP could be completed only with a significant excess of hydrogen peroxide and at reflux temperature, even with an improved catalyst. An excess of oxygen together with kerosene above its flash point constituted an evident fire risk, and the cost of adequate safety measures proved prohibitive on

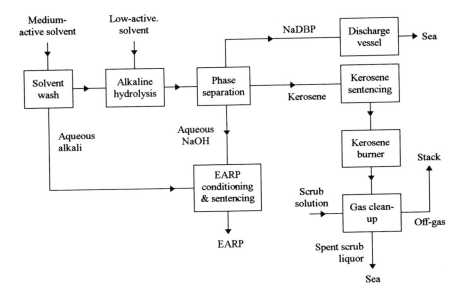

**Fig. 9.5** Solvent destruction process chosen for Sellafield

a large scale. Nevertheless the basic process might still be used to destroy small quantities of other organic waste.

### TBP hydrolysis plus NaDBP oxidation

Combining the best features of two wet chemical processes, namely alkaline hydrolysis of TBP and catalytic oxidation by hydrogen peroxide of the resulting NaDBP, might avoid the drawbacks of both. NaDBP is easier to oxidise than a TBP/OK mixture, being readily aqueous-soluble and so needing only a minimal excess of hydrogen peroxide in a single phase reaction; moreover the flammable kerosene has been largely removed. However, the reaction proved to generate small quantities of highly volatile organic species, which were extremely difficult to remove in conventional condensers and so would have led to very costly plant.

### Chosen Sellafield process[2]

Parallel work demonstrated that NaDBP was non-toxic, and could be discharged to sea if hydrolysis were preceded by an improved wash to minimise residual radioactivity. No downstream plant would then be troubled by the phosphate content. Thus a four-stage process has been evolved and is soon to be put into industrial practice.

The feed is first washed successively with water, 0.25M sodium carbonate (with which uranium forms a soluble complex), and 7.3M sodium hydroxide. This removes the vast majority of the radioactivity and the uranium, which is present in some of the waste solvent and would cause problems in subsequent stages.

The alkaline hydrolysis of the TBP by strong refluxing sodium hydroxide follows, forming the three phases previously mentioned. The activity distribution is > 90% in the NaOH and most of the rest in the NaDBP phase, with less than 1% in the kerosene. Simultaneous distillation of some kerosene and butanol helps the reaction.

The combined very low-active kerosene and butanol streams are fed to a vortex incinerator, chosen because although both phosphorus and activity levels will be low, they might build up slowly in a more conventional ceramic-lined type, gravely hampering eventual replacement.

Finally the alkaline aqueous product is conditioned for feeding to EARP.

### 9.3.4 Differences between processes for oxide and metal fuel

The above processes were developed essentially as adjuncts to the Magnox reprocessing plant, operated since 1964 with successive improvements as the importance of reducing discharges was gradually recognised. The need to minimise waste arisings of all kinds was a major influence on the design of the new plant for oxide fuel, THORP.

Most importantly it demanded a 'salt free' reductant, eventually U(IV) stabilised by hydrazine, to replace ferrous sulphamate in the separation of plutonium from uranium and avoid the limit on waste concentration imposed by the presence of iron salts. (Whilst not strictly 'salt free' in the same sense as, for instance, hydroxylamine which leaves no residues in solution, uranous nitrate is oxidised to U(VI) and follows the product stream with much the same effect.)

An even greater proportion of the active waste isotopes can now be concentrated and subsequently vitrified. Apart from the solvent wash raffinates, in which no satisfactory substitute for sodium carbonate and hydroxide has yet been found, all the aqueous process effluents can be so treated.

A small quantity of boronated water from the inner containers used to transport some of the oxide fuel from the power stations to Sellafield is fed to EARP. Corrosion of the stainless steel or zirconium cladding on oxide fuel is much less than of Magnox alloy, the spread of any contamination is limited by intermediate containers, and the bulk of water from the THORP cooling ponds does not require ion-exchange treatment before discharge.

## 9.4 Solid wastes

Solid wastes include not only those that are intrinsically solid, such as fragments of fuel cladding, but also any radioisotopes and other toxic metals removed as precipitates or sorbates from effluent streams. All need to be encapsulated for safe long-term storage.

(a) High Level Wastes (HLW) contain well over 99% of all the minor actinides and fission products separated by reprocessing. Although initially evaporated

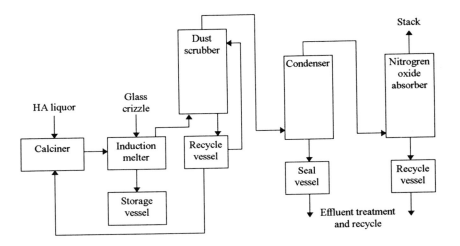

**Fig. 9.6** Sellafield waste vitrification plant (WVP)

and stored as concentrated liquid in high integrity stainless steel tanks as described in Section 9.3, they are later incorporated into glass and stored for eventual disposal as solid.

(b) Intermediate Level Wastes, a wide range appreciably less active than the HLW but still in need of conditioning before safe long-term disposal. Typical streams are shown in Table 9.3.

(c) Low Level Waste, very low-active or merely suspect materials from such sources as the Research and Development and Analytical Laboratories.

### 9.4.1 High level waste vitrification

The highly-active liquid waste concentrate is essentially a solution of metal nitrates. It is converted to mixed oxides and reacted at about 1000–1100 °C with glass-forming materials to form a vitreous product which cools into monolithic blocks.[3] The following details apply to the Sellafield version of the process.

**Preparation of feedstock**

Before vitrification the HLW is fully characterised by chemical analysis, and lithium nitrate is added at about 4 g/litre to inhibit the formation of the refractory oxides of aluminium, chromium and iron during calcination.

The glass forming reagents are added in the form of a granulated or *crizzled* base glass, consisting of particles 0.5 mm to 2 mm in size and of a closely specified composition.

## Calcination

The calciner is a stainless steel tube, sloping at 3% (1.7°) and rotated in an electrically heated furnace. It is operated at temperatures up to 850 °C under reduced pressure (100 mm water gauge), and air-tightness is ensured by graphite sealing rings in each of the end fittings.

The waste solution is introduced into the upper end; flowing down the tube it is progressively evaporated and denitrated to leave the lower end as a finely divided powder (calcine). Caking is prevented by the tumbling action of a heavy iron bar in the calciner tube, and by the thermal decomposition of added reagents which release large volumes of gas in the mass of the calcine. Sucrose used for this purpose has an added advantage as a denitrating agent and in suppressing the volatilisation of ruthenium as $RuO_4$ by destroying nitric acid in the feed.

## Vitrification

The calcine is fused with base glass, introduced through an air lock system, in a directly heated metallic induction furnace at a wall temperature of about 1100 °C. The furnace is designed to produce at least 25 kg of product glass per hour with a typical waste content of 25% by weight. During the dissolution and digestion of the calcine any residual nitrates are decomposed. When the volume of molten material in the melter reaches a specified level, typically after eight hours feeding, the product glass is drained off into a stainless steel product container.

## Product container handling

The product container holds about 400 kg of glass, corresponding to two melter discharges and typically derived from 8 tonnes of Magnox or 2 tonnes of LWR fuel. After filling, it is allowed to cool and is then closed with a welded lid and decontaminated by high pressure water jets. Containers are stacked, up to 10 high, in closed thimble tubes within a store cooled by natural convection of air, where they will remain for at least 50 years.

## Off-gas treatment

The combined off-gas produced in the melter and calciner contains water vapour, nitrogen oxides, entrained dust and some species, primarily ruthenium and caesium, volatilised by the process. Dust entrainment is typically 1% and volatilisation increases the ruthenium content by a further 1%. In order to keep the nitrogen oxides and radioactive emission of the effluent within the authorised discharge limit, off-gas is passed through a dust scrubber column, a tube condenser and a bubble-cap scrubber column for nitrogen oxide removal, then a scrubber column, electrostatics precipitators and high-efficiency filters.

The dust scrubber removes more than 99% of the contamination, and its highly active effluent is recycled to the calciner. The liquid effluent from all remaining stages of off-gas treatment is returned to the site highly-active evaporation plant for combination with HLW solution and eventual recycle to the vitrification plant.

## Glass selection
Borosilicate glass is formulated to satisfy the following criteria:
- maximum waste incorporation;
- ease of manufacture in the process temperature range;
- sufficient reactivity to digest calcine within the process residence time;
- suitable viscosity in both base glass and product;
- durability of the product.

### 9.4.2 Intermediate level waste (ILW)
Cladding fragments and other intermediate-level solid wastes on the Sellafield site have for many years been stored in dry or water-filled silos according to content. Although an acceptable provisional measure it needs constant supervision and so is unsatisfactory for the long term. Current arisings are being encapsulated ready for ultimate disposal, in a form designed to minimise attack by the environment or leakage into it. In due course all previous wastes should be treated likewise.

## Selection of the encapsulation matrix
The matrix for each ILW must be satisfactory in storage, transport, disposal and the encapsulation process itself.[4,5]

Cement has been found to be the most suitable encapsulant in every phase. In particular it is substantially better than other matrices, for example polymers and polymer-modified cements, during the encapsulation process and in disposal. The French process of macerating in hot bitumen is believed to present some difficulties and to be due for replacement when opportune; it has however been installed for medium-active residues in the Japanese plant at Tokai-Mura.

## Choice of cement
Pure ordinary Portland cement (OPC) is unsuitable as encapsulation matrix owing to excessive heat of hydration and consequent thermal stress. Various diluents have been considered, including sand, pulverised fuel ash (PFA) and ground granulated blast furnace slag (BFS). Loaded clinoptilolite from the site ion-exchange plant may also be used.

Sand requires admixtures to produce the rheological properties necessary for solid waste infilling, and has been rejected; formulations are best kept simple, comprising water and blended cement powders without additives. This also increases operational flexibility.

Both BFS/OPC and PFA/OPC have been selected as matrices for the encapsulation of different ILW. The main advantage of BFS is that it is a cementitious material and reacts with water to form a hardened mass. In combination with OPC, it produces a totally hydrated system ultimately better than OPC alone, with lower

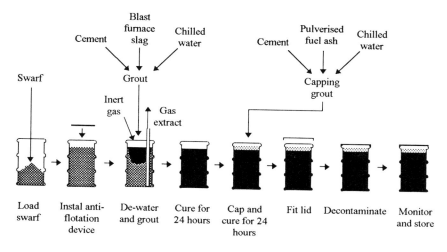

**Fig. 9.7**  Encapsulation of intermediate-level waste in cement

permeability and pore size for a given water content. BFS/OPC has been adopted as the encapsulation matrix for most of the ILW streams.

PFA improves the rheological properties of a cement grout and is therefore used mainly where high fluidity is required.

### Product evaluation

Encapsulated waste must not only be robust enough for immediate handling, but stable at least until final disposal, up to 100 years onward. Typically over 25 properties must be tested—more than 30 are possible—for about 25 waste types. Evaluations cover

- physical and mechanical properties;
- physical and chemical stability;
- radiation stability;
- leach behaviour.

Stability depends on:

- how cement hydrates;
- how the waste interacts with the cement;
- effects of scale-up;
- physical, chemical and radiation-induced degradation processes.

### Cementation practice

Two basic processes are used in the UK for encapsulating ILW, although other options have been considered. Solid wastes are packed into 500 litre stainless steel drums and the voids filled with cement grout; slurries are mixed with grout-forming materials within similar drums by means of a paddle which remains embedded.

The earliest plant opened in 1990 and all are concentrating at present on current waste arisings, but the intention is to retrieve and process historic wastes over the next 15 years. For the time being, the filled drums are stored above ground, but in due course they should be transferred to a deep geological repository (see Chapter 10).

### 9.4.3 Low level waste (LLW)

In the UK, this material, with very low levels of radioactivity, is packed into drums which are compacted and loaded into ISO freight containers. These are placed in concrete vaults immediately below the ground surface, at Drigg which is near Sellafield. When these vaults are full they are covered by an impermeable clay cap. Details are given in Chapter 10.

## 9.5 Aerial effluent

All aerial effluent is derived from ventilation, which in a nuclear plant serves two purposes:
- it maintains a comfortable environment for the operators,
- by setting up pressure differentials between successive levels of containment, it ensures that any air flow across boundaries is towards the region with greater risk or extent of radioactive contamination, and so helps to contain such contamination within the process area.

The latter function clearly requires a forced draught, and that the air expelled particularly from the process areas must be freed from hazardous substances before discharge to the environment.

### 9.5.1 Nature of aerial effluents

Reprocessing is the separation of used fuel from its cladding, followed or accompanied by dissolution in hot nitric acid and subsequent separation into plutonium, uranium and waste streams. Dissolution releases gaseous fission products and a nitrous fume (most of which is recycled); some of the chemical processes have gaseous by-products or require sparging, causing traces of the process liquid to become suspended in the vessel atmosphere; liquid transfers may cause splashing or involve the use of air-lifts, with similar effects; and dry processes, particularly shearing oxide fuel, tend to generate dust of which some may also be suspended. Thus air leaving the process areas is liable to contain:
- radioactive fission gases such as krypton-85;

- dusts and mists rich in fuel or its constituents;
- chemically noxious gases or vapours such as nitrogen oxides.

Levels of all these in discharged air must be kept within strict statutory limits to avoid toxic or radiogenic effects on the public, whether by direct exposure or through the food chain.

## 9.5.2 Treatment technologies

Since aerial effluent includes a wide variety of species, no single technology is suited to all arisings. The three main methods used are wet scrubbing, electrostatic precipitation and filtration.

### Wet scrubbing

Some gases liberated from reprocessing operations are reactive and can be absorbed by a liquid stream. Such gases may be radioactive (eg carbon dioxide containing C-14, iodine vapour containing I-129) or non-radioactive such as oxides of nitrogen ($NO_x$). These can be removed by intimate countercurrent contact with sodium hydroxide solution in a packed tower; $^{14}CO_2$ can be precipitated in the form of barium carbonate, as in the new THORP plant at Sellafield (UK).

A proposed alternative process (IODOX) is based on the oxidation of iodine-bearing species to involatile iodates by hot, highly concentrated nitric acid. However, the concentration necessary for effective oxidation is higher than commonly used in industry, presenting problems of corrosion and fume retention.

The efficiency with which iodine is removed depends strongly upon the form in which it occurs. Both as the element itself and as hydrogen iodide, it is readily absorbed by an alkaline scrub; organic iodides, such as are formed by reaction with the process solvent or with lubricating oils, are not. If these amount to a significant proportion of the total iodine inventory, a final polish with a more effective absorbent such as silver zeolite may have to be considered. (The cost of such materials generally precludes their use as first-line absorbents in a large-scale process.) Partly in order to minimise the formation of organic iodides, the fuel dissolver solution in THORP is sparged with nitrogen oxides which eliminate most of the iodine in absorbable form before it can meet the solvent.

The liquid effluent from an iodine scrubber presents a particular difficulty in effluent management; if it is allowed to mix with other, acidic waste streams, the iodine is likely to be volatilised again. No completely satisfactory way of immobilising iodine is known. On a coastal site the best procedure is considered to be segregation from acidic wastes and discharge to sea, where natural dispersion mechanisms so dilute the iodine that it does not unduly augment the natural radioactivity due to potassium-40.

A column scrubber can also wash out some entrained dust. However, particles are better removed in devices specific for the purpose, such as venturi scrubbers where incoming air saturated with water vapour is expanded to cause condensation by adiabatic chilling. Small particles, which are usually difficult to remove, act as nuclei for condensation and can then be efficiently collected as larger droplets which disentrain readily.

**Filtration**
Filters have been used in the nuclear industry for many years. Developments to improve their efficiency have led to modern glass fibre units which retain over 99.9% of particles, so that the critical requirement is a tight seating. (For comparison, filters in domestic appliances rarely exceed 90% efficiency). These *High Efficiency Particulate in Air* (HEPA) filters each contain about 20 square metres of pleated glass-fibre fabric, arranged in a rectangular or circular case. Particles in the air are forced to take a tortuous path with multiple opportunities for attachment. Large particles are impacted on to the fibres, whilst smaller particles diffuse within the air stream and are drawn to the fibres by electrostatic forces.

Because of the scale of ventilation needed in a major nuclear plant, filters are usually arranged in banks containing perhaps many tens of units. Typically air is drawn through two identical banks in series, with very high overall efficiencies, often as the final *polish* before discharge. After a time, the resistance of the filters may increase and their effectiveness diminish owing to the collected deposits; they are then replaced and become solid waste for disposal.

**Electrostatic precipitators (ESPs)**
The electrostatic forces already mentioned are specifically utilised in an ESP, where very high voltages create the conditions for particles to acquire electrical charge. They then migrate along the electric field to an earthed electrode where they collect. As the air generally bears a mist of water droplets, the collection plate is constantly irrigated, and deposits are washed off into a collection tank for disposal. This system has been used in several plants to supplement other methods, especially where high levels of radioactivity debar routine intervention by operators.

### 9.5.3 Noble gas removal

Chemically unreactive gases, of which Kr-85 is by far the most important with a half-life of 10.7 years, will pass unaffected through all the treatments so far described. Even more than iodine, they resist immobilisation, and in the UK the best current practice is considered to be discharge to the atmosphere. The resulting collective radiation dose to the population is a substantial part of the total due to reprocessing, but still very slight in relation to natural exposure. Nevertheless the position is kept under review, and three separation processes have been considered.

**Activated charcoal**
The gases can be absorbed on a suitable charcoal bed. However, absorption is reversible, and desorption is promoted by a rise in temperature, such as may be

caused by the decay heat of the absorbed gases themselves. The risk of fire, which could release the whole content of the bed almost instantaneously, has virtually ruled out this process.

**Freon absorption**
The noble gases are soluble in certain solvents, most notably the refrigerants collectively known as Freons. A process cycle can be designed with the solvent absorbing the gases from the effluent stream and then distilled to separate them.

**Cryogenic distillation**
The effluent stream could be dried and liquefied, then fractionally distilled. Krypton (b.p. $-152$ °C, m.p. $-157$ °C) remains among the less volatile residues.

After any separation, the problem of disposal would remain. Without reliable immobilisation, a whole inventory or a substantial part of it might be released suddenly, causing a local concentration of radiation dose. Because of their inert nature the gases are unsuited to chemical immobilisation. They might be compressed into cylinders, with the obvious risk of leakage; alternative methods that have been considered are implantation in a metallic matrix, or irreversible absorption in a zeolite.

Separation would incur not only the capital, operating and storage expenses, but also an additional radiation dose to operating personnel that has to be weighed against the very slight reduction to the populace at large. Whether the effort would be worth while remains questionable.

## 9.6 Practices at other sites

At present the principal reprocessing countries besides Britain are France and Russia, although Japan is expanding its capabilities. (Other countries with nuclear power programmes either purchase fuel-management services from abroad or retain their fuel as discharged.) Practices in waste management are generally similar to those already described, although with some significant variations.

This applies even to the small reprocessing plant operated by the UKAEA at Dounreay for fuels from materials-testing and fast reactors. The highly-active raffinates are concentrated by evaporation and then stored as liquids in high-integrity tanks. The preferred route for the concentrates depends at present on the fuel from which they arise:

- for those from MTR fuel, cementation, the plant for which is currently undergoing inactive commisssioning;
- for those from DFR and PFR fuel, cementation or vitrification.

The medium-active aqueous effluents are subjected to a flocculation process which produces a precipitate of ammonium diuranate. (The uranium product, unlike that from THORP, is depleted in U-235 and of relatively little value.) Currently three alternative ways of treating the floc are under consideration: re-dissolving and reprocessing it, cementation, and vitrification.

The low-active effluents are discharged to sea at a pH of about 3–4 following simple filtration. In future however the pH will have to be at least 7, and the consequent active precipitates will need more extensive filtration.

Spent solvent is stored for the present in tanks, pending a decision on a treatment process.

### 9.6.1 France

Highly-active liquid and cladding fragments are respectively vitrified and cemented as in the UK.[6] Medium-active liquids are however directly treated by floc precipitation as in EARP.[7] Because of the higher feed activity, greater decontamination factors are required, and ferric hydroxide is supplemented as carrier by nickel ferrocyanide, titanium sulphate, cobalt sulphide and barium sulphate. Solids are separated under gravity rather than by ultrafiltration, and encapsulated in bitumen rather than cement. Apparently the intention is to redirect some of these medium-active effluent streams to evaporation and vitrification, as is already done with alkaline salt wastes.

Waste solvent treatment also differs. In France the diluent is dodecane, probably the normal hydrocarbon although hydrogenated propylene tetramer has been used and not always clearly differentiated. Unlike kerosene, most of it can be recovered in usable condition by distillation and is recycled, along with some of the TBP. The remainder is currently stored pending the development of a treatment process, for which the Nukem technology (Section 9.3.3.) is believed to be favoured at present.

### 9.6.2 Japan and Russia

Again as much as practicable of the higher-level wastes are evaporated, stored and eventually vitrified.[8,9] Fuel hulls and other wastes with significant activity will be suitably encapsulated for disposal. In Russia, low-level waste goes to land-fill sites.

### 9.6.3 Germany

Although Germany has abandoned its own reprocessing intentions, at least for the time being, it does have fuel reprocessed in France and Britain under contracts that specify return of the fission products. Most will be in vitrified form. A small proportion in lower-level waste would be cemented or bituminised,[10] but to minimise bulk transfers, agreement is being sought to substitute an equivalent amount of high-level material from other sources. This requires the approval not only of the customer but also of the reprocessor's regulatory bodies.

## 9.7 Conclusion

Practices in waste management have evolved greatly over the past two or three decades, with considerable reductions in discharges to the environment. Attempts continue around the world to find cost-effective ways of reducing them still further, and the IAEA has recommended research particularly on chemical precipitation, ion exchange, evaporation, membrane processes, solvent extraction, biological processes, electrochemical processes and treatment of waste organics.[11]

Mere extensions of existing technology could of course further decontaminate effluent streams, but always with diminishing returns and additional secondary wastes. Beyond a certain point, expense on safety measures would be more genuinely effective if applied in completely different areas of conventional hazard. Where this point lies in relation to current practice is a matter for debate.

## 9.8 References

1 R. Ivens. Effluent management—the next step. *Atom*, July/August 1990, p. 20.
2 G. V. Hutson. Selection of a process for a waste solvent treatment plant at Sellafield. *Environmental Protection Bulletin* **19** (1992).
3 W. Smith. Vitrification of Sellafield wastes. *Atom*, March 1985.
4 G. A. Fairhall. Effects of operational variables on the product properties of encapsulated intermediate level wastes. *Conference on Radioactive Waste Management 2*, BNES, Brighton, pp. 79–84 (1989).
5 G. A. Fairhall and J. D. Palmer. The evaluation of properties of immobilised intermediate level wastes. *Radioactive Waste Management and the Nuclear Fuel Cycle*, Vol **9** (1–3), pp. 51–70 (1987).
6 P. Ledermann, P. Miquel, and B. Bouillis. Optimisation of active liquid and solid waste management at the La Hague plant. *RECOD 94*, **2**, session 9A (1994).
7 Chemical precipitation processes for the treatment of aqueous radioactive waste. *IAEA Technical Reports Series No. 337*, Vienna (1992).
8 E. V. Kulikov and N. A. Rabov. Policy and principles of radioactive waste management in Russia. *RECOD 94*, **1**, session 2A (1994).
9 Y. Yamaguchi and A. Yamato. Radioactive waste management in Japan; policy and major issues. *RECOD 94*, **1**, session 2A (1994).
10 P. Bauder and W. Blaser. Management of operational and reprocessing wastes from nuclear power plants in Germany. *RECOD 94*, **1**, session 2A (1994).
11 Advances in technologies for the treatment of low and intermediate level radioactive liquid wastes. *IAEA Technical Reports Series No. 370*, Vienna (1990).

# 10 Disposal of fuel or solid wastes

S. Richardson, P. Curd, and E. J. Kelly
*BNFL, UK Group, Cumbria (S.R. and E.J.K) and Peter Curd Associates, Faringdon, Oxfordshire (P.C.)*

## 10.1 Introduction

Every industry produces waste, some of it highly dangerous. For example certain heavy metals and chlorinated hydrocarbons are toxic, remain so indefinitely, and need to be kept from the human environment effectively for ever. Although radioactive materials and the associated potential dangers are in principle different, decaying through time, the half-lives of some nuclides are so long that similar precautions may be needed. Others are short lived, or present in such low concentrations as to present little hazard, and precautions may then be relaxed accordingly.

Disposal practices need to be matched to the waste itself, which ranges from intact fuel (if not to be reprocessed), concentrated solutions of fission products or their vitrified product down to streams or packages in which radioactivity is scarcely measurable or indeed merely suspected. The essential principle of radioactive waste management is that neither current operations nor their potential consequences should expose any individual to a significant risk of harm. Furthermore, no burden of care that can reasonably be avoided should be placed on descendants who will not benefit from the operations generating the waste.

Particularly for the far distant future, where history suggests that a sustained level of technological competence cannot be totally assured, these requirements imply that dangerous wastes must be proofed against all likely access so long as the danger lasts, with strict precautions against appreciable leakage of radioactive components into the biosphere. This means, among other things, that they must be solidified.

Such precautions are expensive, and for large volumes of scarcely-radioactive waste would be neither practicable nor necessary, so that dispersal into the environment may cause the least overall detriment. Every country concerned has its own legal provisions on what may be discharged, usually based on recommendations by the International Commission on Radiological Protection (ICRP) on acceptable radiation exposures to the general population.[1] In practice, deliberate discharge is limited to liquid and gaseous effluents that cannot realistically be managed in any other way, and even then all reasonably practicable measures must be taken to minimise their radioactive content within absolute limits. Details are given in Chapter 9.

As elsewhere in this book, British practice is taken as an example. It is governed by the Department of the Environment (DoE), Her Majesty's Inspectorate of Pollution (HMIP), and where discharge to the biosphere is concerned, the Ministry of Agriculture, Fisheries and Food (MAFF). In other countries, though similar principles apply, their interpretation may differ.

Apart from dispersal, the practical options are:

- supervised storage above ground (necessarily a temporary measure);
- shallow burial, generally adequate only for low-level waste;
- disposal to a deep repository.

Disposal may also be described as *retrievable* or *irretrievable*. The latter term, although relative, might practically be defined as disposal in a manner that would need a mining operation to recover the material.

## 10.2 Waste categories

Waste is classified broadly as High- (HLW), Intermediate- (ILW) or Low- (LLW) level, according to its radioactivity. Formal definitions are given in Chapter 9, but in brief, HLW is the heat-releasing waste containing the bulk of fission products from irradiated fuel, LLW contains so little that near-surface disposal is considered to pose no unacceptable hazard, and ILW has too much to be classed as LLW but not enough to cause significant heating. Fuel directly disposed without reprocessing would naturally fall into the HLW category, but as some special considerations apply it is described here in a section of its own.

A recent review by the Department of the Environment has suggested that definitions should be more closely relevant to disposal decisions.[2] For example, it may not be wise in practice to dispose all LLW near to the surface. Conversely there is no safety reason why all short-lived ILW need go deep. Likewise in the USA, the practice of categorisation by source rather than content is recognised to have been unduly restrictive in some respects, and is expected to change very soon.

## 10.3 Low-level waste

Low-level waste includes rubble and steelwork from decommissioning and items such as rubber gloves, protective clothing and packaging from the nuclear industry and the medical and industrial use of radioactive materials. UK regulations specify that low-level waste must contain less than 4 GBq of alpha radiation emitting

186  *Disposal of fuel or solid wastes*

**Table 10.1** Maximum annual disposal limits per nuclide

| Radionuclide | Annual limit (TBq/yr) |
|---|---|
| Uranium | 0.6 |
| Radium-226 and Thorium-232 | 0.03 |
| Other alpha | 0.3 |
| Carbon-14 | 0.05 |
| Iodine-129 | 0.05 |
| Tritium | 10 |
| Others * | 15 |
| * In this context 'others' denote: | |
| • beta emitting radionuclides with half lives greater than three months; | • |
| • iron-55; and | • |
| • a constituent of cobalt-60 not exceeding 2 TBq. | • |

material or 12 GBq of beta/gamma emitters per tonne. The evolution of current practice is well illustrated by the Drigg Low Level Waste Disposal Site, owned and operated by British Nuclear Fuels plc (BNFL).

### 10.3.1 Drigg: general description

The site is an engineered disposal facility for waste generated by UK nuclear facilities, such as the fuel reprocessing plant at Sellafield and the power utilities operated by Nuclear Electric and Scottish Nuclear, and also for waste generated by radiochemical activities in hospitals, research, and educational establishments. Additionally, waste is received from Ministry of Defence establishments and from radioactive source manufacturers. Although the annual disposal figures vary, approximately 70% of the waste is derived from the Sellafield site, six kilometres to the north.

The volume of waste disposed to date is around 850 000 m$^3$ and current disposals are of the order of 30 000 m$^3$ per year. The total area of the site is approximately 110 hectares of which 22 hectares at the north end has consent for current and future disposals.

Drigg accepts only solid low-level waste (LLW) for disposal, and has been doing so since 1959. Besides the conditions that the waste must not exceed 4 GBq per tonne of alpha-emitting nuclides or 12 GBq per tonne of all other radionuclides, the site also has limits on the annual disposal of specific radionuclides as illustrated by the current limits in shown in Table 10.1.

### 10.3.2 Geology

The site is on the West Cumbrian coastal plain in the north-west of England, about 1 km east of the Irish Sea coast and 300 m north of the tidal estuary of the River Irt. The elevation of the land varies from 15 to 20 m above mean sea level in the consented disposal area.

The dominant geology is a bedrock of Permo-Triassic St Bees Sandstone, heavily covered by superficial deposits of glacial or other sedimentary origins. A simplification of the stratigraphy would be to interpret the site as number of impermeable clays interlaced with more permeable sand or gravel zones. The lateral coherence of these layers is uncertain although it is known that the clay layers promote the formation of numerous perched aquifers. The St Bees Sandstone underlies these sediments at depths varying from 10 to 50 m. The understanding of the geology is a key area in the future development of risk assessment models to predict the groundwater movement across the site and an in-depth site characterisation programme exists to support the characterisation of the currently disposed areas and to support engineering developments in new disposals.

### 10.3.3 Trench disposal

From 1959 until early in 1995 trench disposal was authorised as the disposal system for LLW. Seven trenches were excavated into one of the more laterally persistent clay horizons, known locally as G5. This stratum has a very low permeability and acts as a primary barrier between the disposal site and the regional aquifer in the St Bees Sandstone. Typical trench dimensions (apart from the last, on a triangular plan to fill a space) are 25 m wide by 700 m long and 8 m deep.

The trenches were filled by tumble tipping over a progressively moving tipping face whilst partially capping the top of the tipping face. In case of spontaneous combustion, firebreaks of earth were introduced along the trenches from 1967 up to 1983 when they were discontinued.

The natural impermeable base in the G5 clay induces a saturated zone in the waste coincidental with the local G6 perched aquifer. In order to avoid the rapid transfer of nuclides from this saturated zone, engineered barriers have been constructed to limit the water infiltrating into the trenches. Since the dominant groundwater flow is towards the south-west, a cut-off wall has been constructed around the north-east boundary of the site. It consists of a narrow trench keyed into the G5 clay and filled with a cement-bentonite slurry which solidified to form a moderately impermeable structure.

An additional source of infiltrating water is rain. The disposal trenches have been covered with an engineered temporary cap, consisting of an earth gradient stabilised with a permeable, reinforcing geotextile membrane. The upper layer of the cap, which is up to 8m deep and comprises approximately 440 000 $m^3$ of earth, contains an impermeable polymer membrane that promotes run-off to catchment drains. It is also vegetated with shrubs to promote stability and limit erosion by water.

At the Drigg site, the annual rainfall is approximately 1100 mm per year. At least half of this is lost in evaporation and plant uptake. A large proportion is run

188  *Disposal of fuel or solid wastes*

**Fig. 10.1**  Low-level waste disposal site, Drigg, Cumbria, UK

off to drains, leaving only a few hundred millimetres to infiltrate into the unsaturated zone of the waste.

### 10.3.4  Vault disposal

Tumble tipping was not acceptable as a satisfactory way of handling LLW from the viewpoints of operator dose, airborne contamination or, simply, appearance. These factors, and prompting by the government Radioactive Waste Management Advisory Committee (RWMAC), led to requirements for a more controlled and aesthetically sound disposal method. Vault 8 as the successor to trenches 1 to 7 is a radical departure from the older methods of disposal (Fig. 10.1), and has many engineered features capable of retaining waste over extended periods. Moreover, improved assaying facilities should better characterise the consigned waste and thus the subsequent environmental impact of the Drigg site.

Where suitable, wastes in 1 $m^3$ boxes or 200 litre drums are compacted with a 5000 tonne press, and reduced in volume by a factor of 4 to 5. The resulting *pucks* are stacked in 15 $m^3$ overpacks resembling half-height ISO containers (as seen on freight ferries) and the spaces filled with a cement grout. Wastes not amenable to compaction will be grouted directly into overpacks. In time the metal components are expected to corrode, but the grout will retain an intrinsic strength preventing collapse.

Vault 8, completed in 1988, has a disposal capacity of 180 000 m$^3$, and 26 000 m$^3$ of concrete with 2600 tonnes of reinforcing steel went into the construction. Large underground drains take the run-off during the operational phase of waste disposal. At one corner where the natural clay is less adequate than elsewhere as an impermeable lower horizon, the geological strata have been excavated and replaced with an engineered sodium bentonite clay.

### 10.3.5 Hydrological provisions

The water flow across the site is managed to prevent radionuclides from entering the local terrestrial environment. Each trench has a leachate drain leading to collection tanks which also take the run-off from Vault 8 and from streams crossing the non-consented area of the site. At high tides, when conditions permit, the tanks are discharged through a marine pipeline into sediments off the coast.

Boreholes and geophysical techniques are used to determine the flow of water beneath the surface of the site and surrounding area. The aim is to develop an understanding of the water balance and calibrate ground-water flow models.

### 10.3.6 Environmental studies

The site has to be monitored for statutory purposes to assure the regulatory bodies that discharges are within approved limits. Additionally, large scale research programmes exist to characterise the behaviour of radionuclides in the site, and to develop appropriate mechanistic interpretations of that behaviour in order to predict the long term radiological impact of the site.

Research topics include:

- solubilities of radionuclides under environmental conditions;
- rates of release where the solubility limit has not been reached;
- sorption of key nuclides by barrier and geological materials from various leachates and ground waters;
- transport or immobilisation of radionuclides in colloidal form;
- the breakdown of cellulosic materials, under both trench and vault conditions, to yield gases and potential complexing agents that might affect the migration of radionuclides;
- details of microbial action, the existence of which is indicated by the evolution of carbon dioxide and methane from venting pipes in the trenches;

- corrosion of metals, important because (i) oxidation products may sorb radionuclides and act as *sinks* or alternatively transport radionuclides as colloids, while (ii) gaseous hydrogen may interact with microbiological cycles;
- environmental science, e.g. chemistry (of both major elements and radionuclides in ground-water, colloidal concentrations and gaseous evolution), larger scale geology and hydrogeology of the site.

## 10.4 Intermediate level waste

Intermediate level waste consists mainly of metal swarf from fuel cans, filters and other wastes from effluent treatments, worn-out contaminated plant and equipment, isotopes used in medical and industrial practices and radioactive materials from the defence programme. It has been suggested that some short-lived constituents might appropriately go with most of the LLW to a shallow disposal site.[2]

### 10.4.1 Principles of disposal

In the UK there is today (1995) the equivalent of 90 000 m$^3$ of ILW in store. The same quantity again will be generated from existing nuclear plants by the turn of the century. To date, it has practically all been held on the surface, and with repackaging and additional stores, this could continue safely for many decades. Long-lived radioactive wastes however demand the right degree of containment for tens of thousands of years; since 1987, the policy of the UK government has been that disposal should generally be in a deep underground repository where inadvertent penetration would be extremely improbable even if the memory of its existence should be lost, and from which seepage into the biosphere would indefinitely remain negligible or within tolerable levels.

Careful engineering can lock up the wastes in a long-lived natural system which has been substantially unchanged for tens of millions of years. There they will pose no significant threat to living things and safety will not depend on man's frail will and abilities to sustain it. The present generations who have exploited nuclear technology can discharge their obligations to the future to clear up after themselves.

In Britain, responsibility for providing and managing such a repository for intermediate level (ILW) and some low level (LLW) solid radioactive waste has been vested in United Kingdom Nirex Limited, jointly owned by the organisations of the civil nuclear industry with the Secretary of State for Trade and Industry holding a special share.

After assessing various possible sites for a repository, Nirex short-listed areas near Dounreay in northern Scotland and Sellafield, BNFL's reprocessing plant in West Cumbria. For the purposes of repository construction the geologies of both looked broadly similar, with initial drilling suggesting the potential to support a safety case; both also have local populations familiar with the nuclear industry.

Clear transport, environmental and cost advantages at Sellafield, where some 60% of the waste for deep disposal arises, dictated that the site be studied in detail. Work is concentrated for the present on establishing whether geological conditions would be suitable—not necessarily the best possible, for which the search might continue indefinitely.

## 10.4.2 Safety considerations

Unlike most areas of the nuclear industry where safety aspects are chiefly concerned with current operations, those of a repository are inherently directed towards the remote future: distant generations must be afforded the same level of radiological protection as people alive today. The principal requirements for post-closure performance are set out by the authorising departments, under the Radioactive Substances Act 1960, in the 1984 publication 'Disposal facilities on land for low and intermediate level radioactive waste—principles for the protection of the human environment', the 'Principles Document.'[3] In particular, 'the appropriate target applicable to a single repository at any time is, therefore, a risk to an individual in a year equivalent to that associated with a dose of 0.1 mSv: about one chance in a million'.

This statement implies a risk/dose relationship (or risk factor) of 0.01/Sv. However, this has since been revised to 0.06/Sv, covering both fatal cancers and serious hereditary disorders, in line with the 1990 ICRP and subsequent recommendations. The target dose limit has therefore been reduced accordingly to about 1% of the average natural background radiation in the UK, which itself varies by about 800% between London and some areas of Cornwall, for example.[4]

Within this limit, exposure from radiation should be As Low As Reasonably Achievable (ALARA), and 'future movement of radioactivity from a facility should not lead to a significant increase in the radioactivity naturally occurring in the general locality of the facility'.

The risk target is very low and applies at any time without limitation. Given that ILW contains a significant inventory of long-lived radionuclides, the disposal concept must provide adequate containment for a long period of time without monitoring or intervention to ensure safety, and be amenable to effective modelling to demonstrate satisfactory performance.

There is substantial international agreement on a concept of deep geological disposal for radioactive waste. The setting provides long term isolation from future human activities, such as resource exploitation, surface erosion arising from climate change, or other potential disruptive effects. The immediate containment is substantial, but is assumed to fail at some time so that radionuclides can eventually escape and dissolve in ground water. A multiple barrier is therefore needed with both engineered and natural components.

The engineering will be simple and robust, but containment will not depend on its indefinite integrity. Rather it will be maintained even after long-term physical and chemical degradation of the barriers towards their equilibrium state, through the establishment of uniform chemical conditions across the repository and the use of minerals with a high sorption capacity for radionuclides, by ion exchange or other mechanisms.

Caverns would be excavated at depth in a stable geological environment. Wastes packaged in steel or concrete containers would be surrounded by a cementitious back-fill to maintain an alkaline environment with a pH of about 10.5. Predictions covering the required periods of time depend on mathematical modelling of leaching and transport processes in pores, and particularly in cracks which could become preferential pathways for groundwater flow. A substantial experimental programme is providing the necessary basic data for the models.

An important property of the cement is its ability to absorb radionuclides, particularly the actinides. Combined with the low solubility of most long-lived radionuclides in alkaline media, this means that even when their primary containment has corroded, their movement through the back-fill will be very much slower than that of the water.

The back-filling material is therefore specified to ensure:

- long term maintenance of alkaline pore water chemistry in order to minimise dissolution of key radionuclides under the prevailing conditions of ground water flow and geochemistry;
- long term maintenance of an extensive active surface for sorption of key radionuclides;
- homogeneous performance through a relatively high permeability and porosity that will prevent localised concentrations of materials in wastes from exhausting the desired chemical conditioning and detracting from containment.

With this concept a *source term*—the effective amount—of dissolved radionuclides in repository pore water can be derived in a manner that is relatively simple, and therefore easily understood.

The repository is assumed to participate in the general movement of ground water through the host rock, so that the *near-field* (repository) source term is susceptible to transport into the geosphere by the many processes, such as advection and diffusion, which control transport of solutes in such systems. The host geological environment is intended to provide a stable setting in which the flow is predictable, with a long pathway and travel time for transport of radionuclides to the biosphere. Because of the sorptive mechanisms already mentioned, the travel time may be very much longer than that of the ground water which itself will probably be millions of years.

### 10.4.3 The investigative programme at Sellafield

The Borrowdale Volcanics Group (BVG) of rocks, the current focus of investigation, were formed some 500 million years ago. They can be found at the surface in the Cumbrian hills but near Sellafield these hard, impermeable rocks dip below layers of sandstone reflecting the geological history of the Lake District.

To date, 18 deep boreholes have been drilled as part of a regional and local study of geology and hydrogeology. The first eight, from which results were published in December 1993,[5] showed that the site continued to hold 'good promise'.

The Science Programme has mapped the boundaries between the rock formations and the main faults that they contain. Evidence of stratigraphic subdivisions of the BVG has been documented.[5] These subdivisions have been correlated between boreholes several kilometres apart. Results to date show that although the BVG is known to contain many fractures, the flow of ground-water is largely restricted to a few of them, that movement is very slow through the bulk rock, and that water in the Potential Repository Zone is ancient, possibly more than 30 000 years old.

The boreholes give a good outline picture of the geology of the area as well as the hydrogeology (how small quantities of water move in the rocks), of which an understanding is crucial as this is the most likely route by which radioactivity could get back to the biosphere. The picture however needs fleshing out with additional data on the rocks and water flow behaviour, to permit a firmer assessment of long-term safety, help engineers decide on the location, design and orientation of a repository and provide data for construction methods. The necessary detailed studies can be carried out only in a rock laboratory, more correctly termed a *Rock Characterisation Facility* or RCF. Many countries including Switzerland, France, Sweden and the United States are planning, or indeed operating, such facilities as part of their geological studies.[6]

Subject to planning permission, work on the RCF is expected to begin in 1996, and by the turn of the century to yield enough information for a decision whether or not to seek permission for a deep repository. On that basis, and allowing a full seven years for construction and further experimentation in the RCF, 2010 AD is the target for first waste emplacement.

### 10.4.4 Safety modelling

Data from the science programme are used in computer models to assess the safety of the repository far into the future. At this stage there are significant uncertainties in both the data and the models, and the range of possible results is correspondingly wide. Illustrative calculations based on models broadly consistent with the updated geological and hydrogeological interpretation indicate however that the Sellafield site has the potential to support a safety case which would satisfy the demanding regulatory target.

Substantial research has also been undertaken on quantities and properties of various types of waste and the performance through time of man-made parts of the repository system. Present estimates suggest that within the hydrogeology at Sellafield, highly alkaline chemical conditions which would strongly inhibit the dissolution of key radionuclides can be sustained in a repository for more than 100 000 years.[7]

### 10.4.5 Repository design concept

The general concept for a repository near Sellafield is a series of caverns excavated in a stable rock formation some 650m below sea level. The preferred form is based on access by twin drift tunnels from the surface near the existing Sellafield works of BNFL to a subterranean receipt centre. From there, spiral tunnels with a rail system would give access for construction and waste transport to the caverns.

Since this concept was first devised, the prospective volumes of waste destined for the repository have come down following an agreement on user-financing principles. This may enable further simplification of the design.

Customers' indications now support an assumption of 300 000 m$^3$ of ILW up to 2060 AD. New technology allows waste producers to reduce arisings and to compact their volume. In addition, they are confident that they can continue to use the Drigg facility for most LLW well into the next century. Only a nominal 100 000 m$^3$ of LLW is now expected to go to the deep repository.

### 10.4.6 Costs of deep disposal

Cost management is second in importance only to concern for safety. Uncertainties in the costings are presently significant but will narrow as the project emerges from the scientific investigation phase into the conventional engineering of a repository. A recent OECD study of estimates in various countries suggests that those in the UK are reasonable.[8]

The Waste Management Review[2] considered the costs of early deep disposal against other options that would be available if (as is currently judged inadvisable) disposal were left to future generations after interim storage at the surface for some decades longer. The DoE's Consultation Document[2] contained a useful summary of a study by government economists of the comparative costs of representative options. That study, for which the industry's figures were independently audited, showed a very clear result. For a wide range of differing assumptions, including variations in the discount rate applied, a 25 or 50 year deferment of final disposal appeared neither to add significant cost nor to yield significant savings.

With a standard annual discount rate of 6% through all time, a 50-year delay could imply a net saving of £100 million (about $150 million). But, the Government's report points out, this is only about 1.5% of the discounted total cost of waste management and well within the margins of error in the total picture which includes decommissioning; waste retrieval, conditioning and packaging; interim storage and disposal. In undiscounted terms the total cost of the programme is assessed by the National Audit Office and AEA at more than £20 billion ($30 billion).[9]

In a report prepared for the Radioactive Waste Management Advisory Committee, the Science Policy Research Unit of Sussex University suggested that an annual discount rate of 4% might be appropriate in assessing nuclear decommissioning options, but falling to 2% after the first 25 years.[10] Advice from the London Business School and from the Centre for Social and Economic Research on the Global Environment points to a rate well below 6% as appropriate to assessing resource costs in radioactive waste management.[11]

Early disposal brings important benefits in terms of reduced risks to the public and to workers in the nuclear industry. The national commitment to sustainable development also accepts the moral argument for not leaving the problem to later generations. This is the view taken by Government and reiterated in the review consultation documents—that 'the Government continues to favour a policy of disposal rather than indefinite storage for intermediate level wastes.'[2]

## 10.5 High level waste

The issues surrounding HLW disposal are complex. For a complete picture we must also consider direct disposal of irradiated fuel (Section 10.6) since the ultimate processing and disposal costs of these waste forms are a crucial part of the economic arguments for and against reprocessing.

### 10.5.1 Disposal policy in the UK

In recent years the policy for HLW in the UK has been directed by the 1982 White Paper 'This Common Inheritance'.[12] This indicated that vitrified waste forms should be stored at the surface for a period of 50 years, to allow decay of the main heat-generating radionuclides, but emphatically not to substitute for eventual disposal. Looking immediately for potential disposal sites would have been inappropriate, but the White Paper also directed policy towards the production of waste forms compatible with disposal, and the continuing development of scientific and technical knowledge that would underpin the eventual site selection.

More recently, the 1995 White Paper 'The Prospects for Nuclear Power in the UK'[13] was published. In addition a 'Review of radioactive waste management policy'[14] was also produced which highlighted that for HLW

> '...disposal to geological formations on land is the favoured option for the long-term management of vitrified HLW once it has been allowed to cool...'

while the UK Government is developing

> '...a research strategy, the aim of which will be to produce a statement of future intent in this area, setting out the decisions to be taken and the milestones to be achieved.'

The development of policies in other countries largely depend on whether reprocessed HLW or spent fuel is the main disposal option. A comparison of strategies employed for all waste forms in other countries is presented later in this chapter (Section 10.7).

### 10.5.2 Waste forms

In Chapter 9 the technology of vitrification for encapsulating HLW streams has been discussed. In the UK, France and some US operations, the major glass waste form is based on a borosilicate composition. Other types of glass are also under consideration including aluminosilicates, aluminophosphates and lead-iron phosphates, to name but a few. Whilst not certain to be an overall improvement on borosilicates, they may possibly find a niche application for certain specialised waste streams.

The behaviour of vitrified waste forms is well understood to the extent of laboratory tests on leaching and long-term durability. When allied to operational experience, this provides substantial evidence supporting models to predict the eventual behaviour. Often the leach tests are aggressive since they are designed to differentiate between various compositions and radionuclide loadings on an experimental time scale. Long-term research programmes exist in the UK and other countries in order to understand better the real processes by which glasses degrade in environmental conditions.

One possible future form for HLW could be ceramic. This area of research is currently being pursued actively by many countries, notably by Australia, the UK, Japan, Russia and the USA. As with vitrified forms, the aim is to maintain a stable structure throughout the high heat-generating period and through to ultimate disposal. Some refractory ceramics may perform better than glass, although this is not guaranteed, and the benefit has to be measured against increased difficulties in manufacture since these ceramics not only require a high temperature but also a high pressure. In many countries active research into glasses and ceramics is developing hand in hand, and with some speculative foresight, a hybrid technology of glass and ceramic could be a future option.

When considering ceramics for the encapsulation of process wastes, and their environmental interactions, we are looking at an end-product that despite some very significant differences is in physical terms not too unlike the ceramic fuel oxides in direct disposal.

### 10.5.3 Principles of disposal

Although no route or site for final disposal of HLW has been agreed, emplacement deep underground seems ultimately inevitable. A key factor is the waste's heat-generating capacity, which remains high for decades owing to the concentration of radioactive fission products, particularly strontium and caesium. To avoid placing an excessive heat load on the repository, the final HLW form therefore requires a period in storage for these radionuclides to decay. Meanwhile the heat is dissipated by natural or forced advection in specially designed buildings.

As a second key factor, the currently-favoured borosilicate glass is particularly prone to attack in alkaline environments. This will largely determine the nature of the repository backfill and preclude cementitious materials. Thus HLW in this form cannot share an ILW repository as currently envisaged, although the requirements for long-term isolation from the biosphere are otherwise similar.

The stainless steel canisters holding the glass are assumed to corrode in time allowing the waste itself to be leached, or on the more pessimistic suppositions to be totally dissolved in whatever ground water may have access to the site. Defence in depth, i.e. a sequence of complementary obstacles to movement as described for direct disposal in Section 10.6.3, is therefore essential. Hydrogeology is a major consideration, along with tolerance for the residual heat release, and both depend largely on the nature of the host rock. Debate continues about the ideal type; the chief contenders are hard crystalline rock such as granite, salt deposits, and clay.

In granite, flow is essentially confined to fissures, and the rock itself, being igneous, would be damaged only by very high temperatures. Fractures are common, however, and since the rock's sorptive capacity for dissolved substances is low, it would offer little impediment to the movement of dissolved radionuclides.

The permanence of a soluble salt deposit implies that the flow of water through the site must have been negligible for geological ages, and barring major upheavals can be expected to remain so. On the other hand, any water occluded within it tends to migrate up the temperature gradient, owing to differential solubility; it would be particularly corrosive and the steel containment of waste possibly short-lived. Salt tends to deform plastically under heat and pressure, and so would collapse around deposited waste, sealing access tunnels which might otherwise provide an easy route for ingress of water.

Clay too has self-sealing properties, a substantial resistance to percolating water, and a considerable ion-exchange capacity that would slow the movement of many dissolved species (Chapter 11). How far these qualities would survive the heating effect of the waste may be a crucial factor in its acceptability.

The eventual choice is likely to vary from country to country according to the actual geological formations, their soundness and accessibility, and of course the projected costs.

## 10.6 Direct disposal

Direct disposal is ultimately the only alternative to reprocessing in the management of fuel discharged from reactors. After long-term storage to allow the fuel to cool, intact or modified fuel assemblies would be packed into disposal canisters, to be permanently sealed in a geological repository. The essential difference from reprocessing is that the remaining fissile material would not be recovered, nor would any comparable amount of associated wastes (LLW, ILW, liquid and aerial) be generated. To minimise these and operating costs, fuel would be processed as little as possible.

### 10.6.1 Long term storage

Long term storage of irradiated oxide fuel is generally accepted as a safe procedure throughout the world. On removal from the reactor core, spent fuel is normally stored in pools at the reactor for a few years to allow the heat and radiation from short-lived nuclides to die away. Thereafter it may be kept in either wet or dry storage, usually at a national site away from the reactor.

An example of wet storage is the CLAB facility in Sweden, which forms part of their direct disposal strategy and is largely based on technology developed at reactors and reprocessing plants.[15] It consists of four large storage ponds for discharged fuel contained in a rock cavern 25 m underground. The temperature, pH and chemistry of the water are tightly controlled and there is regular monitoring for activity release. Knowledge of fuel pin corrosion mechanisms allows the control of conditions to prevent degradation and rupture of the pins. The fuel should therefore still be sound enough to allow handling and packaging for final disposal 40 years hence.

Several dry storage technologies have been developed in recent years to cope with growing world-wide spent fuel arisings. The two principal alternative approaches involve vaults or casks.

Dry storage vaults are normally permanent buildings with concrete to provide shielding and metal storage containers to prevent contamination. Vaults are designed to cool the metal containers housing the fuel by natural convection currents, and maintain an atmosphere which minimises corrosion. The main advantages of such vault systems are that they are safe, reliable, have low maintenance costs and low radiation doses. Examples include the Modular Vault Dry Store (MVDS) from GEC/Scottish Nuclear, NUHOMS from Pacific Nuclear, MACSTOR from AECL/OH in Canada, and the CASCADE facility designed by SGN in France.[15]

Cask storage is based on large, shielded containers which can hold a set amount of fuel. They can be stored in the open, which is sufficient for passive cooling. Casks are simple, portable, comply with safeguards and are resistant to impact or dropping. The German CASTOR V cask has a thick metal wall and can hold 21 intact PWR fuel assemblies. In Canada, concrete casks with steel linings have been developed to store low burn-up CANDU fuel. Cask storage allows flexible fuel management and gives low maintenance costs and doses.

Both vaults and casks have to be located at national storage sites that are guarded and monitored. Their relative economic advantages depend on the timing of expenditure and so on interest rates; vaults require a high initial outlay but are thereafter cheaper to maintain, while casks need be constructed only when required but cost more per unit volume of fuel.

### 10.6.2 Fuel conditioning

The objective of conditioning is to immobilise and package the fuel to facilitate handling, interim storage and final disposal. A number of options are available, so conditioning processes will tend to be specific to the country, depending on national safety regulations, costs, available technology, the type of fuel, and the geology and design of the repository.

The simplest approach is to place intact fuel assemblies in a suitable container, infill the void space and seal. For PWR fuel, the resultant disposal package would be long (some 5 m) and heavy. Moreover the ineffective use of space could be very costly for disposal in a deep HLW repository, and the bulk of the packages could cause handling difficulties underground.

Consolidation of fuel rods, leaving them intact but disassembling the elements, can increase packing densities by a factor of three. A number of systems have been developed world-wide to operate under water or in hot cells e.g. FUEL-PAC (USA), Fuel Master (France), AGR dismantler (BNFL, UK) and the horizontal system (Karlsruhe plant, Germany). Space is more effectively used but the waste package is still long and heavy. There is also a limit to the benefit of increasing fuel packing densities, as the heat output per container is increased and may require wider spacing of packages in the repository.

Disassembly with reduction in the length of fuel rods is being considered in some countries to produce a smaller waste package, e.g. compatible with the vitrified waste canister (1.4 m long). Rod cutting technology is well established in reprocessing plants. This route however increases the complexity of conditioning and gives rise to problems such as the release of volatile fission products (Cs, I, Xe, Kr) when the cladding is broken and a certain amount of fuel crushed.

Recent assessments in Germany and Korea[16] have concluded that the most favourable option is to dispose of intact assemblies in an unshielded container by emplacement through vertical boreholes into a deep repository.

### 10.6.3 Discharged fuel as a waste form

Direct disposal is based, as for separated HLW, on the multi-barrier approach, with release of radionuclides to the biosphere retarded by a number of obstacles, including:

- the depth of the repository;
- its hydrology and geochemistry;
- sorption onto geological material and engineered barriers (e.g. a bentonite back-fill);
- the disposal canister (a corrosion-resistant metal such as copper or stainless steel);
- the fuel rod cladding;
- the fuel matrix itself.

The vast majority of radionuclides (>99%) will be released only if the matrix is dissolved, and leach tests with unirradiated $UO_2$ or actual fuel show that its solubility is very low, especially under the reducing conditions expected in a deep repository. Release rates are of a similar order to those measured with borosilicate glass.

Modelling studies show that the multi-barrier system delays nuclide release for at least l00 000 years, after which long lived nuclides such as I-129, Tc-99, Np-237 and Ra-226 may reach the biosphere. Even then, the predicted individual radiation doses are very low, well below the target of $10^{-6}$ Sv per year.

One added problem, which does not arise with reprocessing waste but which will apply to direct fuel disposal, is the risk of human intrusion at any time in the future to recover the fissile material, either as a civil energy resource or to manufacture weapons; the issue of long-term safeguards has yet to be adequately resolved.

## 10.7 Review of world-wide disposal systems

This section is intended to provide an overview of disposal systems in some selected countries, to provide readers with an insight into radioactive waste disposal systems. While every effort has been taken to ensure that the information is correct, new developments are always under way and the disposal philosophies should be seen as indicative rather than defined.

In order to set the scene, the United States is chosen for a more detailed examination as a country which contrasts markedly in policy and practice with the UK. Thereafter, countries are presented in alphabetical order.

### 10.7.1 USA

Disposal in the USA can be divided into two areas, military and civil. Production of military plutonium inevitably involved fuel reprocessing, which to avoid risks of proliferation was at one time banned for civil material and thereafter not resumed. The two areas therefore have quite different requirements and have engendered two programmes of disposal.

Any military waste containing transuranic (TRU) material is separately defined as ILW. Activity limits are > 100 nCi/g for TRU and 10 nCi/g for LLW; intermediate levels are treated or decontaminated. The practical effect is of categorising by source rather than content. This contrasts with the UK where the LLW is distinguished from ILW purely on activity levels. In the US civil area, where fuel is not currently reprocessed, there is no ILW component.

**Commercial disposal**

At present the individual states are charged with considering their disposal requirements. Under this philosophy collections of states have aligned forming *compacts* to produce nine or ten disposal sites for LLW. Currently the only two fully operational sites are at Richland, Washington State and Barnwell, South Carolina. An additional site exists at Beatty, Nevada. These are taking wastes from outside the compacts.

Total commercial LLW arisings amount to 90 000m³ per year. LLW is further classified as structurally stable (Class A), compatible (Class B) and liquid (Class C) and is treated accordingly.

Irradiated commercial fuel will be disposed with military HLW in the national repository currently under development.

### Military waste

Military LLW is subject to a range of methods including shallow landfill disposal, grout injection and deep well or trench disposal. This is carried out on military sites at Savannah River in South Carolina, at Hanford in Washington State, and at the Idaho National Engineering Laboratory in Idaho.

TRU is to be disposed at the Waste Isolation Pilot Plant (WIPP) in layered salt deposits in New Mexico. The site is still under development and testing, and the TRU is temporarily held in surface stores at sites such as INEL, Idaho.

Disposal of reactor fuels, vitrified HLW and some ILW is currently planned for a repository under investigation at Yucca Mountain, Nevada.[6a] Congress has recently agreed increased appropriations for the project and the target operating date is 2010 AD. Site characterisation studies and the driving of access tunnels for an underground laboratory or *Experimental Studies Facility* are now well under way. The purpose is to determine the suitability of the tuff deposits for the purpose, and in particular the long-term stability of the water table which is projected to be many hundreds of metres below the projected depth of deposition. This is the main site under study; sites at Deaf Smith County, Texas (in salt) and Hanford, Washington State (in basalt) have however also been considered.

### 10.7.2 Argentina

A site for LLW and liquid ILW exists at the Ezeiza Atomic Centre (CAE) 40 km from Buenos Aires. This site uses a combination of trenches and disposal pits and benefits from the soils on the mesa that possess a naturally high ion-exchange potential for radionuclides. Vault disposal methodologies are in the design stages. Sites for ILW and vitrified HLW are currently under investigation with a monolithic repository a prime option, possibly in granite.

### 10.7.3 Belgium

The Belgian programme for LLW to HLW has focused on clays.[6b] In particular most of the research has focused on the Boom clay at the Mol nuclear research establishment. Experimental facilities exist 200 m below ground to investigate the site. Low level waste and short-lived ILW will be disposed in shallow repositories.

### 10.7.4 Brazil

Brazil is currently developing a site at Goia to take wastes from the Goiana accident (uncontrolled contamination by a caesium source). In addition there are five onshore and two offshore sites under investigation to take solid LLW and irradiated fuel.

### 10.7.5 Canada

Canada proposes to dispose of irradiated fuel up to 1000 m deep in the largely gneiss-type rocks of the Canadian shield, but no final decision will be taken until public hearings have taken place in 1996. The complete system would use the IRUS—Intrusion Resistant Underground Structure—concept with multi-layered cement and bentonite backfill. An Underground Research Laboratory has been providing information on the Canadian geology since 1986.[6c]

### 10.7.6 People's Republic of China

Regional repository sites are being set up. Two major sites in the north-west and south of China are expected to be operational in 1997. A further three regional sites are to be constructed.

### 10.7.7 Czech Republic and Slovakia

Disposal sites exist at Hostim, Litomeric and Jachymov. The first two sites are in abandoned limestone workings and are operational, while the last is an abandoned uranium mine without final licence. Two regional sites are being developed including Dukovany (which went operational in 1994) for LLW and ILW, and at Temelin for bitumen, concrete and glass waste forms.

### 10.7.8 Denmark

The nuclear programme is small, but of interest as considering ultra-deep disposal into salt domes at 2500 m.

### 10.7.9 Finland

A 100 m deep repository for ILW (mainly short-lived) and LLW was opened at the Olkiluoto nuclear power station site in 1992. A similar facility is being constructed at the Loviisa power station site and is due to open in 1996.[6d] The disposal of irradiated fuel is under investigation and focuses on the Olkiluoto site.

### 10.7.10 France

As France, like the UK, reprocesses irradiated fuel, the primary waste form for disposal until the HLW has cooled is long-lived ILW. Work on deep disposal is governed by legislation adopted in 1991.[6e] Four potential sites in granite and in deep sedimentary deposits have recently been announced for initial study. Underground laboratories are planned to follow at two of the sites. Depending on the results, and legislative sanction to move formally to final disposal, one of these will be chosen for a repository. HLW, as a vitrified product, is currently stored at the surface.

A shallow landfill site for LLW and ILW is operated at the Centre Manche, Cap de la Hague and has recently developed a vault disposal system for LLW at the Centre de L'Aube.

### 10.7.11 Germany

In Germany, a former iron ore mine at Konrad has been found to be suitable for LLW and ILW, while a site at Gorleben in Lower Saxony is being studied as a potential repository mainly for vitrified HLW but also for irradiated fuel.[17] Numerous shallow surface disposal sites exist, especially in the former East German sector.

### 10.7.12 Japan

As a reprocessing country, Japan will require repository facilities for either vitrified or ceramic HLW. Disposal sites are under investigation in granite and mudstone-sandstone sequences. A commercial LLW disposal site using a shallow vault disposal has begun operations at Rokkasho. There are plans to construct a rock laboratory on the island of Hokkaido as a preliminary step to establishing a deep repository by 2030 AD.

### 10.7.13 Netherlands

No decisions have been taken, although retrievable deep disposal is the main option.

### 10.7.14 Russia

Until recently the Russian disposal systems lacked the control and legislation of those in the West and a range of direct injection, shallow bunker and shallow disposal methods have been used with serious contamination of water systems. The Russian programme is now investigating disposal of HLW at depth in the Mayak region of the Kola peninsula.

### 10.7.15 Spain

Low level and short-lived ILW wastes are being disposed in shallow landfills and vaults at the new repository at El Cabril, while a deep repository for HLW is planned by 2020 AD.[6f]

### 10.7.16 Sweden

A final repository for low-level and short-lived intermediate-level waste has been in operation since 1988, built in granite bedrock about 50 m below the Baltic seabed near the Forsmark nuclear power stations.[6g] Parallel tunnels run from the surface down to the repository area which consists of caverns for LLW and a specially engineered silo surrounded by bentonite clay for the ILW. By 2010 AD about 90 000 m$^3$ of waste will have been accommodated. When the repository has been filled, the tunnels will be sealed with concrete to isolate the waste caverns and to prevent accidental intrusion.

To provide a deep repository for irradiated fuel, initial site investigations are planned at various candidate sites before the end of the century. Consents for repository construction would follow and the target is to commence disposal by about 2080. There is a well-developed experimental programme based at the Stripa underground research laboratory. Meanwhile, the fuel is to be stored in the CLAB facility 25 m underground.

### 10.7.17 Switzerland

The Swiss experimental programme is well established and a site for a short-lived waste repository has been chosen at Wellenberg in Switzerland. Two crystalline rock types are being studied as potential hosts for a deep HLW repository planned some time after the year 2000 AD,[6h] a crystalline underground rock laboratory exists at Grimsel, and LLW/ILW repositories in marl, anhydrite or crystalline rocks are under investigation. However no construction has started.

## 10.8 Conclusion

Disposal practices world-wide are varied and depend on many factors, including the choice between reprocessing and directly disposing of irradiated fuel, the economic status of the country, the development time available for the scientific programme and, not least, the intended geological environment. Although some parallels can be seen between countries meaning to emplace irradiated fuel in crystalline rocks, considerations are largely dominated by local or national factors, safety and economics.

Actual disposal, as distinct from temporary storage however prolonged, has been practised only with the less demanding arisings, but no fundamental obstacle to the disposal of ILW, HLW or intact fuel itself has so far appeared. Precautions against the release of radionuclides from such sources into the environment can be shown in principle to be capable of meeting requirements, although they have yet to be demonstrated in any particular case, and the conditioning necessary for direct disposal is still only conceptual.

The choice between reprocessing and direct disposal strategies is by no means clear-cut on grounds of cost or safety, and countries with nuclear power programmes are divided about equally between them. In terms of the complete life cycle, both options comply with international radiological safety standards. The OECD currently estimates that direct disposal is slightly cheaper, but the difference is less than the variation with different assumptions. In the short term, storage at the surface is cheaper than reprocessing but cannot be considered a permanent solution.

In the future, the choice may be determined not by cost differences, safety or technology, but by the supply of uranium, the commercial uses for plutonium (MOX and FBRs) and the viability of other energy resources.

## 10.9 References

1 *Ionising radiation regulations 1965*. Health and Safety at Work Act.
2 *Review of radioactive waste management policy—preliminary conclusions*. UK Department of the Environment (August 1994).
3 *Disposal facilities on land for low and intermediate level radioactive waste—Principles for the protection of the human environment*. HMSO (1984).
4 J. S. Hughes and M. C. O'Riordan. *Radiation exposure of the UK population*. UK National Radiological Protection Board R-263, ISBN 085951 3645 (1993).
5 *Nirex deep waste repository project—scientific update 1993*. Nirex Report No. 525. (1993).
6 *Going underground—an international perspective of waste management and disposal* compiled by UK Nirex Ltd (September 1994).
 a USA, Office of Civilian Radioactive Waste Management, Department of Energy.
 b Belgium, NIRAS/ONDRAF.
 c Canada, Atomic Energy of Canada Ltd.
 d Finland, TVO and IVO.
 e France, ANDRA.
 f Spain, ENRESA.
 g Sweden, SKB.
 h Switzerland, NAGRA/CEDRA.
7a Nirex Safety Assessment Review Panel Report NSS/G108.
 b Nirex Report **No. 525**.
8 *Costs of high-level waste disposal in geological repositories*. OECD (1993).
9 *The cost of decommissioning nuclear facilities*. National Audit Office, HMSO (May 1993).
10 G. MacKerron, J. Surrey and S. Thomas. *UK nuclear decommissioning policy, time for decision*. Science Policy Research Unit. Commentary by Radioactive Waste Management Advisory Committee (RWMAC), (1994).
11 *The social cost of fuel cycles*. Report to UK Department of Trade and Industry, London Business School Centre for Social and Economic Research on the Global Environment (1992).
12 *This common inheritance*. UK Government White Paper, CMND 8607, HMSO London (1982).
13 *The prospects for nuclear power in the UK*. UK Government White Paper, CM 2860, HMSO London (May 1995).
14 *Review of radioactive waste management policy—final conclusions*. CM 2919, HMSO London (July 1995).

15 Survey of experience with dry storage of spent nuclear fuel and update of wet storage experience. *Technical Report Series No. 290*, IAEA, Vienna (1988).

16 J. H. Whang, H. S. Park, and R. Papp. Scenario study on the direct disposal of spent nuclear fuels. *High-level radioactive waste and spent fuel management proceedings 1989*, pp. 513–517, ASME Nuclear Engineering (October 1989).

17 C. Schütt. Gorleben: technical angles and political wrangles. *Atom*, 436 pp. 18–21 (Oct/Nov 1994).

## 10.10 Further reading

Evaluation of spent fuel as a final waste form. Technical Report Series No. 320, IAEA (1991).

Concepts for the conditioning of spent nuclear fuel for final waste disposal. Technical Report Series No. 345, IAEA (1992).

W. Miller, R. Alexander, N. Chapman, I. McKinley, and J. Smellie. *Natural analogue studies in the geological disposal of radioactive wastes. Studies in environmental science 57*, Elsevier (1994).

N. A. Chapman and I. G. McKinley. *The geological disposal of nuclear waste*. John Wiley and sons. (1987).

# 11 *Environmental radioactivity*

P. D. Wilson*
*BNFL, UK Group, Sellafield, Cumbria*

## 11.1 Introduction

The world environment has always been radioactive, for most of its history probably more so than at present. Nevertheless human actions have redistributed part of that radioactivity and increased exposure to it, as one likely source of occupational hazard that was noted (though not diagnosed) as early as the sixteenth century, in the form of fatal lung cancers among miners.[1] Most of the present concern about radioactivity however relates to the artificial component added since the mid-1940s, first from nuclear weapons and later from civil applications.

While most of the radioactive materials generated by the civil nuclear industry are confined and shielded to minimise external exposure, some do escape or are released into the environment. Whether routine discharges actually cause any identifiable harm is disputed. In the one undoubted disaster, the Chernobyl accident, the escape was large, amounting to 3 or 4% of the radioactive contents,[2] including an estimated 25% of the more volatile constituents; emissions of caesium-137 appeared to be about 5% of the total due to weapon testing.[3] Even so, the known deaths (about thirty or forty) and injuries were due more to direct radiation from the exposed core and remaining molten fuel than to uptake from the cloud of scattered material. Routine discharges are very much lower, but they still pose some risk. Their behaviour has therefore been the subject of extensive study, still continuing, into the pathways and mechanisms by which such materials are dispersed or re-concentrated, and so can reach human beings or accumulate in unexpected ecological niches. As a result, more is probably known about such effects and their consequences than about any other type of material that human kind releases into the environment.

The principal regions concerned are atmospheric, terrestrial, and aquatic (notably marine), although interactions between them are also important. With increasing attention to the long-term disposal of radioactive wastes, scrutiny of the

---
*See note on authorship in preface

terrestrial environment is extending to the considerable depths appropriate for a repository. Despite some common features, there are substantial differences between the regions in the processes that chiefly control the behaviour of radioactive elements, and between those elements themselves according to their chemical nature.

Sources of environmental radioactivity are ubiquitous. Nevertheless the discharges from fuel reprocessing plants, and notably from Sellafield (UK) into the Irish Sea, are rightly or wrongly the object of special public concern. In this almost land-locked portion of the ocean, universal considerations apply with particular force, and the area supplies several specific examples to illustrate general principles.

## 11.2 Atmosphere

Releases from civil nuclear facilities are into the lowest part of the atmosphere, the boundary layer about 1000 m thick within which friction with surface features causes turbulence and mixing. Intentional discharges with significant radioactive content are from stacks to ensure dispersion before the discharged *plume* touches the ground. Their effective height is not necessarily the physical height, but may be increased by the upward kinetic momentum or thermal buoyancy of a discharged air stream, or diminished by downwash effects due to the aerodynamic interference caused by buildings including the stack itself. Such effects must be taken into account in the design.

Once clear of these disturbances, however, the behaviour of discharged material is dominated by the bulk movement of the ambient air. The plume may then move and develop in several different ways according to meteorological conditions: broadening into a cone in smoothly-flowing air, looping through successive convection cells at times of thermal instability, fanning out in a horizontal plane in an inversion, lofting above a horizontal discontinuity in the thermal regime or *fumigating* (gradually dropping its contents) below it. Lateral and vertical dispersion depends on turbulence, which in turn depends on temperature gradients, wind speed and surface roughness. The radioactive content is thus gradually diluted until it ceases to be measurable, normally a few tens of kilometres from the source.

Releases may be essentially gaseous, but despite filtration are liable also to contain or interact with suspended solids if only as the normal dust content of ambient air; the radioactive content may then to some extent be absorbed on to these even if not itself particulate. Particles, especially of the coarser dimensions, tend to settle out of the air stream and so need to be considered specially, although below diameters of about one micrometre, gravity ceases to be significant, while electrostatic or other forms of attraction to solid surfaces become more so. Particles may also be trapped by impaction on vegetation, most probably on the taller growths.

Reactive gases or vapours, such as iodine, can be absorbed by chemical interactions with the ground surface or vegetation.

In some instances, deposition may be dominated by wash-out in rain. A particular example is that from the Chernobyl accident, where within the general increase in dispersion with distance, the distribution of caesium-137 was determined largely by the pattern of rainfall during the passage of the cloud.[3] Its subsequent behaviour depended on the type of terrain on which it fell.

## 11.3 Terrestrial environment

### 11.3.1 Behaviour of radionuclides in soils

Soil, being a porous agglomerate of small particles, has a very large surface area per unit mass, providing a considerable potential for radionuclides to react chemically. Both mineral and organic materials are involved. One of the most important mechanisms is ion exchange, in which fixed negative electrical charges on the solid attract positively charged ions (cations) in solution and can reversibly exchange one type for another without altering the essential structure. Since hydrogen ion is itself positively charged and so competes, besides significantly affecting the speciation of many radionuclides, absorption depends strongly on pH. The mechanism is particularly important in clays and in zeolites, complex alumino-silicates of which the latter group has large channels in the crystal structure that can accommodate water and metallic ions.

To some extent it also occurs in other materials such as the oxides of iron and manganese which often coat the surfaces of soil particles, although here rather different mechanisms such as the formation of hydrogen bonds and eventually oxygen bridges may be more important, especially over long periods of contact. That is particularly so for interactions with most metallic elements other than the alkalis and alkaline earths, since their ionic compounds tend to be hydrolysed at the near-neutral pH values prevalent under natural conditions.

Bulk movement can also occur in soil owing to agricultural practice or the actions of burrowing animals such as earthworms. Over a period of years, even insoluble material may thus be distributed more or less uniformly to the depth affected.

The decay of once-living tissues produces a wide range of humic substances, compounds with molecular weights from a few hundred to about 100 000, and generally acidic owing to multiple carboxylic or phenolic groups; they also tend to act as reducing agents towards higher-valent metal ions.[4] They may be insoluble (humin), soluble under alkaline to weakly acid conditions (humic acids), or soluble at any pH (fulvic acids). Even where soluble, they tend to be strongly absorbed by solid surfaces, whether mineral or themselves organic. The polyacidic character makes them powerful complexing agents for metal ions susceptible to oxygen ligands. In solution the complexes may aid migration of radionuclides by resisting-ion exchange processes, or impede it by being themselves absorbed on solids or coagulated and precipitated. Metal ions, especially if multivalent, can promote coagulation. The effect thus varies from one radioelement to another and according to specific local conditions.

The general effect of absorption processes, even when reversible, is to delay the movement of materials dissolved in percolating water relative to the water itself. The stronger the absorption, which depends not only on pH but for some radionuclides also on the redox potential, the greater is the delay. Reported values of distribution ratio, to be taken as orders of magnitude only, are for strontium, a hundred; for caesium, several thousands; for radium about seven thousand; and for plutonium about a hundred thousand (Ref. 1, p. 89).

Among a few of the more significant radionuclides, the order of mobility, the reverse of the order of absorption, is generally

$$\text{iodine} > \text{strontium} > \text{caesium} > \text{plutonium} \approx \text{americium.}$$

However, according to the nature of the terrain and the duration of contact with individual particles of absorbent, caesium may be the least mobile of all. That deposited after weapon tests in the 1960s has scarcely penetrated below the upper 30 cm of undisturbed soil, and a substantial part of the strontium remains in the same zone.[1] Where deposited as components of actual fuel particles from Chernobyl as in the major contamination zone, these elements must first be leached out before ionic processes can apply to them, on a time scale of years in each case.[2]

Iodine, which generally forms negative rather than positive ions in solution, is not taken up by the cation-exchange mechanisms that favour metallic ions. Strontium in contrast is absorbed by this mechanism, and caesium still more so, while plutonium and americium (like most multivalent metallic elements) tend to hydrolyse into insoluble hydrated oxides that bind by the oxygen bridge mode or by simple adhesion.

Equilibrium may take weeks to be established, if at all, and in any case depends on factors such as temperature, pH, the composition of the soil, and the concentrations of other dissolved species; distribution ratios may vary by orders of magnitude according to specific conditions, which must therefore be defined with any quoted values.[5] Table 11.1 shows distribution ratios for various radionuclides between

**Table 11.1** Distribution ratios for various radionuclides

|  | Clay | Sand | Mix (1) | Mix (2) | Mix (3) |
|---|---|---|---|---|---|
| Cation–exchange capacity (meq/100g) | 2.3 | 0.8 | 3.0 | 2.7 | 2.9 |
| Sr-85 | 50 | 9 | 40 | 30 | 8 |
| Ru-106 | 800 | 5 | 490 | 82 | 34 |
| Cs-137 | 7 500 | 1 700 | 3 800 | 2 600 | 2 800 |
| Th-228 | 24 000 | 280 | 5 800 | 280 | 5 800 |
| U-233 | 46 | 560 | 46 | 900 | 2 200 |
| Pu-239 | 7 600 | 340 | 1 800 | 80 | 32 |

**Table 11.2** Composition of a typical Sellafield ground water

| Species | $Na^+$ | $K^+$ | $Ca^{2+}$ | $Fe^{2+/3+}$ | $Mg^{2+}$ | $Cs^+$ | $Sr^{2+}$ | $Cl^-$ | $NO_3^-$ | $SO_4^{2-}$ | $CO_3^{2-}$ | $PO_4^{3-}$ | C (organic) | pH | Eh (mV) |
|---|---|---|---|---|---|---|---|---|---|---|---|---|---|---|---|
| ppm | 26 | 1 | 38 | 0.2 | 9 | 0.1 | 0.1 | 54 | 12 | 30 | 78 | 0.1 | 0.8 | 5.7 | +200 |
| meq/l | 1.1 | 0.026 | 1.9 | 0.011 | 0.75 | 0.075 | 0.0023 | 1.5 | 0.19 | 0.60 | 2.6 | 0.0031 | – | – | – |

samples of water and soils from the Sellafield area, measured after 14 days' contact; the composition of the water is given in Table 11.2. Departures from systematic variation show the importance of specific affinities, or perhaps of random factors.

Below ground level, mechanisms of sorption and transport are essentially similar to those near the surface, although the movement of water is much restricted by the more compact nature of the environment. Some rocks, particularly of the sedimentary types, are porous, and most being brittle are liable to be more or less cracked and faulted as a result of earth movements, but cracks are often filled with secondary minerals. In such circumstances, water at depth in the Sellafield area is stated to move at rates measured in metres down to less than a centimetre per year, while dissolved substances that interact with the host rock move even more slowly.[6]

### 11.3.2 Uptake by plants and animals

Material deposited on leaves may subsequently be absorbed through the pores and taken into the plant tissue. That deposited on the ground may also be taken up, partly through splashing caused by raindrops, partly by dissolution and absorption through the roots (with the tendency in the same order as in mobility).

Whether adhering to the leaves or incorporated into the plant substance, radionuclides in forage vegetation are liable to be ingested by grazing animals. They may then be transferred through the gut wall and follow the normal processes of incorporation and gradual elimination, or be excreted directly and returned to the ground. For the elements listed above, the fraction absorbed through the gut generally decreases in the order

$$I \approx Cs > Sr > Pu \approx Am,$$

with values ranging from near 100% to about 0.1%.

Once in the animal's systematic circulation (any animal's, including human), iodine tends to concentrate in the thyroid, strontium in bone, plutonium and americium in bone and liver. Caesium is fairly uniformly distributed but favours muscle tissue. To varying degrees all are then subject to clearing processes leading to gradual excretion.

Animals providing food form a link in the pathway to humans, so incorporation into edible tissue and milk is particularly important. Iodine, caesium and strontium transfer readily to milk, which is generally the principal source of human exposure to their radioactive isotopes, although for caesium a secondary route exists through meat. Plutonium and americium do not transfer to milk or accumulate significantly in muscle tissue, but they do accumulate in offal, particularly liver, which is their main pathway to man in the food chain. However, they are generally very much less significant than the natural polonium-210 (alpha emitter, half-life 138.4 days, formed in the decay chain from uranium-238) which follows the same route. It also occurs notably in cigarette smoke.[1]

## 11.4 Seas, rivers, etc.

The sea is particularly important in relation to the reprocessing of nuclear fuel, as the direct recipient of effluents from plants such as La Hague (France) and Sellafield (UK). Discharges at least from the latter have declined sharply since the peak in the 1970s (see Chapter 9) and most of the radioactivity in that area is a relic of past years.

Dispersion in the marine environment is simple in principle but exceedingly complex in practical detail. Initially the material follows the ebb and flow of tidal streams, together with any other currents, but these form intricate local patterns while flows near the surface may differ in direction as well as speed from those lower down owing to the effects of wind and submerged obstacles; in some circumstances, for instance where the range of horizontal movement is restricted, the sea-bed current may be the reverse of that at the surface.

Vertical mixing depends largely on the weather. Wind generates wave motion which for any molecule of water is essentially circular or elliptical, extending to a depth comparable with the distance between crests. To this depth mixing may then be considered complete. In deep water, thermal gradients may create a sharp boundary. In shallow basins such as the Irish Sea this is unlikely, but the influx of river water with its lower density may have a similar effect in calm conditions; in the Baltic it is permanent.[2] Conversely, the sediments borne by rivers in spate may cause their own mixing effects as they settle under gravity.

Radioactive constituents may remain dissolved, be absorbed on sediments whether suspended, mobile or static, or be taken up by marine animal or vegetable organisms; they may enter the food chain by any one or several of these routes. Furthermore, those elements that are precipitated in whole or part as very fine particles tend to accumulate at the sea surface, from which they may become airborne in spray and carried back to land by on-shore winds. Elements such as caesium that are themselves soluble but can be absorbed by solids may behave similarly, and being especially associated with the finest particles travel the furthest.

Just as on land, a variety of chemical interactions including ion exchange and complex formation can bind radionuclides to a solid surface. Although the extent of ion exchange tends to be less, because the substantial concentrations of natural

**Table 11.3** Classification of sediments

| Designation (BS 1377) | Particle diameter (micrometre) |
|---|---|
| Clay | <2 |
| Silt | 2–60 |
| Fine sand | 60–200 |
| Medium sand | 200–600 |
| Coarse sand | 600–2000 |
| Gravels | >2000 |

sodium, potassium and magnesium in sea water compete with the radionuclides for the absorption sites, the mechanism is still important. The finer the particles, the greater the ratio of surface to volume or mass, so the greater the proportion of absorbed material, and the distribution of absorbed radionuclides depends largely on the behaviour of these fine grades. Since, for instance, the finer sediments from the Irish Sea bed tend to migrate under tidal action into estuaries where the flood current is faster than the ebb owing to the conformation of the channels, the highest concentrations of radioactive elements along the affected coasts occur in estuarine mud.

Because the mass of water is vastly greater than that of accessible sediment, however, the latter does not necessarily hold the bulk of any particular radionuclide. Caesium is about a thousand times more concentrated in sediment than in water, yet much remains in solution and moves with the water itself. Plutonium and americium, on the other hand, are still more strongly absorbed and chiefly associated with the sediments. The amounts of all these elements in the Irish Sea are mostly the residue of past decades, but that of caesium has responded more significantly to the great reduction in current levels of discharge.

In the UK, where most industrial nuclear sites are on the coast, river waters are less important receptacles for radionuclides than the sea, but minor discharges do occur, notably from medical or research establishments. In some other countries such as the USA with larger land-masses, major waste-producing centres are located on inland waterways. Similar principles apply, but because of the much lower concentrations of competing ions in solution, uptake of radionuclides by solids tends to be much higher. According to circumstances, this may be advantageous or detrimental. Together with the more restricted opportunity for dilution and the known or possible further uses for the water downstream, it has to be taken into account in determining permissible releases.

## 11.4.1 Sediment behaviour

Sea bed sediment is mixed vertically by a variety of processes, dominated by the activity of burrowing marine animals—a process known as *bioturbation*. Such mixing occurs quite rapidly to a depth of a half to one metre. In areas where sediment accumulates progressively, such as undredged harbours, radionuclides may be incorporated to greater depths simply by the build-up of deposits, and if undisturbed form a historic record of suspended concentrations.

Sediment particles are conventionally classified by particle size (Table 11.3).

The finest sediments settle in areas of low tidal and wave energy, and are typically found in estuaries, on salt marshes (where vegetation is an efficient trap for fine suspended particles) and in harbours. Deposits here are typically of clay or silt.

Beaches are an environment of much higher energy, largely because of wave action which rapidly mixes deposits to a depth of about one metre. Consequently the sediments in these areas are largely sands and gravels, with a much lower overall radioactivity than the silts and clays of estuaries and harbours; indeed, much of their actual content is associated with the few per cent of finer material.

Deposition is not irreversible, and a proportion of the bed or intertidal sediment is continually being recycled to join fresh inputs from rivers in maintaining the mass of material in suspension. Except near the sea bed or under particularly rough conditions, this tends to be at the finer end of the particle range where settling is relatively slow and the concentration of absorbed species particularly high. However strongly it favours the solid, equilibrium between these species and solution is established quickly compared with changes in input; accordingly, this remobilisation is an important route for the return of old contamination to the marine food chain, notably in the Irish Sea for a very small proportion of the plutonium and americium discharged from Sellafield in past decades.

### 11.4.2 Uptake of radionuclides by marine organisms

Living things in the sea can take up radionuclides by direct absorption from the water, by ingestion, filtration or external adherence of contaminated sediment, or together with their food. Equilibration with the environment generally seems fairly rapid, compared with changes in ambient levels buffered by the reservoir of contaminants in existing sediment. Accordingly it is possible to determine a meaningful concentration factor representing the ratio of concentration in the organism to that in the surrounding water, although as in all biological systems there is much variation within and between individuals so that every factor is at best a statistical average. Where the concentration factor is large, the species may serve as an indicator enabling the presence of certain radionuclides to be detected more readily than in the water itself; if the value of the factor is known with any confidence, it may then be used to estimate ambient levels more sensitively than would be possible by direct analysis.

Abnormally high concentrations of radionuclides might harm the organisms themselves, but at the levels of interest there is no evidence that this has occurred, and the main concern is about transmission through the food chain to humans. The important concentration factors are then related to the edible portions of the organism. Thus for strontium-90 in shellfish, the high value in the shell (due to the element's chemical similarity with calcium) has little bearing on human intake, compared with the low value for the flesh. Caesium-137, on the other hand, concentrates mainly in muscle and its uptake by fish is very significant. For this element however the concentration factor in sea water at about 100 is some 50-fold lower than in fresh water owing to competition from the chemically similar and abundant potassium.

Concentrations of radionuclides tend to be higher in sediments than in living organisms, particularly for elements such as plutonium and americium with low solubility and low mobility in the food chain. These are most significant in molluscs that accumulate sedimentary material by filter feeding (e.g. mussels) or by ingestion in the course of grazing (winkles), while the estimated values will depend on how much sediment the animals have taken in, and whether they are cleaned before analysis.

The most important combinations of creature and radionuclide are determined by conditions and habits peculiar to the locality. A knowledge of such circumstances is therefore essential to determining the possible impact of environmental radioactivity. For example, in the 1950s and '60s, the isotope of greatest concern in the northern Irish Sea was ruthenium-106, accumulated by species of seaweed (*porphyra purpurea, p. tenera,* and perhaps *p. umbilicalis,* collectively known as laverweed) that were gathered locally for consumption in South Wales, where some 200 people ate particularly large amounts. When the gatherers retired from business without successors, the emphasis shifted to caesium-137 in fish and actinides in shellfish.[7] Members of the current *critical group* are each assumed to eat 9 kg/year of locally-gathered winkles, 36.5 kg/year of fish and 6 kg/year of crab and lobster.

Similar principles apply to fresh-water species, with the proviso that accumulation factors are liable to be higher than in the sea owing to the dearth of competing ionic substances. Also the discharged materials tend to be, for instance, radiopharmaceuticals or industrial tracers rather than reprocessing wastes, with corresponding differences in the limiting radionuclides. Phosphorus-32 (beta-emitting, half-life 14.3 days) can be of particular concern owing to its metabolic importance.

### 11.4.3 Transfer from sea to land

Tidal inundation of coastal land, whether regular or due to exceptional storm surges, can deposit sediments together with any associated contaminants. Both salt marshes and meadow land occasionally washed by the tide may be used for grazing cattle or sheep and so constitute a route to human consumption. Secondly, as mentioned earlier, material can be resuspended by the action of wind and wave, then transported inland as an aerosol. This mechanism is particularly important for those elements that are themselves insoluble or are absorbed on to fine solids; these tend to concentrate in a micro-layer at the surface where they are much more liable to resuspension than dissolved constituents in the bulk of the water.

## 11.5 Artificial radionuclides of special interest

Radionuclides of particular concern are tritium and isotopes of carbon (specifically C-14 which also occurs naturally), krypton, strontium, technetium, iodine, caesium, plutonium and americium. Ruthenium used to be very important in discharges from Sellafield but is now much less so; not only (for unrelated reasons) is the seaweed which accumulated it no longer gathered locally for human consumption, but a change in the chemical process in 1964 led to greatly reduced amounts in discharged streams, while with relatively short half-lives, the radioisotopes in discharges more than ten years ago have practically all decayed.

The decay properties of these and many other radionuclides are listed in Appendix 4.

### 11.5.1 Tritium

Tritium is the isotope H-3, a soft beta-emitter with a half-life of 12.3 years. Large amounts are produced in nuclear fuel, where it may exist as the element, oxide or (in cladding) zirconium hydride. Once the fuel has been dissolved, the tritium oxide is practically inseparable from the ordinary water, and some preliminary treatments have been suggested to recover it in concentrated form. However, the elemental portion could still escape, the radiotoxicity is low, and dilution in the environment enormous, so these measures have not been generally adopted.

### 11.5.2 Carbon

Carbon contains a small proportion of the weakly beta-emitting isotope C-14 (half-life 5730 years), formed continuously by the effect of cosmic rays on nitrogen in the upper atmosphere and then distributed throughout the lower layers. Taking part in the general metabolic cycle, it is incorporated into living tissue and its subsequent decay forms the basis of the well-known method to determine the age of organic materials.

The reaction forming the natural isotope, i.e. neutron absorption and proton emission, is liable to be reproduced in a nuclear reactor if nitrogenous materials are present; C-14 is also formed by a small proportion of decay reactions. Although its radiotoxicity is low, the reduction in discharges of more harmful radionuclides has greatly increased its relative significance in industrial effluents (see Table 13.4).

### 11.5.3 Krypton

Krypton is one of the *noble* gases, a group with particularly stable electronic configurations in the uncombined atoms which therefore have little or no tendency to enter into chemical combination. The isotope of principal concern is Kr-85 (half-life 10.7 years), a fission product decaying by beta emission to stable rubidium-85. Because of its chemical inertness, it is difficult to trap and immobilise securely in fuel reprocessing plants, which may therefore discharge it to the atmosphere. For the same reason it tends to remain in the natural circulation, giving an external radiation dose to the general population. Although this is slight in comparison with natural sources, krypton is among the most radiologically significant components of routine gaseous discharges from reprocessing plants. It may also escape from damaged fuel elements in a power reactor and appear in the aerial effluent.

### 11.5.4 Strontium

The principal radioisotope, Sr-90, is a pure beta emitter with a short-lived daughter in yttrium-90, emitting much more energetic beta-radiation (Appendix 4). Strontium is one of the alkaline earth metals, like calcium which is essential in bone formation. Consequently Sr-90 entering the body tends to accumulate in the bones where it can irradiate the blood-forming tissues of the marrow. This is its chief ecological significance, but with the half-life of 29 years and high fission yield (Appendix 3), it is an important contributor to the mid-term radioactivity and heat release of stored waste.

Because a principal dietary source of calcium is cow's milk, the main route taken by strontium to man is by deposition on pasture and uptake by cattle. Strontium however is also concentrated by arctic lichens, which are grazed by reindeer and caribou. Lapps and Eskimos thus receive unusually high body-burdens through eating meat from these animals.

### 11.5.5 Technetium

Technetium is a purely artificial element, having no isotope stable enough to remain from primordial times. Environmentally the most important isotope is Tc-99 (beta-emitting, half-life 213 000 years) which is a major fission product. By far the most stable chemical form under aerobic conditions is the pertechnetate anion $TcO_4^-$ (analogous to permanganate), which like iodine is not retained by cation-exchange; despite incompletely reversible absorption through other mechanisms with distribution ratios ranging from under 10 to several hundred,[8] it is highly mobile. In reducing media however, the very much less mobile tetravalent state may become more important. Like iodine, technetium is absorbed by the thyroid, hence its importance as a contaminant.

### 11.5.6 Ruthenium

Ruthenium is one of the platinum group of metals, and a major fission product. The principal isotope environmentally is Ru-106 (half-life 1.02 year), itself only a soft beta-emitter but decaying to rhodium-106 (30 s) which is a hard beta-gamma source. By forming a range of more or less readily interconvertible complex ions in solution, with different affinities for the solvent used in processing irradiated fuel, it tends to spread in significant amounts through all process streams including some destined for discharge as low-active effluent. This behaviour was one of the reasons for the change of process at Sellafield in 1964 in order to restrict the carry-over from the main bulk of fission products. Ruthenium then discharged to sea was largely associated with ferric hydroxide particles (which on current practice would be separated as solid waste), but also as a labile mixture of cationic, anionic and neutral species in solution.[9] The previously-mentioned uptake of ruthenium by edible seaweed might be due to simple adhesion of particulate matter, or to chemical interactions such as cation exchange on the sulphated polysaccharide porphyran forming much of the weed's substance.

Ruthenium also forms a highly volatile tetroxide under oxidising conditions such as exist in the waste vitrification process, and significant amounts emerge in the off-gases from which they must be removed. $RuO_4$ is a powerful oxidant, readily reduced by organic matter of almost any kind to a refractory dioxide, and therefore tending with atmospheric dust to form aerosols that could be inhaled and lodged in the lung, as has happened on occasion.

### 11.5.7 Iodine

The importance of iodine lies in its being volatile, readily taken into the food chain, and accumulated by the thyroid gland particularly in young children. Large intakes of radioiodine can therefore damage the gland or cause eventual cancer. The two most significant isotopes, I-129 and I-131 (beta-gamma emitters with half-lives 15.7 million years and 8 days respectively), are both among the more abundant products of nuclear fission, and I-131 is the most immediately dangerous constituent in any release of material from freshly-discharged fuel. Under these conditions I-129 is less troublesome, since its radioactivity being inversely related to half-life is by comparison negligible. It does however persist, and among the products of the nuclear power industry is one of the chief potential sources of radiation exposure in the extreme future.

In the fuel cycle, problems with I-131 are avoided by delaying reprocessing until the isotope has almost completely decayed. With I-129 that is obviously impossible, and the isotope must either be contained securely or discharged in ways that present no significant hazard, now or in the future. Secure containment is difficult to assure indefinitely, because the element and many of its compounds are volatile, iodide salts are corrosive, and the element might be displaced sooner or later from any solid compound, while once released it is highly mobile. Since the principal pathway to man is by deposition on pasture, consumption by cattle and secretion in milk, discharges to the atmosphere are strictly limited. For a coastal plant there is a practicable alternative in discharge to sea, from which little passes through the food chain. The practice in the UK is therefore to send as much as possible of the iodine by this route.

## 11.5.8 Caesium

Caesium is one of the alkali metal group, like sodium and potassium. Because of the sequence of ionic radii, it resembles the latter more closely, and tends to follow or replace it (particularly where there is any deficiency) in selective biological uptake processes. Free caesium is therefore concentrated by living organisms, though eliminated gradually if the source is removed, with biological half-lives varying from days to months.[1,2] Again because of its size, however, it is strongly bound by materials with ion-exchange properties, which include most soils, with distribution ratios ranging from some hundreds to about 5000 according to conditions; uptake by clays is particularly important.[5] Uptake into the food chain is therefore chiefly from water (for instance by way of fish) and from vegetation on which it has been directly deposited. However, there is an exception in acid, mineral-deficient soils such as occur in upland heath country. On such lands which took a substantial burden of radiocaesium from the Chernobyl accident, owing to heavy rainfall at the time, so much of it has remained circulating between pasture and sheep as to cause restrictions on the movement and sale of the animals to be retained for much longer than originally expected.

Caesium is particularly concentrated by lichens, and so after deposition from fall-out is accumulated like strontium by Lapps and Eskimos.

The important isotopes are caesium-134 (half-life 2 years), Cs-135 (2.3 million years) and Cs-137 (30 years), all beta or beta-gamma emitting fission products with yields among the highest known (Appendix 3). Cs-134 and Cs-137 contribute largely to the decay heat of nuclear fuel and highly-active waste during the first decades or centuries after discharge, while Cs-135 is a significant element of any long-term hazard from waste repositories. In this latter capacity however any real danger is greatly diminished by the strong affinity for ion-exchanging minerals, which would substantially delay its passage compared with percolating ground water.

## 11.5.9 Neptunium

Neptunium-239 is an intermediate in the formation of plutonium from uranium, and the element lies between these two in the actinide sub-series of the Periodic Table (Chapter 1, Fig. 1.1). This isotope is however short-lived (half-life 2.36 days) and unimportant in the environment, in contrast with Np-237 (2.14 million years).

The actinides are formally analogous to the predominantly trivalent lanthanides, but because of the similarity in energy between the outer d and f electron shells, the early members up to plutonium tend to adopt higher valencies, almost as though the main long series were continuing uninterrupted into the succeeding groups.[10] Under environmental conditions the most stable valency of neptunium is probably 5 in the form of the ion $NpO_2^+$ which is soluble and mobile. This, combined with its long half-life and relative abundance in aged wastes derived from irradiated nuclear fuels, makes neptunium the most significant alpha-emitter in terms of potential long-term hazard.

There is evidence however that in the absence of air, neptunium may be reduced to the tetravalent state in which it would be strongly hydrolysed and far less mobile. In this case its presence in buried wastes would pose far less potential hazard than if it remained pentavalent. Moreover, although the element is taken up by vegetation fairly readily, transfer into animal tissue may be far less extensive. Once in the body, it favours bone and liver.

### 11.5.10 Plutonium

Plutonium is likewise one of the actinides with chemical similarities to the main long series of the periodic table. In the *group* valency of 3 it is readily oxidised and as a rule the most stable valency is 4. Ionic compounds of this valency are more easily hydrolysed than any other (in higher oxidation numbers, elements tend to form oxygenated cations of reduced charge such as $PuO_2^+$, with a greatly reduced tendency to attach further ligands), while plutonium hydroxide tends to polymerise irreversibly, and is particularly insoluble with a solubility product considerably less than $10^{-50}$. Consequently plutonium in the environment is generally solid or in a colloidal form that may adhere to other solids,[11] but is still significantly mobile in ground water. Any minute proportion in true solution is liable to be predominantly penta- or hexavalent. With these properties the element has low mobility within the food chain, and the chief cause for concern is generally taken to be inhalation of airborne particles that may lodge in the lung.

Amounts of plutonium estimated as from 3 to 20 tons have been distributed throughout the world by tests on nuclear weapons, and considerable amounts were discharged from fuel reprocessing facilities particularly in the 1970s. The principal isotopes are Pu-238 (half-life 87.7 years), Pu-239 (24 100 years), Pu-240 (6560 years), Pu-241 (14.4 years) and Pu-242 (375 000 years). All are alpha-emitters except Pu-241 which decays by beta-emission to americium-241. In the early nuclear industry Pu-239 predominated, but with increasing irradiation in reactors Pu-240 may be comparable and the higher isotopes considerable.

Plutonium-238 is notable as being used as a heat source in thermoelectric generators for such applications as heart pacemakers and space vehicles. The former could in theory be retrieved when no longer needed, but the practical difficulties and consideration for all concerned in the inevitably distressing circumstances have led to a preference for making the units resistant to cremation. Recovery of satellite power sources is likewise seldom practicable, although for totally different reasons, and these are dispersed in the atmosphere on re-entry.

### 11.5.11 Americium

Americium is the first of the actinides to resemble closely the lanthanide analogues. Its only stable valency is three, and in that it is readily hydrolysed although less so than plutonium(IV). It is thus rather immobile, and its chief environmental importance lies in its alpha decay (half-life 433 years) to neptunium-237.

## 11.6 Natural radioelements

Although artificial radionuclides tend to receive the most attention in discussions of environmental radioactivity, they are greatly outweighed by those of natural origin. These may be primordial, i.e. remaining from the formation of the earth some five billion years ago, or members of the decay chains from such elements (Appendix 2). Some of the more important are now described, most conveniently in reverse order of atomic number. For more detail of decay characteristics, see Appendix 4 or a specialist source document.

### 11.6.1 Uranium

Uranium is the heaviest element known to occur naturally on earth, apart from traces of plutonium. Because of the large size of the atom itself, and its tendency to form even bulkier complex ions, it does not fit readily into the crystal structures formed by other elements. Thus it is liable to be left in the liquid phase during the formation of principal minerals from magma or aqueous solution, until the last stages of solidification; it then forms secondary minerals at the crystal boundaries. Some relatively concentrated ores do exist (Chapter 2), but for the most part the element is widely dispersed in very small concentrations, and is a minor component of all rocks and soils with particular concentrations in granite or similar igneous rocks. Owing to an association with phosphorus, it is a significant by-product or waste of the phosphate fertiliser industry. It is also present at about 3 parts per billion in sea water.

Like other early members of the actinide sub-series, the element usually behaves chemically as though belonging to an uninterrupted long series of the Periodic Table and in this instance to Group 6, with considerable resemblances to tungsten or chromium. In aqueous solution the most important valency by far is 6, represented by the uranyl ion $UO_2^{2+}$ which in the presence of carbonates forms a remarkably stable tricarbonate anion, particularly important in the sea; in minerals U(IV) is also common, sometimes in compounds of mixed valency, and is relatively insoluble. Human exposure is due to dietary intake by way of vegetables or by inhalation of airborne dust such as soil or coal ash, and the element concentrates in bone.

The principal isotopes are U-238 (99.27% by atoms) and U-235 (0.72%), with half-lives of 4470 and 704 million years respectively. Both are alpha-emitters, and a small proportion of U-234 formed by decay through the '$4n + 2$' chain (Appendix 2) is usually present. Recycled uranium also contains U-232 and U-236 (70 years and 246 000 years respectively). The former is a considerable radiation hazard in its own right, while most are so through the members of their decay chains. In high concentrations the element is chemically toxic,[12] but particularly in the presence of the daughter elements, the radiotoxicity is usually a more serious consideration.

### 11.6.2 Thorium

This element is several times more abundant than uranium, similarly widespread but concentrated in minerals such as monazite of which substantial deposits exist in southern India, Brazil and China. The sole natural isotope is Th-232, which with a half-life of 14 billion years (nearly three times the age of the earth) is feebly radioactive in its own right but significant in the much higher specific activity of its daughters, notably radium and its decay products. External radiation doses in the Indian and Brazilian monazite areas are about 5–6 milliSieverts (mSv) per year with some individual exposures over five times as great.

Thorium chemically resembles elements of Group 4, with effectively no other valency than four. It is very insoluble and strongly absorbed by minerals,[5] but its decay products are mobile.

### 11.6.3 Radium

The principal isotopes of radium are Ra-226 (alpha-gamma emitting, half-life 1600 years) and Ra-228 (soft beta-gamma, 5.76 years), decay products of uranium and thorium respectively (Appendix 2). Accordingly the element is associated with its parents, but being chemically analogous to calcium and barium is much more soluble, readily taken into the food chain and incorporated into bone. One curious quirk is its concentration along with barium in Brazil nuts, about three orders of magnitude higher than the weighted average in other foods. Its use in the treatment of cancer, and the damage it can cause when used without precautions against ingestion, have been well known since the early part of the twentieth century.[1] Its own radiotoxicity is greatly enhanced and extended by that of its daughters.

Radium-226 is the most important radionuclide in mining waste. Although its concentration may be only a few times higher than in the undisturbed environment, the treatment of the ore renders it more accessible to leaching processes, and various methods are adopted or proposed to control the releases (Chapter 2). The general situation however has not yet reached the satisfactory stage where no further human intervention is needed to keep seepage within acceptable limits.

### 11.6.4 Radon

As a decay product of uranium and thorium by way of radium, radon is similarly pervasive. Being also a member of the noble gas group, and soluble to a radiologically significant extent in water percolating through the sub-soil, it is not confined to the parent mineral but seeps into the atmosphere, where it may accumulate to substantial concentrations in poorly ventilated buildings and be inhaled by the occupants. Where it has been measured it is responsible for a large proportion of human exposure to radiation, half the average total to the UK population, and in certain granitic areas such as Cornwall very much more. Concentrations are particularly high around deposits of mining waste containing its parent elements.

The main isotopes are Rn-222 and Rn-220 (both alpha-emitters, half-lives 3.8 days and 56 seconds respectively). The latter has little opportunity to escape from any but the shallowest sources and so contributes the less to the radiation dose. The total due to the element is not only from its own radiation, but to daughter elements deposited on surfaces, whether of building interiors or airways in lungs. The daughters are two series of mostly short-lived alpha, beta and gamma emitters, eventually decaying to stable isotopes of lead.

### 11.6.5 Potassium

Potassium is a widely-distributed element, essential to life, copiously applied as fertiliser to arable and horticultural land, and constituting 0.2% of the human body where it is chiefly concentrated in muscle. It naturally contains 0.01% of the energetic beta-gamma-emitting K-40 (half-life 1.28 billion years) which imparts to the element an overall specific activity of 26 Bq/g and to human soft tissues a radiation dose of some 200 microSv/year, about a tenth of the UK average total.

Potassium is also an important constituent of sea water which contains about 10 Bq/litre of K-40. Away from the vicinity of industrial discharges, this is the predominant source of marine radioactivity.

## 11.7 Monitoring and research

In the absence of evidence for other detriment, the chief concern about environmental radioactivity relates to human health. The ideal way to determine its impact might be to establish unequivocally:

(a) how the components of a radioactive discharge became distributed between the various identifiable compartments of the environment;

(b) the radiation doses delivered by the contents of these compartments to people exposed to them;

(c) the effect of these doses.

Like all ideals, it is not realised in practice, as witness the widely divergent estimates of consequences due to any particular event. The environmental compartments are seldom internally uniform and not always sharply delimited; the behaviour of radionuclides within them may have a strong random element owing to unidentified or unquantified influences; patterns of human behaviour—how much time spent handling fishing gear contaminated with radioactive sediments, rates of consumption of foods from affected areas, and so on—vary enormously;

and medical effects of consequent exposure are (rightly) not allowed to reach discernible levels.

To cut this Gordian knot, the practice is to seek out a *critical group* such as the seaweed eaters of Section 11.4.2, who receive a more significant dose than anyone else. (That need not mean *larger*, as a small dose to a sensitive organ may be more significant than a larger dose to the rest of the body.) Provided that this dose is within acceptable bounds, then all is taken to be well. There may of course be arguments on what is *acceptable*, and questions of whether the surveys have missed any important fact, or whether the critical group has been correctly identified. Imperfect as the system may be, however, it is the best available.

### 11.7.1 Exposure

Where current radioactivity is concerned, the task of determining exposure to the critical group falls naturally into two parts, roughly corresponding to topics (a) and (b) above. First, the levels of radioactivity are measured in appropriate forms, such as the concentrations of significant radionuclides in foodstuffs, or the actual radiation levels on frequented beaches and mud-flats.

Where exposure is through food, allowance must be made for factors such as transfer through the gut, affinity for various organs of the body, and elimination rates, all of which are liable to vary between individuals and according to competition from inactive analogues. The actual source of food also needs to be taken into account. For instance, early estimates of exposure due to agricultural produce from the immediate area of Sellafield neglected dilution by purchases from other sources, until a survey of shopping habits showed that the estimates could be substantially reduced.[7]

Estimating future exposures is of course more difficult, and a model of environmental behaviour is needed, more or less elaborate according to circumstances. At the simplest it may be an empirical correlation between discharge rates and observed environmental levels; a degree of sophistication may be added by an allowance for delayed response due to absorption on sediments. At the other extreme, where aerial discharges are concerned, models may exceed in complexity those used in weather forecasting (which may indeed be incorporated), with the saving grace that they do not need to be computed in haste.

Where appropriate a model may be supplemented by experimental discharges of tracers, radioactive or nowadays more probably otherwise, for instance sulphur hexafluoride to follow the effects of atmospheric dispersion on aerial releases from various positions. The results may then be used to refine the model. Wind-tunnel experiments can also provide a valuable input. Besides its predictive application, a model may be used to correlate the results of emission monitoring and environmental monitoring.

Essentially it must cover all the significant radionuclides, the pathways leading to human exposure, the extent of dilution, rejection or concentration factors along the route, and the time taken to traverse it (in a simple hypothetical case, a radionuclide that takes twenty half-lives from release to exposure will be reduced a million-fold from its original activity and can probably be ignored). Often the numerical values are uncertain by orders of magnitude, or the route itself may be in doubt. The practice then is to apply assumptions believed to be pessimistic, and if any exposure is calculated to come within a few orders of magnitude of being significant, then to investigate the crucial areas. Thus the effort of research is concentrated where it is most needed.

This applies particularly to proposed repositories for radioactive waste, where barriers to the movement of radionuclides are deliberately engineered and the site is chosen, among other criteria, for the additional delays that it would naturally impose (Chapter 10). Important considerations in a deep burial site are the nature and extent of absorptive materials around the primary containment, the rate at which water moves through the surrounding rock, the influence of faults or fissures on routes to the surface, the ion-exchange properties of any secondary minerals in the cracks, and the sensitivity to possible geological disturbances.

The uncertainties might appear daunting, given the span of human experience compared with geological time, but for the existence of two ancient sites that provide considerable reassurance.[1] The Oklo uranium deposits in Gabon, West Equatorial Africa, became a natural nuclear reactor some 1.8 billion years ago when repeated inundations provided enough moderation for the then higher proportion of U-235 to sustain a chain reaction; most of the resulting radionuclides of interest have since moved no more than a few metres. Again, a weathered deposit of thorium and lanthanide minerals on the hill Morro de Ferro in Minas Gerais, Brazil, suggests that the analogous plutonium and transplutonium elements in a waste depository, even if exposed by erosion of the overlying rock, would scarcely move from the site.

## 11.7.2 Effects of exposure

Determining the medical effects of exposure to radionuclides derived from the civil power industry suffers from a fundamental difficulty: with certain exceptions they are no different from those of natural radioactivity, and the levels are so much lower than any effect on the general public is lost amid the normal variation. The exceptions concern radionuclides such as iodine-131 that have occasionally been discharged by reactor accidents in unusually large quantities, and are so concentrated by biological processes as to give a selective dose to specific organs.

To be sure, there have been claims of enhanced rates in certain diseases around nuclear installations in both Britain and the USA, but these have involved contentious manipulations of medical statistics and mostly appear to be artefacts of those manipulations. Even the acknowledged cluster of infant leukaemias in Seascale near the UK reprocessing plant at Sellafield was ascribed not to environmental levels but tentatively to much higher occupational radiation exposures received

226 *Environmental radioactivity*

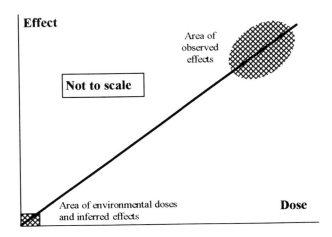

**Fig. 11.1** Extrapolation of observed radiation effects to low doses

by the fathers before conception. In any case the hypothesis seems inconsistent with observations over wider areas.

Thus the effect of normal civil discharges can be estimated only by long extrapolations from observations at much higher doses. In the extreme, radiation can cause severe damage to biological structures leading to unmistakable disease or death; such effects require a certain minimum dose to occur at all, and their severity is directly related to the degree of exposure, which apart from the area around extraordinary events such as nuclear explosions or the thermal explosion of the Chernobyl reactor, greatly exceeds any due to environmental radioactivity.

At normal doses any effect—essentially induction of cancer in various forms—is believed to follow a series of cell changes for which the probability but not the severity depends on the dose level. The form of the relationship cannot be determined precisely, and not at all in the region of practical interest where artificial exposures are orders of magnitude lower than natural background, but is assumed to be a direct proportionality with no threshold (Fig. 11.1). Thus it ignores the cell repair mechanisms that are believed to exist and might break the causative chain. It also neglects the inverse relationship between radiation dose and delay before the onset of resulting disease, a delay that might extend beyond the natural life span of the victim. It does however lead to the statistically convenient prediction that a certain aggregate radiation dose, spread over however many thousands of people, would cause the same number of cancers. Whether this is true cannot be determined, but it accords with the practice of making pessimistic assumptions.

Less conveniently for the nuclear industry, it implies that any level of exposure to a sufficiently large population will cause some harm, and therefore requires that exposures for which it is responsible must be kept as low as reasonably attainable (ALARA). In any case, they must not exceed the maximum recommended by the ICRP, currently 5 mSv to any member of the public over any consecutive 5 years,[13] while the British National Radiological Protection Board recommends a limit of 0.3 mSv per year due to current or future operations of any one site.[14] To put these figures into perspective, the dose to the UK population from all sources (predominantly natural) ranges from 1 to 100 mSv per year with an average of about 2 mSv per year.[15]

## 11.8 Conclusion

Assessing the effects of artificial radioactivity in the environment is a difficult task even for current levels, and for the future even more so. It involves aspects of physics, chemistry, meteorology, limnology, oceanography, geology, hydrology, biology, medicine, and statistics, besides a considerable knowledge of widely varying habits of behaviour and consumption among the public. The practice (or at least the intention) when in doubt has always been to take a pessimistic assumption, and that probably means that any harm due to radioactive emissions has been substantially over-estimated. Continuing research may enable the estimates to be reduced by considerable factors and quite possibly by orders of magnitude.

Whatever the outcome of this research, it will not alter the fact that on the whole, environmental radioactivity is predominantly natural. A sense of proportion is therefore desirable in considering the 0.1% or less due to industrial processes.

## 11.9 References

1  M. Eisenbud. *Environmental radioactivity* (3rd edn). Academic Press, Orlando (1987).

2  F. Warner and R. M. Harrison (eds). *Radioactivity after Chernobyl.* John Wiley and Sons, Chichester (1993).

3  R. S. Cambray *et al.* Observations on radioactivity from the Chernobyl incident. *Nuclear Energy,* **26(2)**, pp. 77–101 (April 1987).

4  G. R. Choppin. Humics and radionuclide migration. *Radiochimica Acta,* **44/45**, pp. 23–8 (1988).

5  J. Bell and T. H. Bates. Distribution coefficients of radionuclides between soils and groundwaters and their dependence on various test parameters. *The Science of the Total Environment,* **69**, pp. 297–317 (1988).

6  Scientific update. Nirex Report no. 525 (1993).

7  S. R. Jones, M. J. Fulker, J. McKeever and T. H. Stewart. Aspects of population exposure consequent on discharges of radionuclides to the environment from the nuclear reprocessing plant at Sellafield in Cumbria. *Radiation Protection Dosimetry* 36, p. 199 (1991).

8  P. J. Hooker, J. M. West and D. J. Noy. Mechanisms of sorption of Np and Tc on argillaceous materials. DoE ref. no. DOE/RW/86/055 (1986).

9  P. D. Wilson. Some aspects of the chemistry of ruthenium in sea water, PG Report 819(W). UKAEA (1968).

10  F. A. Cotton and G. Wilkinson. *Advanced inorganic chemistry* (2nd edn), pp. 1076–81 (1966).

11 S. R. Aston. Evaluation of the chemical forms of plutonium in seawater. *Marine Chemistry*, **8,** pp. 319–325 (1980).
12 S. G. Luxon (ed.). *Hazards in the chemical laboratory,* (5th edn). Royal Society of Chemistry, Cambridge, p. 637 (1992).
13 International Commission on Radiological Protection. ICRP Publication 60, Pergamon Press (1991).
14 Documents of the NRPB, **4,** p. 182 (1993).
15 NRPB, *Living with radiation.* HMSO (1989).

# 12 *Decommissioning nuclear facilities*

S. Buck
*Mallard Consultants, Lamplugh, Cumbria*

## 12.1 Introduction

Nuclear facilities at the end of their useful lives cannot merely be abandoned. The site may be needed for other purposes, amenity must be preserved, and above all else, the public must be protected from any remaining hazards. Decommissioning is a major world-wide task: the IAEA has estimated that some 50 nuclear power plants and over 220 research reactors are already shut down, by the year 2000 a further 40 and 230 respectively will be at least 30 years old,[1a] and supporting facilities, such as the successive British Magnox reprocessing plants, must eventually be included.

Decommissioning is often thought synonymous with dismantling and demolition to *green field* condition. More correctly, it is the whole process of plant management after a facility is taken out of service. Immediate dismantling and demolition is not the only option, or necessarily the best; it may represent a greater initial hazard than, for instance, deferment with interim surveillance.

The chief considerations in the choice of management regime are:

- technical capability;
- the radiological status of the plant;
- the arisings, treatment and disposal of wastes;
- regulatory requirements of the particular nation;
- financial comparisons of technically acceptable options;
- public attitudes to the operation.

The remainder of this chapter addresses these various issues.

## 12.2 Principles

### 12.2.1 Terminology

As codified by the IAEA, nuclear decommissioning comprises three stages, which may be separated in time:[2]
Stage 1—remove fuel or process inventory.

Stage 2—remove all ancillary plant and buildings; seal main plant to minimise external radiation and costs of surveillance.

Stage 3—complete decontamination and demolish to green field status.

A typical interpretation of these stages is that adopted at Sellafield (UK):

- Post-operational clean-out (POCO) completes the final operational campaign, cleaning the plant as thoroughly as possible without physical modifications or abnormal reagents.

- Initial decommissioning—the minimum of work to ensure safety with minimal surveillance and preferably passive maintenance. It may require extensive removal or immobilisation of radioactivity, including some dismantling of active plant within cells.

- Removing remaining plant and machinery, and demolishing virtually non-active buildings, approximating at worst to LLW.

US reactor operators have a choice between alternative approaches with their own terminology:

- Prompt DECON: immediate decontamination or removal of equipment to permit unrestricted use of the site as soon as possible, i.e. immediate progress to IAEA Stage 3.

- Delayed DECON: similar, except that unrestricted use is delayed for 10 to 20 years for spent fuel storage until a repository is provided by USDOE—meanwhile, effectively Stages 1 and 2.

- SAFSTOR: the facility is made and kept safe for ultimate dismantling and decontamination, currently expected to follow a delay of about 60 years for LWRs (see Section 12.3). Again this corresponds to Stage 2.

### 12.2.2 Control and regulation

Countries with nuclear programmes have national operating regulations, generally extended case by case to decommissioning. To aid this evolution as the need expands, the IAEA is planning a Safety Standard covering legal, safety and radiological criteria and requirements, outline strategies and waste management: topics include advance planning, techniques to minimise radiation exposure, and recycling of materials.

Safety Guides covering power, fuel cycle, medical, industrial and research facilities will be derived from the Standard, with more detailed Safety Practice Documents in various areas. Reactors, nuclear facilities and uranium mining and milling are covered in existing IAEA Safety Series documents 52, 74, 85 and 105. In the UK the Nuclear Installations Inspectorate (NII) of the Health and Safety

Executive (HSE) expects operators of nuclear facilities to decommission them in line with Safety Series guides 52 and 105. The derived policy principles are:

- Decommissioning should be regarded as the final phase in the life of a plant, under the same licensing regime until all radioactive matter has been removed.
- NII will continue to regulate activities while there is any possible danger from ionising radiation.
- NII will require an acceptable decommissioning programme and time-scale, particularly for reduction of the radioactive inventory.
- Priority for risk reduction must favour workers and the public, with special attention to rapid decommissioning of plant capable of presenting the higher risks.
- Management of decommissioning wastes must be consistent with local and central government policies and with IAEA and EU guidance.

Several of these principles are embodied in Condition 35 of the licence of each UK nuclear site. A licensee must make adequate arrangements for decommissioning and specify means to achieve an adequately safe state, defining the various decommissioning stages as phases of the project and accounting for risk incurred in any delay. For new plants (e.g. THORP or Sizewell B) an outline decommissioning plan at the design stage is among the pre-licensing requirements.

NII also needs to be convinced that funds and resources will be available for decommissioning, with risk minimisation given due priority.

Finally, for activities deemed particularly significant in safety terms, the HSE requires a demonstration that the dismantling etc. can be carried out safely and under regulation by NII, just as in the construction or modification of plant.

### 12.2.3 Non-nuclear parallels

The requirement for decommissioning is not peculiar to nuclear facilities. It applies equally in the *conventional* chemical and allied industries, where the term is used both of temporary closure for maintenance and modification, and of dismantling and demolition before permanent removal. In all cases, operational plant must be made safe and free from hazardous materials before the start of subsequent work. This may include the installation of new equipment or processes (*revamping*); the plant may be sealed and left secure (*mothballed*) until a decision has been taken about its future; or it may be immediately dismantled and demolished.

To prepare operational plant for decommissioning, it has to be taken off-line, following existing procedures as for maintenance. Then it must be disconnected from services and physically isolated from operational plant. This leaves a non-operational but contaminated plant, maybe on a site with deposits of process materials and wastes which require attention.

Not only current closures are involved. Plants may have ceased production some time before, and perhaps been abandoned, facing current owners or site developers with a task that depends on the type and extent of contamination and on the intended future use. It must be positively managed, from an outline philosophy (defining objectives, hazards and outline methods) through to detailed specification and execution of the project.

All these considerations apply equally to chemical or nuclear plant, where specific problems due to radioactivity, flammability or toxicity must also be taken into account according to the nature of the facility.

## 12.3 Timing and strategy

Where timing is concerned, the chief difference between conventional and nuclear plants is that radioactive contaminants naturally decay, while others generally do not. Depending on the radionuclides involved, dismantling may thus become substantially easier and cheaper after a suitable delay. These considerations are however sometimes overridden by concern for safety in the interim, or by the urgent need for electric power. For such reasons are Russian-designed VVER (pressurised water) reactors shut down in a unified Germany, all nuclear generation ceases in Italy, and Shoreham plant on Long Island was never allowed to reach full power—yet at the time of writing RBMK reactors (including two at Chernobyl) continue to operate as do VVERs throughout Eastern Europe. Thus many subsidiary factors modify straightforward economic calculations on both the lifetime of a facility and the strategy for its subsequent decommissioning.

### 12.3.1 Relevant factors

Apart from special legal or political considerations, decommissioning decisions are based on:

- Issues such as decay patterns of radionuclides, their amount, form and hence potential accident risk, regulations applying to both operational and shut-down mode, and the options for dealing with each category of waste expected.

- Technicalities such as engineering feasibility, containment of activity, protection of operators during deferment or dismantling, any requirement to develop new tools or techniques, and opportunities to apply the experience and knowledge of plant operators.

- Economic comparisons between immediate dismantling plus waste storage, or surveillance and maintenance pending dismantling at some future date. Decommissioning costs tend to fall with declining radiation levels, but regulations may become more restrictive. Accounting conventions (in particular the rate at which distant future costs are discounted, if at all) may be crucial.

- Policy issues such as employment, environmental factors, transport and power supply options, re-use of the building or site, funding routes available for the project and the degree of financial risk.

Strategy will depend largely on the category of plant:

- Where safe containment can be guaranteed indefinitely only at considerable cost (as in facilities handling exposed fuel or plutonium), short-term dismantling is unavoidable.
- Where a few decades' decay in levels of fixed radioactivity simplifies dismantling (as in power reactors), there is a clear technical advantage in postponement.
- Otherwise, as with heat exchangers or fuel storage ponds, a decision depends largely on detailed economic arguments.

### 12.3.2 Cost considerations

Costs and financial provisions vary from country to country, in some will remain uncertain for a time, and have been claimed to be underestimates;[1b] nevertheless much relevant work has already been done and valuable cost information compiled.[3a] Nuclear Electric currently estimates decommissioning costs for Sizewell B at 3% of the value of electricity over its lifetime, or around 1% with discounting; figures for AGR and Magnox stations are rather higher at 3% and 7% respectively.[1c] Confidence in these estimates is derived from comprehensive decommissioning studies.

Minimising the costs of such major engineering projects will clearly require continuing effort. However all can learn from each other, work on reactors will involve much repetition of techniques if not of detail, and experience on earlier partially military facilities in the UK, USA and France is providing much valuable information on:

- cost-effective technology, avoiding expensive approaches specific to the project;
- techniques to minimise primary and secondary wastes;
- designs and options to optimise requirements and share them with related projects;
- subdivision of large projects into phases, completion of each of which progressively reduces technical and cost uncertainties.

Cost issues in timing and strategy may therefore be summarised in three areas:

(a) Significant uncertainties (for instance, in accumulating the necessary funds or providing waste disposal facilities), upon which US regulations place great stress, as befits an industry divided into many companies.

(b) Comparative benefits of delay versus immediate dismantling, subject to legal or political considerations.

(c) Where prompt decommissioning is a technical or legal requirement, the need to arrange the programme for minimal cost.

### 12.3.3 Choice of strategy

National interpretations of these factors differ widely. For instance:

- Japan requires complete decommissioning to green field condition, started immediately after shutdown and completed within 10 years. This follows from very high land values and a disbelief that reductions in radiation exposures and waste disposal problems warrant prolonged deferment.

- At the other extreme Nuclear Electric in the UK has proposed, after external dismantling, to *safe store* Magnox and AGR stations for up to 135 years, expecting dose rates then to be negligible during demolition.

- Intermediate are the US and other major LWR operators where a deferment of 50 to 60 years gains the maximum benefit, short of human access to the reactor vessel which may be impossible even after much longer periods. Consequently the US NRC allows a pause after IAEA Stage 1, but only for power reactors and at most for 60 years.

In the USA, six nuclear plants have so far been shut down permanently for a mixture of technical and economic reasons, coupled in some cases with local public sentiment. They are worth examination:[4]

(a) Shoreham, 800 MW BWR on Long Island: public opposition forced long delays in licensing and construction, then permanent closure after one month at 5% power, and presumably influenced a decision for immediate dismantling.

(b) Rancho Seco 900 MW PWR, California, shut down in 1989 after 14 years of unreliable performance and somewhat related public opposition. It is kept in a safe store condition, partly to hold spent fuel pending provision of a federal repository, partly because insufficient decommissioning funds accumulated during the working life.

(c) Fort St Vrain 330 MW HTGR in Colorado ran only from 1979 to '89 with many problems before being closed down on economic grounds. As the utility's only nuclear station it is being decommissioned immediately, largely to escape liability under the Price-Anderson Act which requires operating or 'safe store' reactor sites to share the costs of an accident at any other commercial site in the country.

(d) San Onofre (Unit 1) 436 MW PWR, California, operated satisfactorily from

1968 to the early 80s but less reliably after an upgrading of safety systems, and closed down permanently in 1992 leaving two reliable 1100 MW units on the same site. Partly because of restricted access, it is in safe store until the other units shut down in 20 years or so and all are decommissioned in a sequence planned to make the best use of specialised equipment and staff training.

(e)   Yankee Rowe 167 MW PWR, Massachusetts, operated from 1960 to 1991 when unconfirmed concerns about pressure vessel embrittlement led to a petition for shutdown, accepted in view of costs for additional analysis. Again safe store and delayed dismantling are proposed.

(f)   Trojan 1095 MW PWR, Oregon, operated from 1976 to 1993, latterly with many steam generator problems. Repairs would have been too expensive in competition with local hydroelectricity, so the plant was shut down. Safe store is necessary until decommissioning funds have been accumulated.

This piecemeal approach by organisations generating power from several sources, and each with only a few nuclear, contrasts with that by larger, often state-owned, utilities in other countries. For example Ontario Hydro, with 20 reactors at 5 stations, consistently allows 30 years for Stages 1 and 2 and a further 10 years for final decommissioning, to be reviewed periodically in the light of changing conditions.

## 12.4  Planning

### 12.4.1  Outlining a project

Decommissioning requires a balance between cost, time-scale, occupational dose and waste arisings, sometimes weighted by particular circumstances as exemplified in Section 12.3. One of these factors may be traded against another, so that for instance a *minimum processing* philosophy reduces manpower and dose but increases the volume and disposal cost of wastes. On the other hand, all may suffer, say, from inadequate design or status information, or benefit from better cleaning or cheaper remote techniques.

The initial concept must therefore weigh these issues whilst identifying areas for subsequent detailed study, such as specific difficulties or major deficiencies in data. Any significant scheme must also assess economically the options for dismantling facilities and re-using buildings or site.

### 12.4.2  Detailed considerations

The first phase is to review plant status, alternative sequences of decontamination, dismantling and demolition, and what is to be done with removed equipment and the building structure.

The status of an operating or recently-closed plant will be well defined and documented, subject usually to some errors on the design, functional competence or inventory. For early facilities, such deficiencies may be serious.

The sequence of plant removal is often constrained by the layout, and by neighbouring facilities with which functional and structural links need careful review, especially where services and access have to be maintained for continuing operation.

Finally the nature, volumes and disposal routes of waste arisings need to be assessed, with associated costs and any constraints such as the dates when disposal facilities will be available. This could prompt a further revision of project strategy, e.g. towards improved decontamination to decategorise waste arisings.

A cost estimate for the project is now possible, followed by details of planning, design and procedure, decontamination and dismantling schedules, resource requirements and safety studies for regulatory approval.

Any major change to the site, especially to its appearance or workforce, may raise significant issues of public or industrial relations. Neighbours of the site may see demolition as an improvement to amenity, a dangerous operation, or a loss of employment, all issues requiring open discussion and reassurance where appropriate.

### 12.4.3 Project interactions

Options should take account of progress throughout the programme, whether to allow for interactions between units or to utilise experience. For instance, deferment may be cheapest for one plant in isolation, but in a larger programme may prevent resource levelling or re-using space for interim waste storage etc. Short-term cost savings must be weighed against *all* future costs and hazards. BNFL has prepared a strategic planning model for decommissioning to minimise lifetime costs and improve their estimation, given complex choices. It uses detailed studies of major Sellafield plants supported by data from completed projects and practical development. Besides scheduling work, it must quickly and reliably analyse the implications of various assumptions and options on facilities to be included, routing of waste streams, manpower levels, regulatory changes etc.

Other tools for the *desk-top* planning capability, emerging especially in France and the UK, include photogrammetry (reconstructing a design image from stereo photographs), simulation of proposed installation and operating regimes, radiation imaging and dose visualisation techniques, and data bases covering tooling, costs, decontamination and a range of decommissioning methods.

## 12.5 Practical techniques

Nuclear plant decommissioning has several aspects, not all necessarily represented in a given project. Within a project tasks may well be phased, in an extreme case with initial wash-out in one part of a plant whilst another is being dismantled or demolished.

### 12.5.1 Plant clean-out

This is usually the first stage, although some upgrading of services may first be needed. It proceeds from a standard end-of-campaign wash out, through a post-operational clean-out (POCO) with standard reagents but increased residence times etc., to *in situ* decontamination with more aggressive reagents, possibly in changed flow patterns.

The first two steps are essentially as on any chemical plant handling flammable or toxic materials, where existing effluent routes remain serviceable. They have long been standard practice at BNFL plants, whether for wet chemistry or dry powder, particularly after early decommissioning experience such as with the co-precipitation plant for mixed oxide fuel. Inadequately cleaned when shut down in 1976, this yielded more than 50 kg of plutonium and uranium oxides during dismantling ten years later.

The last step, *in situ* decontamination, simplifies subsequent dismantling by removing residual active species as far as possible. Since corrosion damage is no longer important, unless components are to be recovered for use elsewhere, conditions may be much more rigorous than for the most stringent previous clean-out. Almost anything is permissible—increases in time, temperature, acidity or concentration, alternating use of acid then alkaline or oxidising then reducing conditions and perhaps completely different reagents. For example, dilute hydrochloric acid, alone or with fluoride, seems a favourite for decontaminating heat exchangers on gas-cooled reactors.

Decontamination however does not eliminate radioactivity but simply moves it somewhere else, and is beneficial only if the contamination is less of a problem in its new home. It must be reviewed in the light of overall cost-benefit, especially in easing waste management. Its costs in labour, reagents, re-routing and effluent-treatment facilities must be weighed against the savings in dismantling man-hours, protective clothing, or remote operations.

Three very different examples are worth describing:

(a) Before gaseous diffusion plant at Capenhurst (UK) was dismantled, thousands of stage units (each with a compressor, cooler and filter) and over 1900 km of pipework, valves etc. needed decontamination. The main technique was fluoridation to convert solid deposits to volatile fluorides which could then be pumped away.

(b) BNFL had to replace a dissolver in the current Magnox reprocessing plant and first to decommission the old unit (Fig. 12.1). By prolonged circulation and agitation with alternately acid or alkaline, reducing or oxidising washes over 4 years, the highly radioactive plant within the dissolver cell was decontaminated by a factor of 10 000, to a level allowing manual engineering work. This saved the costs of remote operation or of building a new dissolver cell.[5]

(c) Access to the reactor building of Three Mile Island Unit 2, essential to evaluate

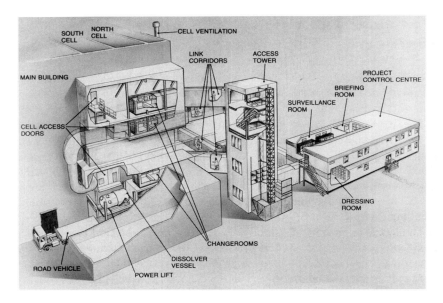

**Fig. 12.1** Decommissioning a Magnox dissolver

damage and work directly on affected plants, was prevented by levels of krypton-85 too high for sustained occupation. After prolonged review, preparation, safety analysis and public consultation, some 46 000 Ci was vented in July 1980, 16 months after the accident.

### 12.5.2 Upgrading of services

Several important services are likely to require upgrading, most frequently ventilation, containment and liquid effluent treatment. For plant essentially of the current generation, this is not because they are inadequate but because they are inappropriate for decommissioning. Differences from normal operations or maintenance need re-routed or increased air-flows and alternative or additional containment.

Possible modifications to ventilation include additional filtering, reversal of flow pattern, temporary containment structures, altered access routes, reduced pressure with respect to the outside atmosphere and dehumidification to minimise corrosion during a delay before dismantling.

Much valuable experience of containment is being gained on earlier plants, sometimes unconnected with the fuel cycle (for instance a radium lumenising facility close to a school, dwellings and railway line,[5a] or the old Windscale pile chimneys—remains of a defunct military installation). They were invariably built to much lower radiological and environmental discharge standards than now apply, and have often been out of use for so long that such equipment as was provided is no longer adequate or even usable. Complete if temporary enclosure of main and ancillary plant may then be necessary.

Existing routes for nuclear wastes (Section 12.6.1) may need upgrading or supplementing to cope with *in situ* decontamination reagents.

### 12.5.3 Dismantling

Following clean-out, the next step is what BNFL has termed *Initial Decommissioning*. Equipment is dismantled, opened or disconnected so that appropriate tools (e.g. a vacuum cleaner) can remove remaining residues, preferably to a level allowing prolonged deferment of final dismantling should this be desirable in overall economic terms.

The technical feasibility of such deferment depends largely on the type of facility and the adequacy of protective measures that can be applied. Where only small amounts of fission products or enriched uranium remain, for example, immobilisation by *tie-down* coatings may permit active safety systems such as ventilation to be shut down.

If a plant has handled highly active materials or large quantities of plutonium, this course is very unlikely. Further dismantling will then continue immediately, with size reduction or removal of whole items until all active plant is cleared or decontaminated.

### 12.5.4 Size reduction

The distinction between dismantling and size reduction is that the former yields components which could be reassembled, whereas in size reduction they are cut into disposable sections.

Size reduction requires decisions on where it is to be done, the tools to be used, and how they are to be deployed, with trade-offs (for instance between cost and time) in each of these interacting areas. Any cutting method may be considered, e.g. Fig. 12.2: the main criteria are safety for operators and environment, and costs in capital, man-hours or waste disposal.

The choices of tool and deployment system interact strongly. For example where radiation restricts but does not entirely prevent entry, a fast manual cutting technique may be justified by the excessive cost of remote operations. Alternatively, tools to be operated remotely may be deployed manually, so saving the costs of the more elaborate remote techniques needed to change, recover or maintain them.

For highly repetitive operations a robotic system may be cost-effective, while in areas of highest radiation fully remote techniques are inescapable (Fig. 12.3), often requiring specific tooling. Even here however costs may be reduced without compromising safety; much reactor dismantling is done under water,[6a] and having many reactors of similar design (as in the French PWR and Canadian CANDU programmes) will allow both development costs and actual tools to be shared between them.

**Fig. 12.2** Dismantling large components during decommissioning

Components may be cut up in place, in local specific facilities, or (capacity permitting) in a central workshop, depending largely on access to the location and compatibility with adjacent continuing operations. For example, in the plutonium plants programme at BNFL Sellafield, *in situ* or immediately adjacent shared facilities are preferable, although expensive items may be transferred from one operation to the next. Local size reduction is usual in reactor schemes because components tend to be large, widely separated and dissimilar, but the Capenhurst diffusion plant, with many similar and transportable units, found benefit in a robotic central cutting and treatment facility.

### 12.5.5 Assessment of radioactivity

Assessment is required to determine the necessary safety measures in all operations, and to plan waste disposal. Residual activity in or under buildings must be measured when a site is to be closed, and wastes as generated must be categorised. Techniques include computation, for instance of neutron activation in reactors, and various measurement systems in process plants.

In plutonium plants it is essential to locate and quantify residues that, with a change of configuration or moderation during dismantling, could cause a criticality incident. BNFL Sellafield has therefore developed a modular inventory

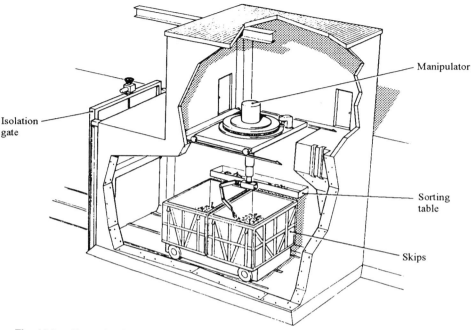

**Fig. 12.3** Example of remote handling

measurement system, using portable gamma and neutron monitors to locate residues and then a re-configurable neutron coincidence system for measurement.

Measurement of residual fission products is rarely so significant. Variants of standard health physics methods, and techniques to predict contact dose rates from external radiation measurements through any shielding, are usually sufficient for operational control. Nevertheless inventory measurement will sometimes be required, particularly under water in fuel ponds, and systems are being developed.

With reactors, activity assessments based on neutronic history suffice for planning, cost estimates, prediction of waste arisings, and regulatory approval of the proposed scheme. Once it is under way, theoretical estimates must be supplemented by analysis of representative samples, since variations in local neutron flux and in the composition of structural materials lead to uncertainties in the induced radioactivity. The particular bugbear is cobalt (activated to the hard $\gamma$-emitter Co-60), ranging typically from 30 to 300 ppm in mild steel or from 180 to 3200 ppm in stainless; tritium and C-14, derived from lithium and nitrogen, dominate graphite activities in short-term dismantling, but Ca-41 and Cl-36 (from Ca-40 and Cl-35) are more important in the longer term for waste disposal, owing to long half-lives.

In the Windscale AGR, for example, Co-60 is determined by gamma measurements, and inactive Co-59 by analysis; the isotopic ratio serves as a neutron indicator for isotopes which cannot be measured easily, such as Cl-36, Ca-41 or Fe-55. Estimates of amount depend on adequate modelling of the components and a reference *library* of material samples.

Large quantities of construction materials contain negligible activity beyond that naturally occurring. It may be checked by high-efficiency γ-ray detectors, or by bulk assay systems based on the radiolysis of freons or similar gases, still under development but already capable of the crucial distinction between ILW and LLW.[7,6b]

### 12.5.6 Site remediation

The final close-down of nuclear sites and associated remedial work resemble those in conventional chemical plants except that radioactivity is easier to detect than many other hazards. Consequently, in the measurement of residual contamination and treatment of the site for re-use, checking progress and compliance with regulations is faster and simpler than for instance with organochlorides. Considerable experience has been gained in decommissioning projects around the world, and techniques continue to be improved.

It must again be stressed that decontamination, which merely moves radioactivity from one location to another, is worth while only if the benefit in free release of the site outweighs the cost of treatment and waste disposal.

No British nuclear site has so far been released for other use, although Nuclear Electric has made proposals for Berkeley.[3b] Experience does exist elsewhere, however, especially for uranium mining and milling activities. US practice is to apply an impermeable multi-layered cap, usually vegetated on top (although at least one is a car park). The concern is always to locate and assess any residual activity below ground and either remove or contain it so that diffusion into water courses or supplies is within acceptable limits. Monitoring of water run-off and of any airborne activity is usually continued for some years, suspended eventually with regulatory consent.

## 12.6 Waste management

The space released for new facilities by decommissioning is valuable, and sometimes a decisive consideration as in Japan; nevertheless much radioactive waste is produced, both primary (structure and contents of the facilities themselves) and secondary (materials used in the decommissioning process). Treating and disposing of it accounts for much of the cost, for example some 30% of the £4.3 billion estimated in 1988 for decommissioning all BNFL facilities over the next 100 years.[6b] Waste management is thus a vital part of the process to minimise overall costs, and arisings (particularly of solids) must be estimated as accurately as possible at the project planning stage.

Although radioactive wastes generated in decommissioning (except of reactors) are generally much less than during normal operations, they must be kept to the minimum that is cost-effective and appropriately allocated to ILW, LLW and *de minimis* categories. Handling should always be integrated between decommissioning projects and with any continuing operations at the site.

Management practices differ for fluid and for solid wastes. Liquids and gases cannot be left unsupervised, since any containment may leak. Sooner or later, after monitoring and any due treatment, they must therefore be discharged to the environment under conditions ensuring that natural dispersion processes reduce the contents to innocuous concentrations. Solids may be disposed to a suitable repository, or if sufficiently low in radioactivity, recycled for limited or unrestricted use. Whatever route is chosen must comply demonstrably with regulatory requirements, allowing for both expected operations and potential mishaps.

Many decommissioning projects have attempted economies by restricting wastes, especially in the more expensive categories. Options can be divided into three groups:

(a) Avoiding or reducing waste production. Means range from simple management methods, such as preventing contamination of packaging, to the control of loose contamination by containment, ventilation or tie-down coatings.

(b) Volume reduction, much studied and often genuinely cost-effective. Techniques include compaction, melting (scrap metal) with or without decontamination, incineration, or simply making the best use of containers by appropriate size reduction and careful packing.

(c) Segregation into clean and contaminated fractions, or decontamination.

## 12.6.1 Gaseous and liquid effluents

Effluents may be residues from the normal process or arise from the dismantling and decontamination itself.

Aerial discharges may be controlled locally, at final discharge or both, e.g. by electrostatic precipitation, cyclone separation or bag filtration near the work area, and HEPA filtration before discharge, always with adequate monitoring.

Liquids must be contained and stored pending sentencing, with account taken of both radioactive and inactive but environmentally-hazardous species, especially heavy metals. Discharge to the environment is usually the most cost-effective practice, but where it is impracticable, the noxious content must be separated and solidified, for instance by evaporation, filtration, ion-exchange or some combination of treatments.

## 12.6.2 Solids

Many items (particularly of building structure or shielding) remain inactive or can be effectively decontaminated. Despite wastes generated in the decommissioning process, such as tools, equipment, protective clothing and residues from effluent treatment, active arisings are only a small proportion of the original total.

It is instructive to compare estimates with actual achievements. The Japanese

**Fig. 12.4** Recovered scrap components awaiting recycling, Capenhurst

JPDR, the whole decommissioning of which has been treated as a source of data, was predicted to yield a total of 30 000 tonnes of wastes, 4000 tonnes radioactive.[8a] By the end of 1992, when everything but a small part of the bioshield had been removed, active waste amounted to some 3000 tonnes plus 400 tonnes generated in the decommissioning operations. Meanwhile, in Germany, of 4000 tonnes of steel from the turbine hall of the Gundremingen BWR, some 2400 tonnes have been free-released, 1500 tonnes recycled in the nuclear industry and less than 100 tonnes stored for disposal. The Niederaichbach GCHWR in Germany is expected to yield 130 000 tonnes total, of which 1700 tonnes will be suitable for low-active recycle and only 1200 tonnes will require final disposal.[1d]

At the Capenhurst diffusion plant in the UK, although contamination was nowhere heavy, the scale of operations with over 2000 process units and 1900 km pipework required careful choice of the most economical disposal route:[6c]

- clean scrap—cell cubicle panels, base-plates, drive motors—was sold on the open market.
- scrap for decontamination—most aluminium and steel plant items—was cleaned below *de minimis* levels then sold on the market (Fig. 12.4).
- items which defied decontamination or economic checking, such as instruments and small pipes, were buried at Drigg.

Many decommissioning projects around the world have released materials from regulatory control into the public domain. Unfortunately, stipulations are seldom published, and differing circumstances hamper comparison. Exemption levels of surface activity, activity per unit mass, and dose exposure have to be derived from dose objectives set by each national regulator, and may depend upon the radionuclides involved. In practice, currently-used values range from 0.37 to 3.7 Bq/cm$^2$ for β,γ-emitters and for uranium, down to an order of magnitude lower for α-emitters not restricted to uranium isotopes.

Some countries set alternative limits depending on whether the intended use is restricted or not. In Germany, where Federal law requires material to be recycled wherever possible and economic, the Commission for Radiological Protection has recommended its fabrication into items for the nuclear industry under controlled conditions.[1d] If this is not feasible material may be released as scrap for remelting (i.e. dilution) if it is below 1 Bq/g and for unrestricted use below 0.1 Bq/g. Contamination at the surface must not exceed, for example 5 Bq/cm$^2$ for Fe-55 or 0.05 Bq/cm$^2$ for Pu-239.

A Japanese study into recycling metals, particularly from the JPDR project, suggests that the decommissioning of their current 40–50 reactors after shut-down beyond 2010 will yield perhaps 40 000 tonnes per year of exempt steel for free recycle plus 30 000 tonnes for melting and controlled recycle as nuclear waste containers, reinforcing bar for concrete structures and the like. Detailed costings show a considerable advantage over disposal unless this is permitted as LLW at up to 37 Bq/g in a simple trench with no barriers etc.

Actual experience is encouraging. Large proportions of scrap may be inactive—over 90% in the Saluggia fuel plant project and more than 3000 tonnes from the high-enrichment stages alone of the Capenhurst diffusion plant. The low-enrichment area of the plant yielded 31 000 tonnes of clean or decontaminated metals, the Rhapsodie fast reactor some 300 tonnes and the Gundremingen BWR 4000 tonnes. Savings on storage and disposal are substantial.

From JPDR, solid wastes were categorised according to activity per unit volume or weight (four levels), type and material.[8a] The information on proportions, activities and attainable packing densities has indeed proved valuable to other decommissioning projects. Some materials, however cannot be economically recycled, and alternative treatments such as incineration, compaction and grouting were investigated.

## 12.6.3 Disposal

Facilities for LLW exist or are being constructed in several countries. For other wastes, although international recommendations may be accepted as basic principles, there are major differences in national interpretation leading to different technical solutions for final disposal sites or repositories.[1a] Where such facilities do not exist, extensive decommissioning must be deferred, limited to safe storage, or (as in the UK) rely on interim storage, duly contained, shielded and guarded against foreseeable hazards.

In the UK, Nirex (United Kingdom Nirex Ltd, established by the major nuclear organisations to construct a deep repository for solid ILW and LLW) is required to find an acceptable site and is investigating one adjacent to BNFL Sellafield, with a view to readiness early in the next century. Nevertheless, most LLW from the UK will still go to the Drigg shallow burial site whilst this remains available. Afterwards, LLW will presumably go to a deep repository.

Interim storage is exemplified by work on the Windscale pile chimneys. They were insulated internally with glass fibre, which was not sealed and so acted as a crude filter during operation and especially the 1957 fire. Thirty-eight years later all but Cs-137 has decayed but the material is still ILW. Packed into 200 litre stainless steel drums, it is housed temporarily in a store formed from the basement foundation voids of one chimney.

## 12.7 Progress

Decommissioning experience around the world, especially in France and the UK, had touched every major stage of the nuclear fuel cycle by the early 1990s. Even that stemming from facilities not strictly in the fuel cycle, or refurbished rather than dismantled, is directly applicable since operations are identical.

### 12.7.1 Achievements

Besides those already described, several others deserve mention:

- Uranium mining and milling—many sites no longer economic have been closed down, decommissioned and their sites restored. Completed examples are Andujar (Spain), Ranstad (Sweden), and Cogema subsidiaries in France, USA, Germany and Canada.[8b,c,d]

- Refining, enrichment and fuel fabrication—several fuel fabrication plants have been extensively modified. At BNFL Springfields, a fluidised-bed reactor converting $UO_3$ to $UF_4$ has been decontaminated and demolished. Smaller fuel facilities at Hanau (Germany) and Saluggia (Italy) have been completely dismantled and the buildings re-used.[8e,6d]

- Reactors and associated facilities—the Shippingport demonstration PWR (72 MWe) was decommissioned to Stage 3 over five years to 1989.[3c] The post-irradiation examination facility at Berkeley, UK, has been completely gutted and refurbished.[5b]

- Reprocessing and waste treatment plants. In 1966 the original Sellafield reprocessing facility was partially dismantled and converted for oxide fuels.[9] The Dounreay reprocessing plant was partially decontaminated and upgraded

**Table 12.1** Reactors well advanced in decommissioning

| Reactor | Power | Type |
|---|---|---|
| Gundremingen | 250 MWe | BWR |
| Niederaichbach | 100 MWe | GCHWR |
| Fort St Vrain | 330 MWe | HTGCR |
| Shoreham | 809 MWe | BWR |
| Windscale | 33 MWe | AGR |

in 1972–75 to handle PFR mixed oxide fuel instead of the previous enriched uranium fuel.[1e] The HA waste vitrification pilot plant PIVER at Marcoule was decontaminated and dismantled during 1986–92.[10]

- Recycling plant for Pu and U—several manufacturing facilities for mixed oxide fuel have been completely decommissioned, usually to Stage 2 for re-use of buildings. Examples include Cadarache, Winfrith and Sellafield.[3d,11,12,1f]

### 12.7.2 Current projects

Other work in hand or about to start will yield considerable further experience by the end of the century. Larger and more diverse plants, up to essentially full-scale, are being dismantled, requiring more elaborate techniques. Those expected to reach IAEA Stage 2 or 3 include power reactors of several types (Table 12.1).

Other facilities to be taken to Stage 2 on a similar time-scale include:

- Sellafield's original full scale reprocessing plant, used for metal and oxide fuels.
- Plutonium facilities at BNFL Sellafield and AEA Harwell (both with remote dismantling).
- The original Sellafield fuel cooling and storage pond, with major refurbishment completed to permit clean-out and decommissioning.

These projects, many full-scale or nearly so, should demonstrate the ability to decommission plants at about the predicted cost.

## 12.8 Future prospects

Besides the power and research reactors already closed down or approaching 30 years of service, the many much younger stations (in France, Japan, Taiwan and Korea for example) must sooner or later be decommissioned. Since decontamination for maintenance, refuelling and replacement of steam generators is already routine for LWRs, and a decontaminated nuclear plant is much the same to demolish as any other large civil engineering structure, then only the dismantling of active components still needs to achieve this established status.

The next few years should bring significant progress on waste disposal and decommissioning time-scales. Despite successive delays to a British deep repository, work continues on the rock characterisation laboratory and on specifying waste packaging. A firm specification will allow dismantling to proceed even if interim storage is needed, while the experience so gained improves estimates of required disposal capacity and its cost. Waste disposal issues in the USA and several European countries should be clarified if not resolved on a similar time-scale, especially as advice becomes available from the IAEA.

Technical advances can also be expected. While existing technology may be capable of decommissioning any present facility, there is ample scope for reducing manpower, radiation doses and waste arisings, while solving challenges specific to certain projects. Gamma cameras to locate radiation sources, high-power laser beams applied through fibre optics to cut up active components, and sophisticated waste assay techniques are all likely and potentially valuable.

### 12.8.1 Costs

Although decommissioning costs are a small proportion of the total in power generation, they should still be minimised. The most obvious opportunity is in a programme involving several similar plants, e.g. the replicated reactor systems of Canada or France and the VVER units in eastern Europe. There will clearly be considerable economies in planning, safety studies, development of equipment and techniques, and training as personnel move from project to project.

Large programmes can still gain some economies of scale even without these *identical twin* benefits. Thus in work on the Windscale pile chimneys, the largely inactive chimney of pile 2 provides a testing ground for equipment for the still contaminated pile 1 chimney, whilst the plutonium plants programme at Sellafield benefits from the progressive evolution of *in situ* assay. Care will be needed to avoid losing these advantages in the increasing commercialisation of the European decommissioning business.

With rising costs of waste disposal, and routes constrained or uncertain, decontaminating materials for recycle as scrap or concrete rubble becomes increasingly attractive, even if the material is restricted in use. The example of Capenhurst (UK) in following this cost-effective route for some 99% of steel, aluminium and building materials has already been described.

### 12.8.2 Design feedback

Experience gained in decommissioning may be expected to influence the design of future plant, with improved access, simplified decontamination and dismantling, and reduced radiation. This last applies especially in reactors where a careful choice of construction materials, in particular avoiding cobalt wherever possible, can contribute substantially. Although increased capital expenditure must be weighed against savings discounted many years onwards, discount rates for such long-term commitments may be low and the extra cost justifiably small.

### 12.8.3 Technical advances

Whether generic or project-specific, development is required more for economic reasons than to fill gaps in existing technology. Regulatory environments are however much more restrictive than when many plants were built or operated, and some projects offer specific technical challenges. Most countries with significant nuclear programmes (notably France, UK and on a massive scale, Japan) include decommissioning development, and international collaboration is extensive both within the European community and more widely under the auspices of OECD/NEA.

Besides areas already mentioned, ideas with substantial prospects include bio-decontamination and using lasers to seal activity on to surfaces for prolonged storage.

### 12.8.4 Strategy speculations

Current timing policies range from immediate dismantling (reactors in Japan, plutonium plants everywhere) through intermediate terms (30 years for CANDU, 60 for LWRs under US NRC regulations) to 135 years for British gas cooled stations. Extending such periods would gain further advantages in cash discounting and sometimes in radioactive decay, subject to a sufficiently robust safety case, tolerable maintenance costs and political acceptability.

Eventually a site must be released from the controls applied to nuclear facilities. An alternative way to minimise low-level waste costs might be to leave materials on site, suitably mounded over and re-vegetated. This is after all common practice with uranium mines; applying it to reactors or other nuclear facilities will no doubt depend on public acceptability, and in turn on the actual definition of low-level waste.

For reactors the ultimate step in this direction is the so-called one-piece decommissioning suggested for Ontario's CANDU stations and for LWRs in Taiwan—lowering a reactor's core and bioshield, minus all removable components, into bedrock at the operating location.[1g] New facilities, especially reactors, might be built underground, given suitable conditions. The idea is not new; in Europe, research or experimental reactors have already been built below ground at Halden, Agesta, Chooz and Lucens, whilst the Swedes and Swiss have long built hydro power stations below ground for technical and environmental reasons. Such a solution would certainly reduce the visual impact of stations and improve public acceptability, during both operation and any safe-store period.

## 12.9 Conclusion

Decommissioning, with its frequently long time-scales, is a prime example of inter-generation responsibilities. By the time a facility is no longer needed, the benefits from its operation have largely been gained or committed, and leaving the remains to be cleaned up by successors who have not directly shared in these benefits is acceptable only if we provide the means.[1b]

The technical means exist to deal with any present civil installation, at costs that can be predicted with some confidence, and the insistence on *design for decommissioning* should simplify future operations. Nevertheless there is scope for reducing radiation doses to operators, waste arisings and costs in general, so continued developments may be expected particularly in the fields of robotics and size-reduction methods.

Financial provision is equally important, though how much depends on the discount rate assumed over long future periods. However it may be invested in the interim, it is essential to ensure that sufficient funds are dedicated to the purpose to cover our duties towards descendants.

## 12.10 References

1 *Decommissioning of nuclear facilities*, IBC, London. (February 93).

    a M. Laraia. Nuclear facility decommissioning—an international perspective.

    b G. MacKerron. Economic and financial implications of decommissioning.

    c S. C. Gordelier. Illustration of an approach to the assessment of decommissioning strategies.

    d M. Mittler. Some experience with the decommissioning of German power reactors.

    e G. Bailey. Refurbishment of the D1206 reprocessing plant.

    f S. Buck. Decommissioning of the dry recovery plant at Sellafield.

    g P. D. Stevens–Guille and N. D. Jayewardene. Decommissioning policies, plans and experience in Canada.

2 IAEA Safety Series No. 52, Vienna (1980).

3 *Decommissioning of nuclear installations*, CEC, Brussels (October 1989).

    a A. Cregut and A. R. Gregory. Decommissioning of nuclear installation in member states, achievements and projects, pp. 5–8.

    b P. B. Woolam. Measurement techniques applicable to residual radioactivity on a decommissioned reactor site, pp. 76–82.

    c J. J. Schreibe. Completion of Shippingport reactor decommissioning, pp. 828–32.

    d A. Caillol. Inventory of gloveboxes (in) dismantling operations, pp. 379–82.

4 T. S. La Guardia. Decommissioning power reactors in the USA, forcing the pace. *Nuclear Engineering International* (August 1993).

5 *ENC '90*, Lyon,.Vol IV (September 1990).

    a M. R. R. Harris *et al.* Removal of Th and Ra contaminated soil from an urban site.

    b A. W. Brant and P. M. Hoogewerf. Decommissioning of PIE caves, Berkeley Nuclear Laboratories.

6 *Nuclear Decommissioning '92*, IMechE, London (February 1992).

    a W. E. Murphie. An overview of the United States decommissioning programme with selected examples of current technology and environmental compliance, pp. 9–15.

b C. L. Walters. Overview of a decommissioning development programme, pp. 149–56.

c D. W. Clements and G. H. C. Begg. Decommissioning of Britain's gaseous diffusion plant. pp. 179–87.

d M. Guidotti and L. Sberze. Decommissioning of a pilot fuel fabrication facility at Saluggia. pp. 173–8.

7 A. Colquhoun. British Nuclear Fuels decommissioning development programme and progress. *International Decontamination and Decommissioning Symposium*, Knoxville, Tennessee (April 1994).

8 *1993 International Conference on Nuclear Waste Management and Environmental Remediation*, Prague. Vol. 2. (September 1993).

a S. Yanagihara *et al.* Systems engineering for decommissioning JPDR—a study of characteristics of decommissioning waste, pp. 423–31.

b J. L. Santiago and M. Sanchez. Decommissioning and waste disposal methods for a uranium mill in Spain, pp. 193–7.

c P. Linder and B. Sundblad. Remediation of a closed down uranium mine in Sweden, pp. 489–99.

d H. Quarch *et al.*,. International developments in uranium mining and mill site remediation, pp. 461–71.

e B. G. Christ and E. L. Wehner. Project Specific selection of decommissioning techniques, pp. 189–92.

9 B. Bailey and A. P. Colquhoun. BNFL decommissioning experience at Sellafield. *ENC '86*, Geneva (June 1986).

10 A. Jouan and S. Roudil. Assainissement final de l'installation prototype de vitrification PIVER—Decontamination de la cellule chaude. *CEC Report* EUR14764F (1993).

11 M. G. A. Pengelly *et al.* Decommissioning of a $MO_2$ fuel fabrication plant at Winfrith Technology Centre. *Joint International Waste Management Conference*, Seoul (October 1991).

12 S. Buck and A. P. Colquhoun. Decommissioning of a mixed oxide fuel fabrication facility. *CEC report* EUR 13057 EN (1990).

## 12.11 Further reading

These respectively update references 6c and 8a above.

I. Mech. E. *Decon '95* (November 1995), D.W. Clements. Decontamination and melting of low-level waste—a complete environmental package, pp. 317–80.

S. Yanagihara *et al.* Data analysis and lessons learned in decommissioning Japan Power Demonstration Reactor. pp. 127–40.

# 13 *Management of safety*

A. C. Fryer, M. Merry, C. Sunman, and A. E. Waterhouse
*BNFL, UK Group, Sellafield, Cumbria (except M.M.—THORP Division)*

## 13.1 Introduction

The successful management of safety in any organisation is through the development of a Safety Culture appropriate to that organisation and the function it serves. For the nuclear industry the definition of the term *Safety Culture* proposed by the International Nuclear Safety Advisory Group[1] is:

> 'Safety Culture is that assembly of characteristics and attitudes in organisations and individuals which establishes that, as an overriding priority, nuclear plant safety issues receive the attention warranted by their significance.'

This definition emphasises that Safety Culture is about attitudes as well as management structures, policies, and directives. Statements of policy and managerial directives are of little effect if not supported by appropriate attitudes throughout the organisation. Safety Culture is therefore intangible, though it can (and does) lead to tangible effects. It is simpler to consider these two major components of Safety Culture separately, though recognising that they are intimately linked. External influences are also discussed.

### 13.1.1 Managerial aspects of Safety Culture

Managerial structures assign responsibilities to individuals, and one of the functions of each manager in such a structure must be to ensure that all individuals understand their own responsibilities, those of their immediate colleagues, of their management unit, and of other units within the organisation. Formal responsibility for plant operational safety always rests with the managerial unit operating the plant. Hence it is in the interests of that unit to ensure that other organisational units who are expected to contribute to the safe operation of the plant, do so satisfactorily. For this reason it is standard practice, within the UK nuclear industry, to have a rigorous and systematic vetting before approval of any modification to plant or process.

As established practice, this vetting process ensures that measures are taken to minimise both routine and accidental radiation doses to the public and to the operators. Sometimes it may be possible to install equipment which reduces the routine dose (e.g. by shielding) or minimises exposure to unusual levels (e.g. by alarms). Where relatively high background radiation is the problem, operators' working times may be controlled. Such controls require the advice of Health Physicists regarding levels of direct dose and contamination. Where the potential exists for accidental doses, to either the public or the operator, a risk assessment is conducted as an aid to deciding if, or where, additional controls should be applied (see Sections 13.2 and 13.3).

Any group of people who have to work together will develop a set of values regarding how they behave towards each other and towards outsiders. Such values may be generated by the work team or by the managers, and may be either openly declared or unofficial. A poor Safety Culture is perhaps more likely to develop where the official values for a group do not match the actual values. It is very easy for the official values to become a managerial illusion. However, ensuring that there is only one set of values for the team and that these are suitable rests with the team management.

Individuals who lack the technical expertise for the tasks they are expected to do, can severely impair the safe operation of any plant or process. Hence selection of personnel plays an important part in development of a good Safety Culture, as does training.

If tasks and roles are not clearly defined then management can have little confidence in their understanding of the processes in their charge. Therefore, for any nuclear facility, a series of documents is prepared to clarify the processes: such document systems may include instructions, quality plans, and guidance notes and are subject to periodic review to ensure their continued accuracy and relevance.

It would be easy to fall into the trap of paying incentive bonuses which are entirely based on plant throughput, even if this is seen to conflict with safe practice. Such incentives schemes must therefore reward safe performance as well as productivity.

To support the technical expertise and plant knowledge of personnel within an organisation it is normal practice to encourage the work of professional scientific and technical societies (e.g. the Royal Society of Chemistry). Similarly, the circulation of periodicals which report on accidents, and events which contribute to them, serves to heighten awareness of potential faults in plant design or process. More locally, the publication of *in-house* newsletters or bulletins which address issues relevant to a particular site or organisation, can be extremely useful in ensuring that lessons are learned from past failures.

The treatment of individuals who have made mistakes can be an important influence on the safe operation of a plant. Individuals need encouragement to look for imperfections in their own performance, and where appropriate, help should be given to achieve that improvement. It is dangerous for the cause of a potential hazard to remain hidden, and for this reason, nobody should be pushed into the role of *victim*, or *scapegoat*. It is far better to involve actively those persons associated with the fault in establishing ways of avoiding such faults in future than to apportion blame. This is not to say that disciplinary measures should not be taken but rather that a balance has to be struck, so that individuals are not more afraid of disciplinary action than they are of the actual hazard.

To assure managers that their plant and processes are operating as intended, and in particular safely, it is necessary to audit their areas of responsibility. Such audits need to be conducted internally within the organisation, and there is also much benefit from external auditors who can bring a fresh view on the workings of the organisation. Serious consideration must be given to auditors' reports, to make the exercise worth while.

### 13.1.2 Attitudinal aspects of Safety Culture

The development of a questioning attitude among all members of an organisation is essential if that organisation is to develop a good Safety Culture. The sort of questions which each individual should ask themselves before commencing a task could be:

- What is the task?
- What are my responsibilities?
- What are the hazards?
- Who is at risk?
- How do I perform the task while avoiding the hazards?
- Who do I need to inform about the task, or obtain permission from?
- When is it safe to do the task?
- Where is the task to be performed?

These questions, of which there are many equally valid variants, can provide a rigorous and prudent approach to any task.

Sometimes a questioning attitude can be viewed with suspicion, particularly among middle managers. This is, to some extent, because of the way in which information is presented from one group of people to another. Constructive comments from one group of workers or individual can be viewed by another as the airing of a *grievance*.

From a management perspective, however, is this the entire answer to the question of how to establish and maintain attitudes among the workforce which lead to a good Safety Culture? For example, as a means of understanding the individual attitudes which contribute to safety, BNFL invited Professor Terence Lees of Aberdeen University to survey attitudes among the workforce at its Sellafield site. The most important single contributor towards safety was found to be trust;

trust in the individual's immediate line management, and trust in the individual's colleagues. The less trust, the greater is the potential for accidents.

### 13.1.3 External influences

#### Regulators
External regulatory bodies can have a significant effect upon the safety culture within an organisation.[2] Four features of such relationships are worthy of particular note:

- The Regulator as an organisation has a strong effect, depending on its role, powers, and the way it operates. This relationship is relatively slow to establish or change.

- Regulators as individuals may be more visible to the operator, but their effect (which is not usually long-lasting) depends on their personal attitudes, skills and conduct.

- The regulatory framework (e.g. the licensing arrangements) has a strong effect, and changes to the framework can take a long time to develop and implement.

- So long as the Regulator's requirements can be seen to be reasonable and prudent, they have a positive effect on Safety Culture.

#### Local environment
The local environment in which an organisation operates can influence the attitudes of both the managed and their managers, and thereby affect the Safety Culture within the organisation. Such issues as unemployment, worker mobility, the strength of the local economy, pressure groups, and the fortunes and working practices of other local industries (e.g. fishing, farming, tourism and manufacturing), may each make a subtle contribution to Safety Culture.

#### Pressure groups
The attitude of workers and managers to pressure groups, whose influence may be seen as ill-informed political interference, must have an effect on morale. Managed prudently, it should be possible for pressure groups to have a positive effect on Safety Culture. In the extreme opposite case, the workforce may develop a *siege mentality* which is unlikely to enhance safety.

## 13.2 Risk assessment

Any public discussion on the subject of nuclear facilities invariably raises questions of safety, commonly in emotive and inaccurate terms. A nuclear engineer, professionally concerned about the issue, needs to be more precise, and to measure the *safety* of various design options, operational modes etc., in a manner that allows meaningful comparison of their acceptability. The tool used is Risk Assessment, variously described as Probabilistic Risk Assessment, Safety Assessment and Hazard Analysis, the outline of which is described below.

### 13.2.1 Risk

The concept of *risk* combines both the likely damage due to an event—the consequence; and the likelihood that it will happen—the frequency. In familiar terms, the consequences of being involved in an air crash are very high but the frequency is relatively low, whilst for a car crash the opposite is true; most travellers would consider flying the safer option. Many variables may be relevant, for instance the model of plane or vehicle, the type, length and conditions of the journey, and the standard of maintenance. A test pilot is at greater risk than a holiday-maker, a travelling salesman than a teacher.

Another fundamental concept is that of *probability*—the likelihood that something will occur. To say that the probability of an event is 1 in 10 000 years does not mean that 10 000 years must elapse before the event occurs, or that it will occur once every 10 000 years, but that the chances against the coincidence of all the factors necessary for it to occur in any one year are 10 000 to 1. In Britain it is highly unlikely (a million to one) that any individual will be hit by lightning, although this is no consolation to the four people per year on average who die by this means.

### 13.2.2 Plant design

The fundamental safeguard is the design of the plant or process itself. Safety aspects must be considered at every stage. A *passive* system is always favoured, e.g. enough shielding round a source of radiation to prevent any possibility of exceeding permitted dose levels. Systems are designed as far as possible to be fail-safe, so that loss of power or any other necessary service sends the device or process to a safe condition; if that is impracticable, there must be time for back-up arrangements to take effect before a dangerous situation arises.

In practice it is rarely possible to design without *active* safety devices—things which act to restore the safe condition, but are capable of failing to act correctly. The response to such failure must therefore be considered.

### 13.2.3 Fault identification

The first stage of safety assessment is the fault identification, to be followed by an examination of consequences. If we are driving a car in heavy rain and the windscreen wipers fail we have impaired visibility, and the driver must take action to reach a safe condition—slow down or stop. The effects of such a failure will depend on the weather, light level, type of road, speed etc., and all these factors must be considered.

The standard technique used to predict fault events is Hazard and Operability study—HAZOP, essentially a structured approach to fault identification. It is carried out by a mixed team of specialists, all familiar with different aspects of the plant or process. A standard set of guide words, relating to plant deviations—e.g. high flow, no flow etc.—is systematically applied to each section of the plant or process. The team discusses the likely outcome of each deviation, stimulating and building on each others' ideas. A full-scale study will cover each line and vessel on an engineering line diagram for a continuous operation, and each sequence or process step for a batch process or specific task.

Whilst HAZOP is a near-universal technique with general applicability, it may be supported or substituted by other techniques. With complex electronic or mechanical equipment, a Failure Modes Effect Analysis (FMEA) may be applied. In some systems it is more appropriate to construct Event Trees by considering the possibilities of all conceivable events and their effects in combination.

All techniques rely heavily on the experience and working knowledge of the team. They will generate a wide range of fault conditions, which are then logically grouped for hazard analysis. For a car journey such faults would include a range of *operator* errors (bad driving), *mechanical* faults such as brake failure or incorrect instrument readings, and *external* factors, e.g. other road users, adverse weather conditions.

### 13.2.4 Risk assessment

Once generated, a set of potential fault conditions must be analysed to determine their consequences. The first stage is to construct a model of the event. A good working knowledge of the plant or process, including whatever operational history may be available from internal or external sources, is essential here too so that relevant scenarios are developed. For our example of the car such a model will be simple, but for a complex nuclear plant it may involve a maze of intricate and related events.

Certain conventions are adopted in this model, making it easier for the analyst to decide on the scope and for plant operators to understand the resulting analysis. One is to develop a bounding case including all the lesser events, rather than try to analyse the maze of small complex events. To this end, in the example of the car journey, we would assume total brake failure to cover all losses of brake efficiency, or in a fuel-handling plant, dropping a flask to cover all mishandling events etc.

Analysts are encouraged to use models that are realistic and pragmatic, though with reasonably pessimistic data and scenarios. These assessments are intended to assist the plant operators in identifying important equipment etc., and excessive pessimism could cause means of mitigation to be overlooked.

It is also normal convention to analyse a fault sequence up to the point where controlled recovery is possible, at the natural conclusion of the event. Thus in general if terminating an event requires an immediate or sustained effort, then the assumption is made that no intervention occurs. For example in a fire, the analysts would assume no intervention and an uncontrolled burn. If the required action was evacuation, for example in response to a fire or alarms, it would be assumed to occur with a probability derived from experience, albeit after a time delay. However the case of failure to evacuate would also be studied. Subsequent recovery under controlled conditions is not included in the assessment.

Other conventions are concerned in detail with the scope of each assessment and each safety case. These are essential to ensure that on completion all the analyses and safety case boundaries fit together without omissions or areas of uncertainty between them.

With this information the analyst can now develop the accident scenario and progress to the next stage of the assessment.

### 13.2.5 Consequences

This is to calculate the consequence, that is the result of the fault occurring. In a nuclear facility the fault sequence will be developed to determine where and what the effect could be, including the effect of correct or feasibly incorrect intervention. Malicious events, such as sabotage, deliberate maloperation or terrorist attack, are excluded.

The scenario that has been developed may be complex. A maloperation on the plant may lead to an effect on the workers involved, while a release of radioactive material may also affect workers in the surrounding area and consequences must be calculated for them. An incident within the plant containment may result in an abnormal discharge from an engineered route, e.g. the stack, a coincident failure of the ventilation clean-up system would compound the event, and the effect upon members of the public must be calculated. With an external release of activity, some type of standard dispersion model will be used, such as in the United Kingdom has been developed by the National Radiation Protection Board (NRPB).[3]

Each site or facility has developed its own specific version which will take account of the local geographical and weather features. This site-specific model will be capable of varying the nuclides involved, and their physical form; the height of the release and its duration; and differing weather conditions which may occur during the release.

Similarly based models are available for liquid releases, again site-specific to account for local conditions of discharge to riverine, marine or estuarine environments.

Where a postulated release occurs within a plant, then release fractions and decontamination factors are applied as appropriate. These are derived from specific experimental work, plant experience and extensive databases which draw on a wide range of published sources world-wide.

Having determined the size of release we must calculate its effect. How this is done will depend on the type of release and its route back to the affected group. An aerial release will be inhaled; if it is large, deposition on vegetation and subsequent consumption must also be considered. For all routes, therefore, the potential pathway of the release must be determined and the most critical group (the people most substantially affected) identified. The critical group will depend on the type of release supposed, for example the local fish eaters after a release to sea, or young children living within a few kilometres of an aerial release.

Standard models exist to specify these pathways and to quantify the committed dose per unit of activity released (Chapter 11). In this way the appropriate consequence for each type of release and each potential subject will be calculated, always with inbuilt pessimistic assumptions.

The ultimate aim of the assessment is to compare the results with some set of criteria. To this end *cut off* points are established, estimates of consequence below which no further assessment is carried out. Such thresholds are normally numerical values of predicted dose uptake, with account taken of frequency, derived from criteria which are discussed in more detail later.

If, however, a significant consequence has been calculated then a frequency must now be calculated in order to assess the risk.

## 13.2.6 Frequency

To calculate the numerical frequency of a *top* (maximum compound) event we need values for its components. For example the likelihood that failing windscreen wipers will occur on an rainy night, leading to loss of visibility at high speed, will depend on the reliability of the wipers (in turn depending on design, quality of construction and maintenance), the frequency of rainy nights, and the proportion of time spent in fast driving. We also need to know which component of this failure is *failure in operation* as opposed to *failure on demand*—due, for instance, to shedding a blade or to a faulty switch respectively. Once again such data are specific to the task in hand; the value for an overnight motorway delivery service in the north is much higher than that for a metropolitan office worker.

Failure in operation is usually caused by wear or damage to the equipment itself, and less common than failure on demand, more probably due to a failure of control system or power supply: since the causes are different, so are the frequencies. Both must be taken into account, for example in a ventilation system where a running fan may break down and the back-up fail to start.

Reliability data must be specific to the type of failure, and operators collect them from their own plant over a period, usually including down-time for maintenance without distinction from actual failure. Similarly all similar failures tend to be recorded together and an overall figure is used, once again ensuring pessimism.

In general, top event descriptions are complex and a fault tree must be generated,

260  *Management of safety*

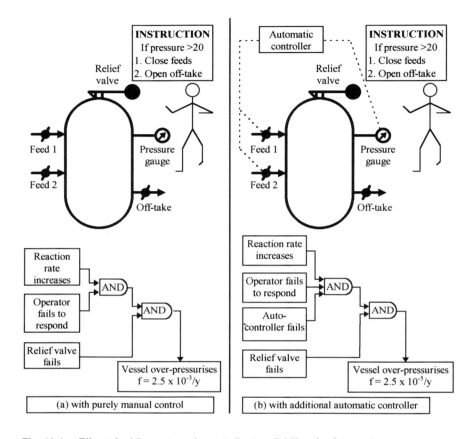

**Fig. 13.1** Effect of adding automatic controller to reliability of safety systems

using logic (Boolean algebra) to combine the various frequencies and probabilities to give a result for the combined event. An example is shown in Fig. 13.1. As noted earlier, some components of this fault tree will be human errors, which in actual incidents have proved to be dominant. Analysis of the fault trees can supply useful insight into the control of an operation or plant.

The examples in Fig. 13.1 show a simple chemical reaction in a vessel. In the example (a), we are relying on an operator to observe a maloperation and take corrective action. The probability is that he will achieve this on nine out of ten occasions. In the example of (b), an automatic controller is added to measure relevant parameters, and respond to rectify the situation if the operator fails to do so; it is ten times more reliable than the human being and the combination reduces the probability of system failure by two orders of magnitude.

We must also take account of common-cause failures. In the example above, both the human operator and the instrument controller may be observing the same pressure gauge. If this gauge is faulty and does not register an excessive rise, neither will respond and the protection system will fail. Systems are therefore designed with *redundancy* (in this case, several pressure gauges) or *diversity* (the operator and controller look at different parameters, or the controller collects several types of measurement). Each solution is tailored to the specific problem.

The fault tree can be used, as seen simply above, to identify for improvement points of weakness and particular elements to which the outcome is especially sensitive. The effects of data variations are examined in tests for excessive pessimism. The sensitivity of estimated risk or frequency to such variations is analysed whenever:

- the primary risk target (see below) is approached within an order of magnitude by a single fault (this analysis covers both consequences and frequency);
- the frequency limit for certain criteria is approached within an order of magnitude by a single fault.

This information allows the assessor to identify features of the model which will require special attention on the plant in order to maintain the integrity of the safety case. Examples are:

- equipment—a gamma alarm;
- a requirement—a proof test on a key instrument;
- an operation—the order in which a set of valves is opened.

The final result of all these analyses is an estimate of the probability that the top event will occur.

### 13.2.7 Criteria

We now need to judge the significance of such results by means of *criteria* against which to compare them.

It is convention not to declare calculated frequencies of less than $10^{-6}$/year (i.e. probabilities of less than one in a million years). Because of the uncertainties and averaging used throughout the process, lower values are meaningless.

If an event is highly undesirable, e.g. death, then our criterion might be that it must occur less than $10^{-6}$/year, with the implication that in practice it is most unlikely to occur at all.

If the event is less severe, for example a release of activity resulting in an abnormal radiation dose to the work-force, a higher frequency will be acceptable. The criteria will become a ladder of increasing dose and decreasing frequency, i.e. we are controlling the risk. For instance,

$N$.mSv $\quad 10^{-2}$/year
10 $N$.mSv $\quad 10^{-4}$/year
100 $N$.mSv $\quad 10^{-6}$/year

In the nuclear industry, consequence and risk targets have been in use for more than a decade, with considerable development and refinement to take account of advances in methodology. Consequence criteria are given in terms of a consequence threshold occurring at an acceptable frequency level. They include both:

- Primary criteria, which limit the summed mortality risk for the most exposed member of the general public.

- Secondary criteria, which account for public perception, and in particular aversion from certain consequences. These criteria limit the frequency of such events to levels below what would be necessary on grounds of objective risk.

### 13.2.8 Safe envelope

It is at this final stage that the assessment is reviewed to determine the need for any Rules or to identify any Key Devices. A safe envelope can then be defined, that is, a set of conditions within which the plant can be assured of safe operation, so far as is humanly possible. In our earlier example of the car this has been done on a generic basis with rules specifying

- maintenance requirements for Key Devices, to be carried out by properly trained and authorised personnel at specified intervals—the MOT test;

- operating conditions—the various speed limits etc.;

- standards of equipment—lighting etc.;

- training and certification of operators—the driving test and licence;

- provision for checking compliance—powers of police to inspect and enforce.

All this, *mutatis mutandis*, will apply to the nuclear plant, with the Rules typically specifying the safe envelope of the plant's operation, and the Key Devices being the equipment and instrumentation necessary to maintain a safe condition.

We would designate an Operating Rule only where a failure to maintain the condition would on its own result in a breach of a primary risk criterion or a designated secondary criterion for either the public or the workers. It is important that such rules are stated in a way that the plant operators can demonstrably comply with them, e.g. in relation to Fig. 13.1, 'If the pressure exceeds 20 psig both feed valves must be closed and the off-take valve opened'.

This definition, based on the numerical analysis, is not exclusive. It is always open to the plant operator to specify any additional rules deemed necessary.

Similarly Key Devices or Safety Mechanisms are also identified by the hazard analysis. This is done in two ways: either they support an Operating Rule that has been identified in the analysis, or they are *key instruments* necessary to keep the estimated fault frequency within the target of a safety criterion.

It is also essential that the assessment is kept up to date. Any changes to the plant, process or conditions must be assessed for their effect on the overall case before they are implemented; maintenance regimes must be established and sustained, and operators must be trained and certified. The case must be reviewed as operating experience is gained or new information comes from other sources.

## 13.3 Criticality control

While the preceding section of this chapter is concerned with risk assessment generally, criticality control is a risk management issue unique to the nuclear industry and deserves more detailed consideration.

A critical mass reaction occurs when sufficient fissile material is assembled in a conformation that allows at least one neutron from any fission to cause another fission. Criticality control, also referred to as nuclear safety, can be broadly defined as the prevention of uncontrolled critical mass reactions. The physics of criticality is discussed more fully in Chapter 5 (Power reactors). It is of importance to note the scale of fault scenarios addressed here: no system studied for nuclear reactors, nuclear fuel storage, or nuclear reprocessing facilities could produce the effects of even the earliest atomic bombs. Nevertheless, a critical event could lead to lethal radiation levels in its vicinity. That is the extreme case. In practice, well contained and shielded installations would significantly mitigate the consequences should an uncontrolled criticality event occur.

There are two situations where criticality control must be exercised:

- systems which are normally sub-critical (such as reprocessing plants which handle fissile material), and which must not be allowed to become critical;
- systems which are normally critical (ie. reactors) and which must not be allowed to become super-critical, with uncontrolled increases in power levels.

There are two unique problems associated with criticality control. The first is that there is no obvious external mechanism (e.g. chemical state or temperature change) which initiates a criticality. It simply happens when enough fissile material reaches a favourable configuration. The second problem is that there is no practical method for measuring the approach to an unexpected criticality. Instrumentation is useful only in recognising that an event has happened, and to a limited extent, for measuring the quantities of fissile material present in a location.

It is important to note the distinction between *fissile* nuclides and *fissionable* nuclides. The term *fissile* refers to nuclides such as U-235 which can undergo fission with neutrons of any energy. The term *fissionable* refers to nuclides such as U-238 which will undergo fission only with high energy (fast) neutrons.

### 13.3.1 Fault identification

As with radiological safety assessment (Section 13.2) one of the methods of identifying potential criticality hazards is the technique of HAZOP. With criticality control however it is assessor training which provides the most important input to fault identification. This is because assessors need to be aware of all the factors which can affect criticality, before they can identify the potential for an uncontrolled criticality.

### 13.3.2 Factors that affect criticality

Although there are always appreciable numbers of neutrons in the atmosphere, it is possible to control the neutron economy of a system containing fissile material. The factors which allow control of the neutron economy are listed and discussed below. In each case, for clarity, it is assumed that only the one factor being discussed is changed.

**Mass**
If there are only a few fissile atoms in a given location, most of the neutrons from those which undergo fission will be able to escape from the system. If there is a significant mass of fissile material then a neutron will have a high probability of being captured and causing fission.

**Density**
It follows from the discussion of mass that the density of nuclei is important. The closer packed the nuclei are, then the greater is the likelihood of collisions with neutrons and therefore the higher the potential for a sustained chain reaction.

**Geometry**
The shape of systems containing fissile material can be designed so that the probability of collisions is significantly reduced. For example, a system may be required for operational reasons to be a flat slab. The slab can be made sufficiently thin so that it makes no difference how much fissile material is present in the system. A sufficiently high fraction of neutrons will escape, and a criticality cannot occur.

**Interaction**
Fissile material in one container cannot distinguish between neutrons generated within itself and neutrons generated from another container of fissile material. Storage of two or more containers of fissile material close to one another can therefore increase the potential for criticality.

**Moderation**
Neutrons which move relatively slowly through fissile material have a greater potential for colliding with a fissile nucleus than a neutron which is moving more

rapidly. The slowing down of neutrons, known as moderation, is used to produce controlled criticality in nuclear reactors. The simplest method of slowing neutrons down is to increase the potential for them to strike non-fissile and non-absorbing nuclei with a small atomic mass. Such collisions may be regarded as elastic impacts between spheres, and the nearer the masses of the colliding spheres, the more effective is the transfer of momentum. It follows that the most effective nuclei in slowing neutrons down are the smaller ones, hydrogen (having a single proton nucleus) being very effective in this respect. This is why water is considered to be one of the most dangerous substances to associate with fissile materials without due precautions.

### Concentration

Even where fissile material is dispersed in a moderator (e.g. dissolved in water) it is possible for the concentration of fissile material to be so low that criticality is impossible, because neutrons are more likely to be captured by non-fissile than by fissile nuclei. With U-235 dissolved in ordinary water this occurs at a H/U-235 ratio of about 2200 (a concentration of about 10 g per litre).

### Poisons

A poison, in the context of criticality control or prevention, is a material which will absorb neutrons, thus reducing their chance of colliding with fissile nuclei. Hence the potential for criticality can be controlled by mixing poisons with the fissionable material. Most substances, even strong moderators like hydrogen, have some poisoning effect. The most effective poisons are considered to be boron and cadmium.

### Reflection

If a moderating material is placed around a piece of fissile material, some of the neutrons which would have escaped will be reflected back into the fissile material, perhaps with their energy altered. This process is called reflection. The effect is similar to that of interaction, though reflection is usually much stronger. The properties which make a material a good moderator also make for a good reflector. However, a good reflector is not necessarily a good moderator (e.g. iron is a reflector but also a poison).

### Enrichment

Natural uranium contains the fissile isotope U-235 at an enrichment of about 0.7%, the rest being predominantly U-238 which is not fissile. The U-238 has a variety of effects. It is slightly fissionable and may undergo fission as a result of collision with fast but not with slow neutrons. It may also absorb neutrons without undergoing fission. Hence, the probability of fission resulting from neutron capture by U-238 nuclei tends to be low, particularly in moderated systems. Uranium-238 also acts as a reflector, as a separator between the U-235 nuclei affecting the fissile density of the system, and has some effect as a poison. Below an enrichment of 5% U-235, U-238 becomes increasingly effective in reducing the potential for criticality until, at an enrichment slightly above that of natural uranium, criticality appears to be impossible without special conditions.

**Inhomogeneity**

A homogeneous system is one in which the nuclides are evenly distributed around the system, for example in a solution. An inhomogeneous system is one in which the fissile material is physically separate from the other parts of the system, for example in a reactor where the fissile material in rods or pins is physically separate from the moderator. At high U-235 enrichment a homogeneous uranium salt solution is more reactive than is any comparable inhomogeneous system. At enrichment close to natural uranium however an inhomogeneous system may be more reactive than a comparable homogeneous one.

### 13.3.3 Control principles

It is generally accepted that nuclear safety should depend as little as possible on decisions or actions of personnel, particularly for routine activities. Hence processes are designed as far as practicable with geometrically safe equipment (tanks, pipes etc.). Where this cannot be done, and particularly where human actions are required, the plant operation is governed by the so-called *double contingency* rule:

> 'Nuclear safety controls are so established that the simultaneous occurrence of at least two independent and unlikely events is necessary for criticality to become possible.'

For example, the introduction of a *safe mass* into a container where a *safe mass* is already present, may be regarded as one unlikely event. Introducing a third *safe mass* into the container, or a change to one of the other factors which could affect nuclear safety, could be the second unlikely event. This would not mean that, if both the unlikely events did occur together, a criticality would actually happen, because many other factors which tend to preserve safety would not have been allowed for. In short, the calculations are deliberately pessimistic. It does however mean that a criticality could occur if the factors not allowed for (the pessimisms) were inadequate, or if the two unlikely and unrelated events did occur.

Unlike most of the nuclear industry, which adopts this *double contingency* approach alone, BNFL does apply probabilistic risk assessment (PRA) to nuclear safety assessment. Within BNFL, PRA is used to *benchmark* double contingency arguments to confirm that the assessed risk is within site risk criteria. Additionally, PRA has been found to be of value when there are multiple initiating events for a single contingency. It provides confidence that the event is sufficiently infrequent to be regarded as a contingency and is not part of the normal variations of plant

**Table 13.1** Radiological design criteria

| Potential Hazard | Criterion | |
| --- | --- | --- |
| | Average | Maximum |
| External radiation to personnel | 5 mSv/y | 15 mSv/y |
| Internal radiation from air contamination | 1–3% DWL | 10% DWL |
| Total radiation dose uptake | | 15 mSv/y |
| Aerial effluent discharges | Less than 50 microsieverts to critical group | |
| Liquid effluent discharges | Less than 50 microsieverts to critical group. Less than 74 GBq alpha per year and 56 TBq beta per year (as measured by 'total beta-5' method) | |
| Radiation dose to the public | Total from Sellafield site not to exceed 500 microsieverts per year. | |

operation. A further benefit has been found where the assessor is faced with a single event which might directly result in a criticality (e.g., dropping a fuel flask down a hoist well), where the double contingency approach tends to be inadequate.

## 13.4 THORP: a specific example

THORP, the THermal Oxide Reprocessing Plant recently built by British Nuclear Fuels plc at Sellafield in Cumbria, is an integrated facility with a design throughput of 1200 tonnes per year of irradiated fuel from light-water and advanced gas-cooled reactors. It comprises two main parts, the Head End and Chemical Separation (Chapter 7), linked to several waste treatment plants on the same site besides themselves incorporating equipment to abate pollution.

### 13.4.1 Design safety principles

A fundamental objective is to ensure that the safety of no one within the plant, on the site or in the vicinity will be prejudiced by normal operations or fault conditions. Key criteria are listed in Table 13.1.

Early in the design of THORP, reducing radioactive discharges was recognised as important. One way to achieve this aim would be to evaporate aqueous wastes to small volume for storage and further treatment. In this respect the flowsheet of the Magnox plant was unsuitable since it used ferrous sulphamate as reductant to separate plutonium from uranium; the resulting ferric and sulphate ions in the waste streams limited the evaporation factors achievable. A salt-free process was therefore adopted for THORP, while separating uranium and plutonium at an earlier stage than in the Magnox plant reduced the amount of plutonium-contaminated waste.

At the Windscale Inquiry of 1977, a commitment was given to limit the radiation dose to the most exposed member of the public, from all operations on

## 268  Management of safety

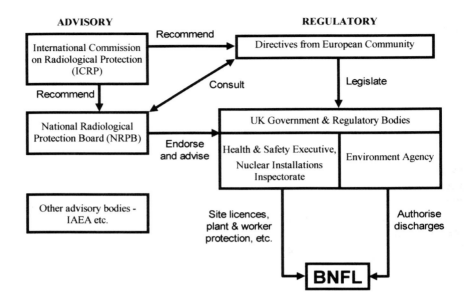

**Fig. 13.2** Advisory and legislative framework for radiological issues in relation to BNFL

the Sellafield site, to less than 10% of the limit recommended at the time by the ICRP. Of this limit, 50 microsieverts per year ($\mu$Sv/y) were allowed to THORP, and the plant includes instrumentation and control arrangements to ensure that discharges remain well within both design and statutory limits. Non-radioactive toxic or otherwise health-endangering materials are controlled by containment or protective equipment to comply with the Control of Substances Hazardous to Health Regulations 1988 (COSHH).

### 13.4.2 Advisory and regulatory framework

All British nuclear installations are regulated to maintain both nuclear and radiological safety, principally by the Health and Safety at Work Act 1974 and the Nuclear Installations Act 1965, with specific requirements in the Ionising Radiation Regulations 1985. Radioactive wastes are governed by provisions of the Radioactive Substances Act (1993). The formal network of advice, instruction and consultation is outlined in Fig. 13.2. In practice there is some feedback which ensures that regulation, while strict, is not unreasonable.

### 13.4.3 Environmental aspects

A key concept in radiological protection for the public is that of *critical group*. This is a small group of people (in one instance elsewhere, a single person whose daily consumption of a lobster meant a uniquely high intake of Zn-65 accumulated by

**Table 13.2** Contributions of various radionuclides to critical–group dose

| Nuclide | Contribution (%)—aerial | Contribution (%)—liquid |
|---|---|---|
| H-3 (tritium) | 0.7 | 0.8 |
| C-14 | 4.7 | 2.2 |
| Co-60 | – | 13.6 |
| Kr-85 | 4.0 | – |
| Sr-90 | 5.4 | 4.5 |
| Tc-99 | – | 0.4 |
| Ru-106 | 7.4 | 70.0 |
| I-129 | 74.5 | – |
| Cs-137 | 0.5 | 5.0 |
| Ce-144 | – | 1.0 |
| Pu (alpha) | 1.5 | 1.6 |
| Am-241 | 1.3 | 0.1 |
| Other | – | 0.8 |
| | 100.0 | 100.0 |

such animals) whose residence, occupation or extreme eating habits make them the most highly-exposed members of the public. For nuclear sites in England the critical group is identified through surveys undertaken by the Ministry of Agriculture, Fisheries and Food (MAFF). MAFF and BNFL conduct extensive environmental monitoring programmes involving thousands of samples per year of air, water, vegetation and foodstuffs. Annual summaries of results are published.

There are distinct critical groups, without common membership, for the aerial and liquid discharges from Sellafield. Their estimated radiation doses due to THORP operating at its design throughput are respectively 22 and 23 µSv/y, about 1% of the average natural background in the UK and well within the constraint of 300 µSv/y recommended by the National Radiological Protection Board (NRPB) for present and future discharges from a single site. When all discharges—present, future and historic—are considered, the limit recommended by the ICRP is 1000 µSv/y. If all effluent discharges from the Sellafield site were at the full authorised limits, the resulting doses to critical groups would be 214 µSv/y (aerial) and 165 µSv/y (liquid). The contributions of various nuclides to the totals are shown in Table 13.2.

**Table 13.3** Collective doses due to THORP

| | Man.sieverts | | | | | |
|---|---|---|---|---|---|---|
| | Liquid 500 years | Infinity | Aerial 500 years | Infinity | Total 500 years | Infinity |
| UK | 0.3 | 3.4 | 2.5 | 2.8 | 2.8 | 6.2 |
| Europe | 1 | 40 | 16 | 20 | 17 | 60 |
| World | 7 | 560 | 163 | 220 | 170 | 780 |

**Table 13.4** Nuclide contribution to world collective dose

| Nuclide | Contribution (%) | | | |
|---|---|---|---|---|
| | Liquid 500 years | Infinity | Aerial 500 years | Infinity |
| C-14 | 80 | 10 | 4 | 20 |
| Kr-85 | 0 | 0 | 95 | 70 |
| I-129 | 4 | 89 | 1 | 10 |
| Other | 16 | 1 | 0.4 | 0.3 |

The contributions of various nuclides to the critical group and general doses are shown in Tables 13.3 and 13.4.

There has been some concern about the effects of krypton-85, a chemically inert gaseous fission product that is not removed from aerial effluents by the treatment provided in THORP (nor by any other chemical treatment). Although it is proportionally a major contributor to radiation doses, the average dose to an individual in the world population is only about 0.02 µSv, roughly equivalent to that from eating a bowl of cereal. Nevertheless developments in technology are kept under review, and the design allows for future installation of removal equipment if a cost-effective process becomes available.

Although radiation is undoubtedly a potential hazard to health, the risks should not be exaggerated. Nor should they be supposed to arise solely from the nuclear industry; some comparisons are listed in Table 13.5

Besides radioactive materials, other components of waste discharges are subject to regulatory control. Some process reagents or impurities inevitably find their way into effluents, and those with a potentially significant impact on the environment need the consent of the National Rivers Authority, Her Majesty's Inspectors of Pollution and local authorities. For some, Environmental Quality Standards (EQS) have been set; the insignificant proportions estimated to come from THORP, all less than 1% and some much less, are listed in Table 13.6. In issuing authorisations for these discharges, the regulatory bodies concluded that human health, the food chain and the environment in general would all be effectively protected.

No significant risk to health should be accepted without good reason. Nevertheless, incremental risks due to the operation of THORP are slight compared with

**Table 13.5** Comparative doses (µSv) from various sources

| | |
|---|---|
| From eating one bag (135 g) of Brazil nuts (Section 11.6.3) | 10 |
| From a single chest X-ray | 20 |
| Annual maximum from THORP to local residents | 25 |
| Annual average to a coal miner | 1200 |
| Annual average to BNFL employees (including non–radiation workers) | 2100 |
| Annual average to residents of Cornwall | 7800 |

**Table 13.6** Controlled non-radioactive discharges from THORP

| Element | Discharge from THORP (kg/y) | Concentration outside mixing zone (µg/l) | EQS (µg/l) | THORP contribution as % EQS |
|---|---|---|---|---|
| Boron | 15 250 | 26.5 | 7 000 | 0.38 |
| Chromium | 56 | 0.10 | 15 | 0.67 |
| Iron | 60 | 0.105 | 1 000 | 0.01 |
| Nickel | 24 | 0.04 | 30 | 0.13 |
| Copper | 3 | 0.005 | 5 | 0.10 |
| Cadmium | 0.3 | 0.0005 | 2.5 | 0.02 |
| Lead | 10 | 0.015 | 25 | 0.06 |

others that are willingly accepted. The justification of risk must be related to the benefits gained, and such issues are outside the scope of this review.

## 13.5 References

1 *Safety Culture: a report by the International Nuclear Safety Advisory Group.* Safety Series No. 75-INSAG-4, IAEA, Vienna (1991).

2 E J Polmear. The Regulator as an influence on safety culture in a nuclear installation. *Proceedings of the international topical meeting on safety culture in nuclear installations*, Paper R.N. 940664, American Nuclear Society, Austria Local Section, ISBN 3-9500255-3-7 (1995).

3 R. H. Clarke. *A model for short and medium-range dispersion of radionuclides released to atmosphere.* NRPB-R91, Harwell, Didcot, Oxon.

# 14 Future perspectives

P. D. Wilson
*BNFL, UK Group, Sellafield, Cumbria*

## 14.1 Introduction

The future of the nuclear industry depends on many questions, philosophical or political as much as technical. They are typically whether nuclear power is a fundamentally misguided attempt at maintaining a high-energy economy in a world that must learn to live within renewable means, or rather a way of extending the real benefits of such an economy to countries and centuries that would otherwise lack adequate supplies of other fuels; whether fission—tapping the fossilised energy of ancient supernovae—is the right way forward, or a dead end into which we have been diverted away from reproducing the power-houses of the stars by means of controlled fusion; whether fast reactors are the key to full utilisation of uranium reserves, or an exercise in technological hubris inviting a terrible nemesis; whether nuclear wastes represent a peril waiting to be visited upon our descendants, or the necessary, duly-controlled and acceptable side effect of making provision for the needs of those descendants.

Such questions have no clear-cut answer and will no doubt be debated for many years. This book however aims to provide some of the factual background, not the debate itself. Previous chapters have dealt chiefly with the current situation, which in many ways demands improvement to meet present difficulties or future challenges. Expectations of what is to come differ widely, but for the present let it be supposed that provision is required to maintain some form of nuclear power well into the next millennium; that fusion meanwhile remains technically or economically impracticable (as appears in fact very likely); and that fossil hydrocarbons can still be used as fuel, albeit to a diminishing extent and at generally increasing prices as they become scarcer.

## 14.2 Aims of the industry

The aims may then be summarised as:

- a secure and adequate energy supply;

- safety (for present and future populations) and environmental friendliness at least comparable with the best that coal and oil provide;
- political acceptability of all operations;
- minimal consumption of strategic resources;
- the simplest practicable facilities, with provision for their care and eventual dismantling when redundant;
- costs as low as may be consistent with other objectives.

How to meet them should ideally be determined by an energy policy based on a considered balance between immediate and longer-term interests. In practice it is likely to depend largely on economics. The results and indeed the basis of comparison between nuclear and other sources, established or alternative, are matters of controversy unsuited to discussion here, although touched briefly in Chapter 1; what is beyond doubt is the intense pressure to reduce costs throughout the fuel cycle.

At the same time there is also a campaign to minimise the impact of the industry, in particular the effect of its waste products, on the health and safety of populations, present and future. The present population of course includes the work-force of the industry. How present and future safety may be balanced is again subject to debate, as is the seriousness or reality of grounds for anxiety about either, but the pressure will undoubtedly continue.

One objection to current practice in the industry is that separated fissile material, in particular plutonium, might be diverted from civil to military or terrorist use. Any proposal offering improved security against such diversion will to that extent have an advantage, and some have indeed been promoted on exactly those grounds.[1]

Unless the more immediate concerns are satisfied, the strategic aims could be jeopardised. The industry therefore has to achieve or maintain:

- greater overall economy;
- acceptable occupational radiation doses, individual and aggregate;
- low routine discharges of radioactive effluents to the environment;
- reduced volumes and adequate containment of radioactive wastes intended for indefinite storage or disposal;
- security against diversion of nuclear materials;
- assurance against the consequences of misjudgement or accident.

There is nothing new in any of these objectives, but as circumstances develop, the means will tend to alter even though the ends remain the same.

## 14.2.1 Economy

Improvements are needed across the whole cycle. Up to now the usual sequence, subject to avoiding utterly impracticable requirements, has been to optimise a power reactor design, devise a fuel, then a process for the discharged fuel and finally a means to dispose of process wastes.

Such a division of the overall problem has two great virtues: it restricts the range of parameters to be considered at one time, and attends to problems in order of cost. On the other hand the costs of treating fuel discharged from a reactor are by no means negligible in the overall cycle, as witness the debate over whether it is preferable to reprocess the fuel, dispose of it directly, or store it temporarily pending a decision—an argument admittedly complicated by political issues, but largely economic. Similarly, the treatment of waste streams, particularly since the tightening of discharge limits, accounts in aggregate for a large fraction of the total reprocessing cost. There is thus a great incentive to eliminate as many as possible of such expenses at source.

The most obvious scope for economies lies in reprocessing, even though it is far from the most expensive part of the cycle. Currently both uranium and plutonium are highly purified, with fission product levels reduced by factors of a million or more, to simplify storage and re-enrichment or refabrication into reactor fuel. Such fastidiousness is quite unrelated to the performance of the fuel once in a reactor, which would quickly generate levels of impurity higher by orders of magnitude.

Relaxed specifications for recycled uranium and plutonium could allow processes to be simplified greatly, for instance eliminating purification cycles and associated waste streams. Benefits must be weighed against having to handle incompletely decontaminated products remotely, but this may be needed in any case for highly-irradiated plutonium on account of its intrinsic radioactivity. It also provides a measure of security against diversion to illicit use. As a corollary, remote or robotic handling equipment will be increasingly needed, and this is one of the secondary developments in the industry.[2]

On the other hand, uranium intended for re-enrichment by current processes must be highly purified. Future processes, for instance based on laser excitation, may be less demanding, or enhanced U-235 content may prove entirely replaceable by plutonium. In time this could altogether eliminate the need for isotopic enrichment, and with it the most expensive stage of fuel manufacture, but would require changes in reactor design to accommodate the differences in neutronic properties. Proposals for such designs have in fact been presented and may be expected to attract more attention as a means of utilising stocks of plutonium.[3]

If plutonium is to be mixed with uranium in refabricated fuel, completely separating the two elements during reprocessing appears wasteful of effort. However far the uranium stream needed to be purified, the plutonium product could be allowed to retain about twice its weight of uranium which indeed would simplify subsequent formation into a homogeneous mixture. High-security storage space need not be greatly increased, since current provisions are determined more by the need to disperse heat from the shorter-lived plutonium isotopes than by the actual bulk of material. Storage requirements may also be limited by the desirability of rapidly recycling both uranium and plutonium to restrict the ingrowth of troublesome daughter elements.

Whatever the specification, the main bulk of the fission products must be separated from the useful products, and so long as all conditions are met, the more cheaply the better. However, regulatory authorities must be satisfied that processes will always remain safe, not only under design conditions but under any that might credibly arise through accident or maloperation. One risk peculiar to nuclear operations is that of an unplanned criticality, which although self-limiting (by forcing apart the interacting masses) and unlikely to be violently disruptive would cause an intense burst of radiation. With increasing proportions of plutonium, such a risk would be hard to eliminate from squat mixer-settlers of the kind used in the Magnox reprocessing plant at Sellafield; this is one reason why the new plant (THORP) uses tall, narrow pulsed columns throughout the plutonium path. Unfortunately they require tall buildings, and the need to guard against seismic disturbance accounts for much of the capital cost.

The reasons for moving away from the low-profile mixer-settlers of the earlier Magnox processing plant were sound, but the most crucial could be met at least as well by centrifugal contactors—essentially mixer-settlers with the residence time reduced by accelerated settling—needing relatively little headroom. The civil engineering costs could then be greatly reduced.

So far the suggested changes are not fundamental to the traditional processes based on solvent extraction. One objection to such processes is that the fuel has first to be dissolved in a large volume of an aqueous reagent, which after the recovery of valuable materials must be treated as waste while the uranium and plutonium products are eventually reconverted to solids. Avoiding aqueous stages has obvious attractions, and alternative pyrochemical processes based on chemical or electrochemical reactions between metals and molten-salt media have been proposed. However, the conversion of conventional fuel from oxide and back again has been assessed as a grave drawback, and unless a more direct reaction with the oxide and a subsequent recovery route can be satisfactorily developed, processes based on the volatility of uranium and plutonium hexafluorides may offer better prospects despite needing their own conversion steps.

The drawback would not apply to metallic fuels, as in the American Integral Fast Reactor (IFR) scheme.[4] Here an alloy-fuelled reactor is closely coupled to a reprocessing plant where uranium, plutonium and the bulk of fission products are electrochemically separated in a molten salt bath, and the heavy metals re-cast into fuel elements. Given the initial charge and a supply of depleted uranium, the system could be self-sufficient in fissile material, or might continue to accept plutonium generated in thermal reactors.

Further economies might be achieved through a still closer association between reactor and fuel processing. At present, fuel elements have an elaborate structure necessary for service in the reactor but expensive to build and an impediment to reprocessing. Arrangements permitting a simpler structure might reduce overall costs without excessive penalty in performance. In this sense the ideal would be to have no structure at all. That implies eliminating the cladding, which would detract from the principle of defence in depth against leakage; however, in various accidents that have actually released significant amounts of fission products to the environment or might have done so, fuel cladding has been either irrelevant or (through its own chemical reactivity) a contributory cause, so the principle may not be held sacrosanct.

The ultimate structureless fuel is fluid. Several kinds have been proposed, and could perhaps provide the best opportunity for reducing overall costs. Aqueous solutions might be expected to couple most readily with conventional processes, but nitrate salts suitable in that respect are radiolytically unstable. A slurry might serve if sustained suspension could be assured.

Systems based on molten salts have shown promise in trials but proceeded no further, partly perhaps because cost estimates for the development programme were too low to be taken seriously.[5] They are claimed to be very safe (power irregularities are self-correcting, and coolant cannot separate spontaneously from heat source) and very flexible since the required reactivity could be made up from any suitable combination of fissile and fertile nuclides. A proportion of the flow would be continuously bled off, stripped of deleterious fission products, replenished in fissile or fertile material, and returned to the core. Nothing but waste need leave the site, and fissile products would not be separated, countering fears of proliferation.

Salt as coolant would allow temperatures to be considerably increased without the risk of boiling. This is desirable for the sake of thermodynamic efficiency, mostly around 33% in current nuclear stations compared with about 50% in modern fossil-fuelled systems. The same reason, with the possibility of using process heat directly, lies behind the Chinese interest in high-temperature gas-cooled reactors, where a bed of fuel balls is cooled by a stream of helium (Chapter 5). An alternative way to increase overall efficiency is to utilise the waste heat, but opportunities for doing so are limited, particularly while safety considerations require nuclear stations to be distanced from centres of population that might benefit from combined heat and power schemes. One general difficulty with these is the cost of controlling and metering the heat flow.

In more conventional systems, a smaller but useful increase in temperature might be permitted by a change of cladding material, say from zirconium alloy to stainless steel with its lower reactivity towards the fuel substance, despite its rather greater neutron absorption.

Improvements in running economy may also come from increased duration between fuel changes. The trend is already well established in all types of reactor, and will no doubt continue. Besides reducing down-time it cuts the cost of storage or reprocessing per unit of generation.

Most of the cost of nuclear generation however lies in the capital outlay on the reactor system. Whatever basic type of reactor is chosen, much unproductive cost can be avoided by standardisation; design charges are minimised, and once the safety of the generic type is established, licensing proceedings can in theory be confined to local issues for individual cases. Moreover solutions to common problems can be applied directly.[6]

Other suggested economies are based on using prefabricated components as far as possible and minimising assembly work on site. That currently implies relatively small modules which in aggregate would consume considerably more material and effort than an equivalent single unit. On occasion this may be acceptable, for example where a portion of the whole output would suffice for some years and the modular system would permit an early return without the full capital outlay. It also has the advantage that in case of breakdown, the loss of capacity would be limited to the affected module; on the other hand, the statistical probability of breakdown rises by the same factor as the reduction in lost capacity.

Optimising over the whole cycle multiplies the number of considerations to be treated together; the case of product specifications shows how they can ramify even in a small area of the overall scheme, but help may be available in the form of computerised decision-support systems. These require every aspect of a proposed scheme to be scored against every appropriate criterion, suitably weighted. They do not take over the responsibility for final decisions from human managers. They can however help to ensure that relevant factors are not overlooked, and that where they conflict, their relative importance is assessed consciously. The human element comes into the choice, assessment and weighting of these factors, and results may thus depend strongly on the viewpoint.

### 14.2.2 Operational dose reduction

Radiation doses arise largely during maintenance: apart from some glove-box operations, likely to be phased out, routine tasks are performed behind as much shielding as necessary. Reduction thus depends on minimising mechanical equipment within the shielding and improving reliability. At present the most frequent breakdowns are in the heavy mechanical equipment used to strip the cladding from fuel elements or chop the elements into short pieces for leaching. One line of attack is therefore to eliminate this step by directly dissolving or disintegrating the cladding, before or together with the fuel substance: either way the cladding residues would eventually be combined with the fission products as highly-active waste. Means might be chemical or electrolytic.

Other mechanical operations that might be eliminated or streamlined are the loading and discharge of fuel dissolver baskets. Several schemes are being investigated or actually used to replace batch by continuous operation (Chapter 7). One such is the French rotary dissolver, in which baskets are formed as sectors of a wheel rotating in a trough of nitric acid; another is the Japanese system adopted for the Recycle Equipment Test Facility (RETF) at Tokai, where sheared fuel is to be moved horizontally by an Archimedean screw against a countercurrent of acid.[7] Attempts to escape the maintenance requirements imposed by mechanical movement are represented by fluidic devices such as the pulsed-vee system in which fuel sections are to be moved through a series of angled pipes against the general liquid flow by a periodic pressure pulse.

Wherever practicable, items liable to need replacement or maintenance, such as motors, are placed outside the shielding with drive shafts penetrating into the operating cell. This reduces but does not entirely eliminate doses when work has to be done on them, and the tendency is to avoid mechanical devices as far as possible. Air-lifts and steam ejectors have long been preferred to pumps for moving active liquors, and mechanical pumps are being displaced by fluidic types; in THORP, air-pulsed columns replace the stirred mixer-settlers of the Magnox plant. A move to centrifugal contactors, as in RETF,[7] would reverse this trend, although the possibility of supplying the necessary energy through the impetus of the liquor itself is being investigated (Chapter 7). Separating devices of the cyclone type have so far proved less effective than their mechanical counterparts, but development continues.

### 14.2.3 Reduced discharges

Doses to the public may be short- or long-term. Short-term effects are due to effluents, mainly from reprocessing plants; ideally all contaminants should be stripped out before discharge, but in practice separations can never be absolute. Great efforts and expense have already been devoted to decontaminating waste streams by various means, with diminishing returns; a more profitable line is to eliminate such streams so far as possible at source.

Streams that cannot be so eliminated are those derived directly from the fuel itself, that is the bulk fission products and the cladding. The ideal here would be to separate them cleanly and immobilise them totally in a minimal volume, but again separation is imperfect and some fission products cannot of their nature be absolutely immobilised.

For instance krypton-85 is a noble gas forming no unreactive compounds. Its half-life of 10.7 years requires a decay period of 35 years for every tenfold reduction in beta-gamma activity, and there is some dispute whether overall safety would be better served by an attempt at containment with the risk of a concentrated local exposure in case of failure, or by dispersing the gas immediately with an individually slight but universal effect. The current view within the UK industry is on balance that any greater risk in dispersion is too slight to warrant the high cost of separation from gaseous effluents, but the latest discharge authorisations require continuing reappraisal of the position.

A rather similar argument applies to iodine-129, although its long half-life (15.7 million years) and radiotoxiciy make it a candidate, along with technetium-99 and transuranic elements, for nuclear incineration. Whether the long-term benefits of such a practice would really outweigh the extra short-term risks, secondary waste production, difficulty and expense of separation is another moot point.

Whatever additions to natural radioactivity may be tolerable, methods of limiting them may well be improved. Waste minimisation as a basic objective of process design should help. To this extent advantages will accrue to methods producing minimal fluids, such as the American pyrochemical techniques or those based on fluoride volatility. Failing such innovations, a serious attack can be expected on the daunting complexity of primary and secondary waste treatments in current plant.

### 14.2.4 Reduction and containment of stored wastes

Long-term radiation hazards to the public lie partly in accumulated traces of long-lived radionuclides in routine discharges to the environment, partly in the possibility of escaped high-level wastes reaching the biosphere. Bulk fission products have been stored for decades at reprocessing sites as aqueous solutions or suspensions, concentrated to minimise volumes. While tolerable as a temporary measure, this cannot be contemplated as a permanent method since it needs constant surveillance to maintain conditions and deal with leaks. As has often been pointed out, these wastes will remain potentially dangerous for many lifetimes, and to avoid placing an unacceptable burden of care on succeeding generations, methods of permanent disposal are needed.

The route currently favoured in Britain is to incorporate the waste in blocks of glass to be buried deep underground once the heat release has decayed, although other forms such as synthetic insoluble minerals are not ruled out should they eventually prove preferable (Chapter 10, Section 5.2).

One topic much discussed at present is whether the minor actinides (neptunium and the transplutonium elements formed increasingly with rising irradiation of fuel) and long-lived, mobile fission products should be disposed with the rest or separately. If any separation is to be undertaken, then other groups such as the heat releasers and noble metals might also be considered. It would be inappropriate to go into the full argument here, but some points of it should be mentioned:

- Indefinite containment cannot be guaranteed for nuclides with half-lives beyond about a thousand years.

- The physiological affinities of iodine-129, technetium-99 and caesium-135 make them more radiotoxic than the purely radiological properties suggest, while if released to the environment the elements would be mobile (apart from the tendency of caesium to be strongly held by ion-exchanging minerals).

- For a period of decades to a century the storage volume will be determined by the need to disperse heat from the caesium-137 and strontium-90, both with half-lives of about 30 years.

- Some fission products, such as the platinum-group metals, have low or negligible radioactivity after a relatively short cooling period, and if recovered would be commercially or strategically valuable.
- The net cost of waste management might thus be minimised if the various groups were separated and treated according to their several requirements: long-lived nuclides incinerated or placed in a repository designed for maximum durability, Cs-137 and Sr-90 placed in a heat-dispersing store that need not last for more than a few centuries, commercially valuable products decontaminated and sold, the remainder disposed in a repository with a relatively undemanding specification.
- Nuclear incineration of the minor actinides would yield fission energy estimated to add some 5% to the value of the source fuel.
- On the other hand, separations would inevitably be incomplete and generate extra contaminated wastes; methods so far known for achieving them are less than satisfactory or non-existent, they are liable to be complex and expensive, and would involve additional radiation exposure to operators possibly outweighing any long-term benefit.

The only advantage of merely separating the long-lived radionuclides would be to reduce the volume to be contained for geological ages. As this alone is unlikely to warrant the effort of development and implementation, separation is usually considered together with incineration, i.e. transformation by fission or neutron absorption into short-lived nuclides. The combined operation is commonly summarised as partition and transmutation, P&T, and possible methods are being examined in several centres. The current effort stems largely from the Japanese OMEGA initiative (Options Making Extra Gains from Actinides)[8] undertaken partly in the spirit of intellectual challenge besides the hope of tangible benefit, and from the French SPIN programme.[9] More detail is given in Section 14.3.3.

A radical suggestion for putting high-level wastes permanently out of circulation is to utilise their decay heat to melt rock around a depository.[10] Provided that they were denser than any melt that might be met, they would then bury themselves deeper than could be achieved by direct excavation. Such a scheme has its attractions; for instance, it could virtually rule out the risk of accidental intrusion by succeeding civilisations, and might well be cheaper than conventional methods. The risks are of meeting barriers of dense or infusible rock, of thermosyphoning that might bring contaminated country rock back to accessible levels, and of vulcanism or tectonic movements that could return the waste itself to the surface. Whether regulators would be ready to countenance the abandonment of direct control implied in the method has not been tested.

### 14.2.5 Security

The nuclear industry was born in a military programme and will never be allowed to forget it. Whatever the rights and wrongs of possessing nuclear weapons, and there are cogent arguments on both sides, the technology and materials must by common consent be kept out of irresponsible hands.

Civil plutonium differs from the military grade in having large proportions of isotopes other than Pu-239. Nevertheless, a somewhat unpredictable explosive device could be made from it, and even a failed nuclear explosion would be an extremely serious event. The stocks of plutonium already separated are therefore regarded with disquiet in some quarters, and steps are likely to render at least new arisings as invulnerable as practicable to diversion.

In that respect plutonium is safest in a reactor. Up to the 1970s the expectation in the industry was that fast breeder reactors, using fuel with 20–30% plutonium, would be needed within a few decades to make the best use of increasingly scarce and expensive uranium resources. Nuclear power programmes however failed to expand as predicted, so that uranium is still abundant and cheap, and in the West may remain so until well into the twenty-first century. The situation will probably be different in the Far East (see Section 14.3.1), where some of the gravest current issues of security have arisen.

Although fast reactors were first intended to extend the production of plutonium, they are now paradoxically suggested for use as burners rather than breeders. Meanwhile some plutonium is already in use instead of uranium-235 as the fissile component of thermal reactor fuels (Chapter 8). Mixing it with say twice its own amount of uranium at source would render the product less attractive for theft; the more direct route of separating only an excess of uranium during reprocessing would also have the technical merits previously mentioned, and such practices are likely to become standard. Irradiating refabricated fuel slightly or leaving in a small proportion of fission products would be even more effective as a protective measure.

The greatest risk of theft appears to arise while materials are in transit, and may be minimised by bringing complementary operations together on island sites. A complex comprising fast reactor, reprocessing and refabricating facilities, once primed with an initial fuel charge, could in principle continue thereafter with no inputs but depleted uranium and no outputs but power and wastes. Problems in setting up new nuclear sites may encourage this type of development to maximise the use of those already licensed.

### 14.2.6 Safety assurance

Where routine activities of low individual impact are compared with events that are highly improbable but would be extremely serious if they did occur, the public perception of relative risks does not match the simple formula of impact times probability; for instance, hundreds or thousands of deaths, a few at a time in road accidents, arouse less public concern than single events with much lower overall casualties. It may be understandable on the analogy that a social community, like a human body considered as an organised assembly of cells, can tolerate many isolated deaths more readily than the loss of a smaller number in a compact mass. Whether right or wrong, such a view leads to grave suspicion of the chance, however remote, of a major nuclear accident.

Any process inherently immune to dangerous failure would thus have the advantage over an equivalent relying on potentially fallible control measures. The preference for passive rather than engineered safety features, already accepted in reactor design, is likely to spread increasingly throughout the fuel cycle.

The greatest risk is of breaching containment and so allowing radioactive gases or aerosols to escape. Such an event might occur in some kinds of reactor through a runaway reaction (as at Chernobyl, where a misguided experiment, on a reactor inherently prone to instability, set up a positive feedback system with safety mechanisms disabled; see Chapter 5 for the source of instability), or more generally through a failure to remove decay heat from a shut-down core—a loss-of-coolant accident or LOCA.

A negative temperature coefficient of reactivity promotes stability by cutting the rate of reaction automatically if the fuel temperature rises. Nothing however can affect the generation of decay heat once the reactor is shut down; core cooling in case of pump failure is then most obviously secured if natural convection limits the temperature to safe levels. This is one merit of pool-type sodium-cooled fast reactors, which also have the virtue of needing no pressurisation to prevent the coolant from boiling.

A risk peculiar to water-cooled reactors with zirconium-clad fuel is that an uncontrolled temperature rise could reach the point where more heat was generated by reaction with steam than by radioactive decay,

$$Zr + 2H_2O \rightarrow ZrO_2 + 2H_2,$$

with the consequent risk of a hydrogen explosion, as apparently happened at Chernobyl. This is a further reason for replacing zirconium by the less reactive stainless steel.

In reprocessing, as has already been mentioned, concentrations of fissile material must never under any credible conditions reach a critical level, and a great deal of effort has been needed to show that they cannot. One advantage of the intensified process options now being investigated is of allowing high throughputs within dimensions too small for criticality, whatever the fissile content—a welcome instance of safety and economy going together.

Contactors of the centrifugal type share with conventional mixer-settlers the merit that in closely resembling theoretical stages of equilibrium between solvent and aqueous phases, the units readily allow performance under unusual conditions to be predicted theoretically. (In this they contrast with columns, where continuous mass transfer depends on never-completed approaches towards equilibrium under the influence of many interacting factors.) The relative ease and reliability of simulations greatly simplifies demonstrations of safety even during maloperations, if not by showing unsafe conditions to be impossible, then by identifying those combinations of circumstance that must be checked experimentally.

This is one aspect of the increasing extent to which computing systems are used in the industry. Comprehensive modelling may also reveal unexpected features of proposed schemes more readily than when segments are treated individually; it can be used to aid the choice between alternative lines of development that at first sight have comparable prospects distinguishable only by expensive and time-consuming experimental programmes. On a molecular scale, it may allow new reagents to be tailor-made for particular applications.

Where safety is not inherent, computer-based information systems will be increasingly important in risk assessment and control. At the conceptual stage they will be needed to ensure that known hazards and their interactions are not overlooked in the design of a plant. Simulators may be used to confirm flowsheet performance and stability, and later to train operators. During operation computers may diagnose malfunctions and advise a rational course of remedial action, or where appropriate implement it automatically, so avoiding the problem at Three Mile Island where operators were overwhelmed by a confusion of different alarms. As the most dangerous situations are liable to arise from unexpected coincidental failures, the effective validation of such control systems and of test methods for them will be crucial to ensure a correct response to rare events.

## 14.3 Special topics

### 14.3.1 Reactor systems

Currently, the most common reactors for power generation are cooled by water, whether boiling (BWR) or pressurised (PWR). With their lineal descendants, these are likely to dominate the market for at least some decades. (There is little sign of continued interest in the AGR, appreciably more expensive to build per unit of output despite its higher thermodynamic efficiency.) One significant modification, mooted but not yet put into practice, would allow a complete core load of plutonium-enriched fuel rather than the 30% or so that is the current maximum.[3] If this provision became universal, there would be no further need for isotopic enrichment of uranium, the most expensive stage of conventional fuel manufacture.

Barring a total collapse of the nuclear fission industry, the present abundance of uranium cannot continue indefinitely. Once the importance of husbanding resources is generally recognised, some type of fast reactor will be needed to utilise stocks of plutonium and depleted uranium, and that may come sooner in the Far East than in western Europe.[11]

India and China are apparently set to exhaust their estimated coal reserves in 30 to 50 years. Neither country has much uranium or the means to purchase it from abroad; if either were to acquire the wherewithal, it could transform the balance of the world market, but in any case both have every incentive to make the best possible use of uranium from whatever source. So has Japan, which relies heavily on imported fuels and is perhaps the last country to have shown any serious interest in the rather unpromising possibility of utilising the natural 3 ppb uranium in sea water.[12] Breeder reactors are therefore likely to become economically important earlier than in the West, and Japan is already building up plutonium stocks for that purpose.

India, with its large monazite deposits, could opt instead or as well for the thorium-uranium breeder cycle, in which thorium-232 is transmuted by thermal neutrons into fissile uranium-233. The system has attracted interest for many years, but in the West has been found rather disappointing, besides the sensitivity to risks of weapon proliferation that caused development to be abandoned in Germany. A similar cycle has however been suggested in connection with the incineration of actinides (Section 14.3.3.).

In the uranium-plutonium cycle, one form of fast-neutron breeder reactor has become the standard for consideration, essentially comprising:

- a core of mixed uranium and plutonium oxides;
- a blanket or breeder zone of uranium oxide alone;
- stainless steel cladding for the fuel;
- a pool of molten sodium as primary coolant;
- control gear, containment etc.

Such a system, despite its many advantages, has various drawbacks:

- The structure is complex and expensive. There is a suggestion that serial ordering of fast reactors would overcome the problem of cost, but that is so far a matter of opinion.
- The ratio of two light atoms to one heavy in the fuel substance limits the concentration of fissile or fertile material, softens the neutron spectrum and so, by reducing the effective neutron yield per fission, restricts the breeding ratio.
- The temperature and thermodynamic efficiency are limited by the relatively low boiling point of sodium (883 °C) and the necessary drop across heat exchangers.
- Metallic sodium can react with the fuel substance if the cladding is breached. With water, reaction is violent, and while small leaks from the steam circuit seal themselves, no substantial ingress may be permitted. Air must likewise be excluded.
- Although moderate fluctuations in power level are automatically limited by the negative temperature coefficient of reactivity, disturbances large enough to start sodium boiling would reach a region of positive feedback and consequent instability.

To improve fuel performance, a change from oxide to carbide was suggested some years ago. Reprocessing would however have been impaired by the formation of troublesome organic compounds on dissolution, and the idea fell out of favour. Nitride may be feasible, and has the merit of a high thermal conductivity that might allow fatter fuel pins with a consequently simplified sub-assembly structure. A disadvantage is that the transmutation of nitrogen would yield potentially embarrassing amounts of carbon-14. A metallic alloy with zirconium is proposed for the Integral Fast Reactor (IFR), with its totally different processing route; the cladding would remain stainless steel.

If helium were used as coolant, only a gas phase would be possible and no problems due to voidage could arise. Temperature could be as high as the structure allowed. The coolant might directly operate a gas turbine, avoiding the complexity, proneness to leakage and thermal inefficiency of a heat-exchanger; on the other hand, there would be some risk of contaminating the turbine with fission products released from defective fuel elements.

A more radical departure would be some variety of molten-salt system, with the rigid fuel replaced by a fluid circulating between reactor, steam generator and a purifying plant. As mentioned before, this offers perhaps the greatest scope for reducing cycle costs, while the inherently negative temperature and void coefficients, ensuring a reduction of power in case of overheating, promise particularly good safety characteristics. Operating temperatures could be higher than with sodium coolant, and ingress of water would cause no violent reaction, although depending on its extent it would act as a moderator and presumably enhance the corrosiveness of the salt.

### 14.3.2 Process developments

At present, as explained in Chapter 7, all commercial reprocessing depends on some form of Purex flowsheet, with uranium and plutonium separated from fission products by extraction into diluted tributyl phosphate (TBP), and from each other by reductive back-washing of plutonium: to meet specifications, both streams undergo cycles of purification. The process is well established, safe and effective, and might continue indefinitely with little essential change but for the need to cut costs and rationalise waste management.

To meet these needs, the approach might be either to modify the Purex process, with a good prospect of moderate gains, or to seek a fundamentally different alternative which might yield more substantial benefits, though with a lower probability of success. Both lines are being followed.

**Improved Purex**

The prime target is a single-cycle process, whether achieved by improving performance to meet specification without further purification, by relaxing the specification to match current performance, or by a compromise between the two. If specifications are to be relaxed, then a mixed uranium-plutonium product is likely to replace pure plutonium, for reasons already described.

The extractant might also be changed, as diluted TBP is excellent but not perfect; the chief drawbacks are that its decomposition products hamper separation, the wash liquors used to remove them form a rather intractable waste stream, and the solvent itself when degraded beyond further use needs elaborate processing in the course of disposal (Chapter 9).

An ideal extractant, when appropriately diluted if necessary, would have the following properties:

- high distribution ratios for uranyl nitrate and plutonium tetranitrate at moderate acidities;
- a strong dependence of distribution ratio on acidity (or some other easily adjusted parameter) to facilitate backwashing;
- affinities for uranium and plutonium sufficiently different to allow adequate separation without the use of complexing or reducing agents;
- a high loading capacity without splitting into light and dense phases (third-phase formation);
- negligible extractive power for fission products or other impurities;
- suitable density and low viscosity to aid disengagement after mixing with an aqueous solution;
- low toxicity;
- a flash-point comfortably higher than any process temperature;
- high stability to chemical or radiolytic degradation;
- innocuous degradation products;
- easy disposal when no longer useful (distinction from TBP, which resists chemical attack and produces a mist of phosphoric acid when burned).

As usual, the ideal probably does not exist, and the best compromise between conflicting requirements may well depend upon particular circumstances. Some compounds such as substituted amides of dicarboxylic acids show possibilities,[13] with promising extractive properties and no residue on combustion, but on the other hand a rather strong tendency to form a third phase.

Solvent properties depend on both extractant and diluent; a relatively sensitive diluent may protect an extractant from radiolysis, without itself yielding such troublesome degradation products. Again, the more stable diluents (such as the straight-chain as against the branched hydrocarbons) have a lower capacity for extracted metal complexes and so tend to form third phases which interfere with operation.[14] There may perhaps be scope for some tailoring to introduce a controlled degree of sensitivity in the diluent or in a region of the extractant molecule remote from the ligand function, for instance at the far end of a long alkyl group. Studies of molecular design may help here.

The extent of experience with TBP tends to discourage any move to replace it, and an alternative would have to offer convincing advantages to be considered at all seriously. A less drastic change would be a modification of the solvent purification routines. The current sequence of acid and alkaline washes produces an aqueous waste too radioactive to discharge directly, too salt-laden to vitrify with the bulk of fission products or evaporate to very small bulk, and inconvenient to store. The obvious alternative to sodium alkalis, ammonia, has the unfortunate drawback that the ammonium nitrate formed on contact with residual nitric acid could be violently unstable if evaporated to dryness. More suitable means are still being sought, and the French development programme is committed to finding an alternative.

Since decontamination factors are limited by entrainment as well as true solvent extraction of impurities, the choice of equipment is also important; the ideal would promote adequate mixing without over-fine dispersions, then separate the phases rapidly and completely; it would have no moving parts and accommodate high throughputs in small physical dimensions. Possibilities have been mentioned in Section 14.2.1.

The preliminary and finishing processes offer more scope for improvement, particularly if batch operations can be made continuous, to reduce handling and smooth out intermittently large flows, particularly of effluents. Three models of continuous dissolver have been described in Section 14.2.1.

Shearing might be replaced by chemical decladding or total dissolution, although as cladding is intended to resist corrosion, reagents that would effectively attack it while leaving equipment intact are hard to find. Anodic dissolution offers more promise than purely chemical attack; the problem here is to maintain electrical contact with disintegrating fragments without causing troublesome stray currents elsewhere. Simply melting off the cladding would have the attractions of recovering material that might be recycled for instance as waste containers, but would require possibly hazardous temperatures and risk a reaction between the molten metal and fuel substance.

Current finishing processes are geared to historic requirements and not ideally suited to the kind of product likely to be needed in future. Uranium directly denitrated, plutonium precipitated as oxalate and calcined to oxide, both need further processing to adjust powder properties before refabrication into fuel. The extra decontamination afforded by the oxalate route may be unnecessary with a revised specification, while its very selectivity is ill adapted to producing the mixed uranium-plutonium product probably required. Some form of direct denitration is a likely replacement, given a solution to the tendency of hydrated nitrates to fuse and cake at operating temperatures.

**Novel processes**
Crystallisation has been suggested, but by itself cannot give adequate recovery, while the required solid-liquid separations are inconvenient under active conditions. The more plausible alternatives to the Purex process are essentially *dry*

routes, using pyrochemical techniques and avoiding aqueous dissolution with the consequent waste disposal problems.

Probably the simplest is the *Airox* process, in which fuel is powdered by successive cycles of oxidation and reduction at 400–600 °C which drive off volatile impurities.[15] The residue is refabricated directly for use in reactors of lower fissile requirement, such as CANDU, or with added fissile material (maybe from redundant military stocks) to restore its initial reactivity. Such fuel is self-protected against theft, but can presumably be recycled only a limited number of times, maybe only once, before the accumulation of neutron poisons renders it useless.

For a more thorough decontamination, the fluoride volatility process shows promise, particularly now that fluorinating agents such as xenon difluoride allow operation under milder conditions than hitherto.[16] Unfortunately some of the fission product fluorides, e.g. those of caesium and technetium, are also volatile and cause difficulties in meeting current specifications. Disposing of wastes could also present problems which are still being examined, and a defluorination step may be needed.

For metallic fuel, as proposed for the IFR, the process is dissolution in molten cadmium and electrolytic transfer to successive cathodes through a mixture of lithium and potassium chlorides at 500 °C.[4] The combination of pyrochemical temperatures with intense radioactivity tends to arouse alarm, but has become more familiar with the vitrification of high-level wastes and should not be an insuperable obstacle.

### 14.3.3 Nuclear incineration

One constant objection to nuclear power is that some wastes will be dangerous for millions of years, and that no containment can be trusted to last so long. Protesters are unconvinced by the counter-argument that in other industries, permanently dangerous toxic wastes are discarded with less precaution and fewer complaints; the only complete answer is to convert the offending radionuclides into short-lived products. Whether doing so would provide any real benefit is disputed and outside the scope of this discussion.

The questions first arose in connection with the so-called minor actinides, the transuranic elements other than plutonium formed on absorption of neutrons by uranium or through decay of the primary products. The principal concern is with neptunium-237, the 2.1 million year, alpha-emitting daughter of americium-241, although as fuel irradiation increases, other transplutonium elements such as curium become more important.

A detailed examination concluded that any hazard posed by such elements was too slight to warrant the enormous effort that would be needed to overcome it. Japanese organisations have however financed a research programme aimed at combining the supposed safety improvement (recognised as inadequate grounds by itself) with an energy bonus, estimated at about 5%, from the fission of minor actinides; the initiative has the acronym OMEGA, Options Making Extra Gains from Actinides, and aims also to provide rising technologists with a worthy challenge.[8,17] Contracts have been placed with several European and American as well as Japanese research teams.

The operation has two distinct aspects, separation of the minor actinides from the discharged fuel and subsequent conversion to short-lived species, commonly designated partition and transmutation (P&T). Currently the minor actinides follow the bulk of the fission products, so partition means separation from the highly-active waste and inevitably increasing the number of contaminated streams—hence the objection that it could create more risks in the short term than it would ultimately save.

## Partition

One approach has developed from efforts to extract transuranic elements from American military wastes, and so simplify disposal of the remainder. The problem is that the transplutonium actinides chemically resemble the lanthanide elements comprising about a quarter to a third of the total fission products (one peak of fission yield covers mass numbers 130 to 150), and no known extractant distinguishes adequately between them.

Octyl(phenyl)-$N,N$-diisobutylcarbamoylmethylphosphine oxide (CMPO), one of the most promising used in the extensively-developed TRUEX process,[18] extracts both groups besides zirconium, palladium, technetium and some iron. Further separation can be achieved by selective complexing with various reagents either in a progressive backwash or a second cycle (the TALSPEAK process), and such a combined scheme was reported in 1987 to be almost ready for deployment at some sites. Tests on its use for highly active liquors have also started.

CMPO has drawbacks, for instance needing TBP as a co-solvent to counter its tendency to form a third phase. Synthesis is difficult and expensive, while ultimate disposal would be hampered as with TBP by the presence of phosphorus and consequent unsuitability for incineration. Concern for disposal has prompted a search for solvents composed solely of lighter elements, for use in both main and ancillary processes. Substituted malonamides show particular promise, and are easily synthesised, but their effectiveness has yet to be demonstrated on more than laboratory scale.[13]

In so far as partition is directed towards long-term safety, removing the minor actinides from high-level waste is only half the task: fission products such as iodine-129 and technetium would remain. Proposed comprehensive schemes of waste treatment include the separation of these, and of the main heat-releasing radionuclides caesium-137 and strontium-90 which limit the extent to which high-level waste can be concentrated for disposal.

Caesium and strontium have traditionally been separated from solution by ion exchange, and mineral-based exchangers (more stable to radiation than pure organics) may serve for this purpose. Alternatively, solvent extraction might offer better prospects; the most probable extractants are of the crown ether type (not yet commercially available), or according to Russian work now taken up and adapted in the USA, substituted cobalt dicarbollides based on a carbon-boron skeleton.[19,20]

If so much is to be done to highly-active waste, objections to the recovery of strategically valuable metals such as rhodium and palladium may lose much of their force. Specific extractants have been suggested to separate their salts from aqueous solution; the metals themselves might be separated from vitrified waste, for example by contact with molten lead, possibly after a cooling period to allow decay of Rh-102.

Extraction from highly-active waste might be avoided if the nuclides concerned were suitably directed in the main process. In the Purex system this is readily feasible only for neptunium, which currently goes to waste but with some modification of process chemistry could be made to follow the plutonium stream. This would remove the principal immediate source of long-term hazard from the waste, but a comparable amount would eventually grow in from americium.

Neptunium is said to blend well with uranium and plutonium whether as metal or dioxide, and serves as a burnable poison—at the start of irradiation it absorbs neutrons without contributing to fission, towards the end of the cycle it has been sufficiently transmuted to make a substantial contribution and thus compensates for the depletion of initially fissile material. Thus it reduces the range of control adjustment that would otherwise be necessary.

So far only aqueous systems have been considered. In the IFR process, actinides automatically follow plutonium; so do part of the lanthanides, but decontamination is apparently sufficient to avoid difficulties. Again, fuel for the molten salt type might in principle be processed to remove only the relatively short-lived fission products, with everything else recycled.

**Transmutation**

In this context, transmutation is the conversion of long-lived radionuclides to products decaying relatively fast into stable species, whether by absorption of neutrons or by fission. The latter option is of course available only to actinides, which may undergo fission directly or absorb neutrons until a fissile nuclide is formed.

Absorption is generally favoured at low neutron energies, and with most of the minor actinides fission occurs only above a certain threshold energy beyond which it becomes increasingly probable. An incinerator intended to cause their most direct destruction should thus have a high neutron energy (hard spectrum); for fission products, lower energies are preferable.

Proposed incinerators come in two classes, reactors and accelerator-based systems. From what has gone before it follows that thermal reactors are inefficient consumers of actinides higher than neptunium, including the even-numbered plutonium isotopes. To burn these requires a fast reactor, preferably with a harder neutron spectrum than in present types which would tend to accumulate curium. Thus interest is increasing in fuels other than oxide, in which the high proportion of light atoms softens the spectrum; nitride and metal are the likely candidates.

In accelerator systems, a proton beam is caused to impinge either directly on the transmutation target, or on a heavy metal (lead, bismuth or tungsten) from which it generates a dense neutron shower.[21] The metal target could be surrounded by a sub-critical fissile-fertile array to multiply the neutron flux by an order of magnitude. The array might be divided by a moderator into regions of hard and soft neutron flux, respectively consuming actinides and fission products. To avoid generating more transuranic actinides from the assembly, it might be based on the thorium-uranium cycle, which could thus return to favour; this however would not be as proliferation-resistant as sometimes claimed, since uranium-233 could be used instead of plutonium-239 for military as well as civil purposes.

All these systems are calculated to offer a substantial surplus of power beyond that needed for the accelerator itself. The most obvious obstacles are the difficulty of separating the wastes, and the capital cost of the radio-frequency power supply.[22]

### 14.3.4 Waste management

Process wastes are fission or transmutation products, cladding and operational materials, currently to be vitrified if high-level, cemented if intermediate, packaged for disposal if low-level, or discharged to the environment as liquid or aerial effluent. Decommissioning waste at various levels of radioactivity must also be taken into account.

One route to cleaner technology is to utilise unwanted material rather than discarding it. For instance, cladding might be recovered for recycle, if not in the original form then as containment for other wastes. As a variant on this approach, Zircaloy could be oxidised to form a matrix in which fission products were incorporated. Nitric acid is already recovered from evaporated aqueous streams, while the condensate might also be recycled as the dilutant for acid make-up or the liquid component of a cement grout, and so on.

Recycling however tends to be expensive, to return unwanted impurities to the process, and to increase radiation doses to operators; benefits in waste management must be considered in that light. Enthusiasm for recycling must therefore be tempered with realism, and it will generally be better to avoid generating a waste stream than to find ingenious uses for it.

This principle may apply particularly in the treatment of wastes themselves, where progressive restrictions on effluent discharges from existing plants have been met by additional processing with secondary wastes generated at each step. Ideally, all wastes should either be innocuous enough for uncontrolled release to the environment, or amenable to one common treatment, and an approach to this ideal is likely to be a major preoccupation of the next few decades.

## 14.4 Conclusion

The future of nuclear power depends on improving public acceptability and in particular on cutting costs. Both in turn depend on aiming to meet the actual or expected needs of customers, with a minimum of constraints due to internal demarcations which may owe more to historic requirements than to the current situation. This requires closer co-operation between different segments of the industry, even if they cannot be organically integrated.

Certain objectives may not be fully attainable but could serve as ideal targets. The most obvious are:

- reactors, of both thermal and fast-neutron types, with simplified totally plutonium-enriched fuel and a minimum of disposable structural elements;
- close operational if not structural integration between reactors and reprocessing plant;
- reprocessing to meet relaxed product specifications in a single cycle; wastes reduced as closely as possible to a single highly-active stream, with all else eliminated at source, harmless or recycled.

The cost of development work is so high that international co-operation is almost certain to increase. With the collapse of the Soviet system, contacts with the former USSR are increasing, and to some extent the same applies to China. Japan has long been an increasingly constructive business partner of western suppliers and has taken the lead in the topic of actinide recycle. Other eastern countries such as Korea and Taiwan are eager to expand their civil nuclear capabilities.

In the USA the nuclear industry has seemed for years to be in the doldrums, largely owing to concern over the possible misapplication of materials. Nevertheless, work on new systems has never completely stopped, and advanced designs for both thermal and fast reactor systems are still being proposed. The latter of course imply reprocessing to separate plutonium, and anxiety to prevent its diversion is one of the stated reasons for pursuing integrated reactor and reprocessing facilities from which fissile material never emerges. The need to dispose of somewhat intractable residues from the early military programme has already led to the development of techniques that could be applied elsewhere—the recycling of minor actinides is the obvious example—and may bring more. Conversely, developments on waste management in Europe are likely to be applied in America.

Whether justified or not, concern over stocks of plutonium is not confined to the USA. Ironically in view of the reasons originally advanced for them, fast reactors may gain acceptance as a means of consuming rather than enhancing these stocks. That at least should mean that the technology is still available when, as may reasonably be assumed, their importance to the conservation of resources is generally recognised.

To the extent that they offer similar services in a limited market, nuclear suppliers are in competition with each other. On the other hand, a misfortune to one is liable to rebound on all, and they have many problems in common. The next few decades are therefore likely to see an increase in rather cautious collaboration with jealous reservation of intellectual property rights.

It promises to be an interesting time.

## 14.5 References

1 K. K. S. Pillay. Safeguards and non-proliferation aspects of a dry fuel recycling technology. *Global '93: Future nuclear systems; emerging fuel cycles and waste disposal options*, pp. 715–21, Seattle (1993).

2 P. G. Wood. Robot design—do not forget the operator. *Nuclear Engineering International* **40**(495), p. 50 (October 1995).

3 M. W. Crump and E. P. Flynn. System 80 + plutonium disposition capability. *Global '93*, pp. 817–24.

4 J. J. Laidler, J. E. Battles, W. E. Miller and E. C. Gay. Development of IFR pyroprocessing technology. *Global '93*, pp. 1061–5.

5 H. G. MacPherson. The molten salt reactor adventure. *Nuclear Science and Engineering*, **90**, pp. 374–80 (1985).

6 T. Price. *Political electricity*. Chapter 6, Oxford University Press (1990).

7 J. Nedderman. Reprocessing fast reactor fuel. *Nuclear Engineering International* **40**, pp. 46–7 (September 1995).

8 C. Newman. International programs related to transmutation of transuranics. *EPRI report NP-7265s* (1991).

9 H. Rouyer. Séparation et transmutation des actinides: le programme SPIN. *Revue Genérale Nucléaire* **5**, pp. 407–13 (1992).

10 V. A. Kashcheev, A. S. Nikiforov, P. P. Poluéktov and A. S. Polyakov, *Atomic Energy*, **73**, pp. 735–9 (trans. Plenum Publishing Corporation) (1992).

11 Nuclear Engineering International **40**(495), p. 4 (October 1995).

12 M. Kanno. Extraction of uranium from sea-water. *Nuclear Power and its Fuel Cycle*, Vol. 2, pp. 431–43, IAEA Vienna (1977).

13 C. Cuillerdier, C. Musikas and P. Hoel. New extractants for the treatment of nuclear waste solutions. *Extraction '90*: I. Chem. E. Symposium Series No. 119, pp. 47–59 (1990).

14 P. D. Wilson and J. K. Smith. Third-phase formation by plutonium(IV) in 30% TBP with various diluents. *Extraction '87:* I. Chem. E. Symposium Series No. 103, pp. 67–74 (1987).

15 H. Feinroth, J. Guon and D. Majumdar. An overview of the AIROX process and its potential for nuclear fuel recycle. *Global '93*, pp. 705–8.

16 J. J. Smit and E. R. Els. A pilot plant for the fluorination of $UO_2$ to $UF_6$ using $XeF_2$. *South African Journal of Science* **84** pp. 456–7 (1988).

17 T. Inoue *et al.*. Development of partitioning and transmutation technology for long-lived nuclides. *Nuclear Technology* **93**, pp. 206–19 (1991).

18 W. W. Schulz and E. P. Horwitz. The TRUEX process: removal/recovery of TRU elements from acidic waste solutions. *Extraction '87*: I. Chem.E. Symposium Series No. 103, pp. 245–62 (1987).

19 Feasibility of separation and utilisation of caesium and strontium from high-level liquid waste. IAEA Technical Reports Series No. 356, Appendix, Vienna (1993).
20 R. L. Miller, A. B. Pinkerton, P. K. Hurlburt and K. D. Abney. Extraction of caesium and strontium into hydrocarbon solvents using tetra-C-alkyl cobalt dicarbollide. *Solvent Extraction and Ion Exchange*, **13,** pp. 813–27 (1995).
21 W. C. Sailor, C. A. Beard, F. Venneri and J. W. Davidson. Comparison of acelerator-based with reactor-based nuclear waste transmutation schemes. *Progress in Nuclear Energy* **28,** pp. 359–90 (1994).
22 R. A. Krakowski. Accelerator transmutation of waste economics. *Nuclear Technology* **10,** p. 295 (June 1995).

# *Glossary*

Actinides—the 14 elements following actinium in the periodic table, analogous to the lanthanides although the early members have closer chemical similarities to transition metals of the following groups (e.g. uranium compares with tungsten). All beyond uranium are essentially artificial, although traces of plutonium occur in nature.
Activation—making artificially radioactive, normally by absorption of neutrons, as in the partial conversion of Co-59 in steel to Co-60.
ADU—ammonium diuranate, $(NH_4)_2U_2O_7$.
Advection—the bulk motion of fluid.
AGR—Advanced Gas-cooled reactor—a follow-on to the British Magnox type, using enriched oxide fuel to permit smaller core size and higher operating temperatures, retaining graphite moderator and carbon dioxide coolant.
Air-lift—a transfer device in which liquor in one limb of a U-tube is lightened by injection of air and so moved to a higher level than the other.
Alpha radiation—emission of helium nuclei (alpha particles) comprising two protons and two neutrons each.
Americium—the next (artificial) element after plutonium; atomic number 95.
Annealing—heat treatment to relieve stresses due to welding, cold working etc.
Atomic number—the number of protons in an atomic nucleus, characteristic of an element.
Attrition—general wear and tear, or fracture (of ion-exchange resins etc.) into particles too small for retention.
Autoclave—a device for heating materials under pressure to a higher temperature than the atmospheric boiling point of the liquid content.
Backfilling—packing an underground excavation with solid material.
Backwashing—reversing a solvent extraction by contacting the extract with a fresh aqueous solution of appropriate composition. Sometimes used in ion-exchange practice of a reverse flow to remove broken particles or impurities.
Batch process—in which operations are performed successively on a quantity of material as a whole (distinction from *continuous* where material flows through the successive stages).
Becquerel—S.I. unit of radioactivity, one disintegration per second.
Bench—a level in open-cast mining.

Beneficiation—preliminary treatment of ore to concentrate valuable components.

Bentonite—a clay mineral with strong ion-exchange properties.

Beta radiation—normally emission of energetic electrons (beta particles). If denoted by $\beta^+$ (beta plus) indicates emission of a positron instead of an electron.

Binding energy—the energy holding together the protons and neutrons in a nucleus, equivalent to the difference in mass between the whole and the sum of the individual parts. Greatest for elements around iron, so the lightest nuclei can in principle release energy by fusing and the heaviest by fissioning.

Biocide—a chemical used to prevent or destroy growth of algae etc.

Biosphere—the sum of living things on Earth.

Blanket—in a breeder reactor, an array of fertile material surrounding the core to utilise escaping neutrons in generating fresh fissile nuclei.

Breeder reactor—in which more fissile material is generated by neutron-induced transmutation of fertile nuclei than is consumed by fission.

Buffer storage—for material passing between two process areas, to prevent interruptions in one from interfering with operation of the other.

Burnable poison—element which strongly absorbs neutrons but in doing so is transmuted into a weak absorber. Used to limit the reactivity of fresh reactor fuel and so reduce the required range of control by other means.

Burn-up—the proportion of heavy-metal atoms fissioned in nuclear fuel. See also *irradiation*.

BWR—Boiling Water Reactor—uses as coolant water which is allowed to boil and drive the turbines directly.

Calcining—strong heating, e.g. to a point at which recrystallisation or solid-state chemical reactions occur.

CANDU reactor—a Canadian design cooled by heavy water and capable of operating on natural uranium. Used on a few sites outside Canada, notably in India, Korea and Romania.

Caro's acid—monopersulphuric acid, $H_2SO_5$

Cascade—a series of similar processes in which the output of one forms the input to the next.

Catalyst—substance which promotes a chemical reaction without being consumed in it.

Centrifugal contactor—a form of mixer-settler in which the separation of phases is centrifugally enhanced.

Chain reaction—series of similar reaction steps in which a product from each initiates the next. In a nuclear chain reaction, a neutron released by fission goes on to cause a further fission, and so on.

Cladding—a sheath to separate the fuel substance and fission products from the coolant stream.

Clinoptilolite—a natural mineral ion-exchanger with particular affinity for caesium.
Commissioning—bringing a reactor or other plant into use through a gradual approach to normal operation so as to discover any problems before they become hazardous.
Complex—a chemical species formed (usually in solution) by association between others each capable of separate existence.
Contamination—the presence of some undesirable substance in a material or on a surface.
Control rod—neutron-absorbing material that can be inserted into or withdrawn from a reactor to maintain the neutron flux at the required level.
Conversion—of uranium oxide into hexafluoride, or vice versa.
Coolant—fluid circulated through a reactor to transfer heat from the fuel to its destination.
Cooling—a period of delay following discharge of fuel from a reactor, allowing much of the initial radioactivity to decay.
Core—the region of a reactor containing the fuel. In a breeder reactor the term is normally restricted to the *driver* fuel, in distinction from a *breeder* zone of fertile material.
Cosmic radiation—produced by highly energetic particles from the sun or beyond, interacting with the atmosphere.
Countercurrent—mode of operation in which (for instance) an aqueous solution and an immiscible organic solvent flow in opposite directions through contacting equipment.
Criticality—a fission chain reaction proceeding at a steady or increasing rate. In a reactor, the normal operating condition; elsewhere an accident to be strictly avoided.
Cross section (nuclear)—the probability of interaction for instance between a nucleus and a neutron flux, measured in barns ($10^{-24}$ cm$^2$).
Crud—fine particulate matter formed in a process and tending to accumulate for instance at solvent-aqueous interfaces. (Alleged etymology 'Chalk River Unidentified Deposit')
Curie—the pre-S.I. unit of radioactivity, $3.7 \times 10^{10}$ disintegrations per second, equivalent to 1 gram of radium.
Curium—the next (artificial) element beyond americium; atomic number 96.
Cycle—in solvent extraction, the sequence of extraction, backwashing (stripping) and solvent purification; in ion exchange, the sequence of loading and regeneration; of nuclear fuel, the history from mining and milling, through fuel manufacture and irradiation in a reactor, to disposal or reprocessing and return of fissile or fertile material to a manufacturing plant.
Cyclone—a device in which material carried by a fluid stream is separated centrifugally when the flow is constrained to a helical path.
Daughter—a nuclide formed by spontaneous decay of another (the parent).

Debonding (of fuel pellets)—destruction by heat of adhesive material incorporated to maintain the shape between pressing and sintering.

Decay heat—energy released by the radioactivity of fission products in reactor fuel after fission has ceased. In power reactors it requires a continuing flow of coolant to prevent damage to the fuel itself or to the reactor structure.

Decommissioning—the process of dismantling redundant or obsolete plant, together with associated operations.

Decontamination—removing an undesirable component from a product etc.

Decontamination factor (DF)—the proportion of contaminant in a desired component before a separation process, divided by the same proportion afterwards.

Demineralisation—removing salts from solution, usually by exchange of cations and anions respectively for hydrogen and hydroxyl ions.

Deuterium—the hydrogen isotope with mass number 2 (one proton and one neutron in the nucleus).

*de minimis*—level (of contamination) beneath regulatory attention (*de minimis non curat lex*).

Differential contactor—solvent extraction equipment (e.g. a pulsed column) in which material is transferred continuously between countercurrently-flowing solvent and aqueous pases, rather than in discrete equilibration stages.

Diluent—a nominally inert carrier for the extractant in a solvent extraction process, added to improve characteristics e.g. by reducing density or viscosity. C.f. modifier.

Direct disposal—the consignment of discharged fuel in its entirety (apart perhaps from appendages) to a permanent repository. Contrast reprocessing.

Disproportionation—the spontaneous conversion of a compound to a mixture of higher and lower oxidation states with the same overall mean oxidation number.

Distribution ratio—in solvent extraction, the concentration of an element or compound in the organic phase divided by that in the aqueous phase when both are in equilibrium. In ion exchange, the equivalent ratio of concentrations in absorbent and solution.

Doppler coefficient—in a reactor, the effect of temperature on the probability of fission. A negative value is important for stable operation.

Effluent—a liquid or gaseous discharge of waste.

Ejector—a transfer device depending on the reduction in pressure as a fluid (usually steam or compressed air) is passed at high speed through a venturi.

Electrolysis—process depending on the passage of electric current through a conducting *electrolyte*.

Electron—particle with a mass of 1/1836 on the atomic scale (where the common form of hydrogen atom is 1.0079), carrying unit negative charge.

Enrichment—an artificial increase in the proportion of one isotope of an element (usually uranium) by partial separation from others, leaving depleted *tails*.

Entrainment—carriage of fine particles or droplets of extraneous material in a process stream.

Ever-safe (against criticality)—of such dimensions that a self-sustaining fission chain reaction cannot occur because too many neutrons escape.

Extractant—the effective component in a solvent extraction process.

FBR—fast breeder reactor—with fissions sustained by fast neutrons, and producing at least as much fissile material as it consumes by transmutation of fertile nuclides.

Fertile—capable of absorbing neutrons to form a fissile material, e.g. U-238 which is thus converted by way of the short-lived U-239 and Np-239 to Pu-239.

Fines—material reduced to smaller dimensions than is intended.

Finishing—converting separated uranium or plutonium into forms suitable for utilisation or storage.

Fissile—capable of undergoing nuclear fission induced by thermal neutrons (c.f. *fissionable*—subject to fission only when induced by neutrons of higher energy).

Fission—the splitting of a heavy atomic nucleus into two very roughly equal fragments plus a few neutrons and sometimes a light nucleus such as of tritium.

Fission products—the main fragments from nuclear fission or products of their decay, whether radioactive or stable.

Fission yield—the proportion of total fission products represented by a particular nuclide or decay series; has two peaks around mass number 90–100 and 130–140 respectively, depending in detail on both the fissioning nucleus and the neutron spectrum.

Flask—shielded container for highly radioactive material.

Floc—a voluminous precipitate, e.g. of ferric hydroxide.

Flowsheet—a diagram showing a sequence of processes in outline with the nature and amount of materials passing through them, but no detail of the actual plant.

Fluidised bed—a mass of solid particles kept in suspension by an up-current of gas or liquid.

Fluoride volatility (process)—the formation and distillation of uranium (and perhaps plutonium) hexafluoride as a means of separation from other materials that do not form volatile fluorides.

Free-issue—(material) available at nominally zero cost, e.g. as a by-product of operations undertaken for other purposes.

Fusion (nuclear)—the union of two light nuclei (for instance of hydrogen isotopes, forming helium) with release of energy; the process that powers the stars once gravitational collapse of precursor material has sufficiently raised the density and temperature.

300   *Glossary*

Gamma radiation—electromagnetic radiation, similar to X-rays but higher in energy.
Gangue—worthless material associated with valuable minerals.
Geotextile—a strong, permeable fabric used to stabilise earth slopes.
Glove box—an enclosed space, commonly under reduced pressure to prevent escape of contents, in which manual operations can be performed through flexible gauntlets without breaking containment.
Graphite—the most stable crystalline form of carbon under normal conditions.
Green-field (site)—never used for building, or restored to that state.
Grout—a cement-based semi-fluid used to fill spaces and harden within them.
GW—gigawatt, $10^9$ watts—in reactor ratings may be distinguished as electrical (e) or thermal (th), related by the efficiency of generation (commonly a factor of about 3).
Half-life—the time taken for half of any amount of a specified radionuclide to decay.
Head end—part of a reprocessing plant separating the fuel substance from the cladding and converting it into a form (conventionally a solution in nitric acid) suitable for the chemical separation.
Heavy water—deuterium oxide, in which the hydrogen of ordinary water is replaced by the heavier isotope. Used as a moderator because of its particularly low neutron absorption.
Helium—the second chemical element, with a nucleus comprising two protons and two neutrons.
Hex (colloquial)—uranium hexafluoride, a volatile compound used in isotopically enriching uranium by gas diffusion or centrifuge processes.
HLW—High Level Waste—essentially the bulk of fission products, with associated material.
Hulls—sections of cladding from which the fuel substance has been leached.
Hydrocarbon—compound of carbon and hydrogen alone, as in the predominant constituents of coal and petroleum. Refined products such as kerosene may be complex mixtures of them.
Hydrogeology—the behaviour of water within rock formations.
Hydrolysis—decomposition by water with insertion of hydrogen and hydroxyl ions.
IDR—Integrated Dry Route—from uranium hexafluoride to dioxide.
ILW—Intermediate Level Waste—materials such as fuel cladding that have gained a substantial measure of radioactivity by contact with the fuel proper or by exposure to the neutron flux, but do not generate significant heat levels.
Incineration (nuclear)—the conversion of radioactive wastes to inactive or shorter-lived forms by transmutation in a reactor or other device.
Ion—an atom or group of combined atoms carrying an electric charge because of a surplus or deficiency of electrons.

Ion exchange—process in which ions held electrostatically in a usually solid material are exchanged for others of a different kind in solution.
Ionise—to convert to charged particles, usually by removal of electrons.
Interrogation (neutron)—estimation of fissile content by the response to a flux of neutrons.
Irradiation—in general, exposure to ionising radiation; of fuel, use in a reactor, as commonly measured by the gross energy in GWd/t (gigawatt-days per tonne) obtained from it.
Isotope—one of perhaps several varieties of an element with different numbers of neutrons per nucleus. As a rule their behaviour is virtually identical except where it depends directly on the nuclear mass. A few elements, notably fluorine, comprise only one isotope in nature.
Labile—subject to rapid change.
Lanthanides—the 14 *rare earth* elements following lanthanum in the periodic table. Having electronic configurations differing essentially in an inner shell, they are remarkably similar chemically.
Leaching—dissolution of some part from a solid mixture.
Lixiviation—deliberate leaching.
LLW—Low Level Waste—lightly-contaminated rubbish or process residues that can be stored or discarded with relatively simple measures to protect the environment from harm.
LMR—liquid metal reactor—cooled by molten sodium or other metal; usually if not always of the fast-neutron type.
LOCA—loss-of-coolant accident—in which a reactor fuel tends to overheat owing to the continuing decay of fission products.
LWR—Light Water Reactor—uses ordinary water as both moderator and coolant, whether boiling or pressurised.
Magnox—an alloy principally of magnesium, used to clad uranium metal fuel; also the fuel as a whole or the type of reactor using it.
Mass number—the number of protons plus neutrons in an atomic nucleus—distinguishes different isotopes of an element.
Masterblend—a mixture of uranium and plutonium oxides with a proportion of the latter equal to or higher than any eventually required, and therefore capable of matching all needs by dilution with additional uranium.
MEB—multi-element bottle—container to prevent the spread of contamination from irradiated fuel elements during transport or storage.
Micron—one millionth of a metre.
Military-grade plutonium—Pu-239 with much less of the other isotopes than is produced in power reactors.
Milling (ore)—a process or series of processes to separate valuable and worthless materials.
Minor actinides—neptunium and the transplutonium elements formed directly or indirectly in irradiated nuclear fuel.

Mixed oxide fuel (MOX)—contains plutonium instead of an enhanced U-235 content as the main fissile material.

Mixer-settler—equipment for solvent extraction in which the two phases are mixed (ideally reaching equilibrium and in practice closely approaching it) and allowed to settle at each stage before passing to the next in either direction.

Moderator—a material used to slow down neutrons by elastic collisions; should have mass number and absorption cross section as low as is compatible with other requirements.

Modifier—component added to a solvent extraction system, e.g. to prevent separation of extracted species from diluent.

Monazite—an ore of thorium and rare earth elements.

Monel—a corrosion-resistant alloy.

Monitoring—checking for the presence or level of contamination etc.

MOX—Mixed OXide fuel (uranium and plutonium).

MTHM—Metric Ton of Heavy Metal (uranium plus plutonium, elemental content regardless of chemical form).

Neptunium—the first transuranic element, atomic number 93.

Neutron—neutral particle with a mass of 1.008665 on the atomic scale. Unstable in isolation with a half-life of 10.4 min, decaying to a proton and beta-particle.

Neutron interrogation—checking fissile content by the response to an applied neutron source.

Neutron spectrum—the energy distribution in the neutron flux of a reactor.

Noble gases—helium, neon, argon, krypton, xenon and radon, so called because of the resistance particularly of the lighter members to entering into chemical combination.

Nucleon—a constituent of the atomic nucleus.

Nuclide—a particular isotope.

Packing fraction—a measure of the mass lost on combining nucleons into atomic nuclei.

Paraffin—a class of hydrocarbons with fully-hydrogenated chains of carbon atoms.

Parent—see daughter.

Partition—separation, commonly of plutonium from uranium or of minor actinides etc. from high-level waste.

Partition coefficient—the ratio between the concentration of a particular molecular species in a solvent and that in the aqueous phase at equilibrium (c.f. the distribution ratio, which is concerned with all forms of the element in question regardless of combination).

Passivation—formation on a metal of an oxide coat that prevents further reaction.

Passive safety—state in which any change towards unsafe conditions is self-correcting without the intervention of operators or engineered devices.

Perched aquifer—porous rock holding water at a higher level than the main water table but supported by an impermeable stratum.

Periodic table—a tabulation of the chemical elements, arranged in horizontal series by atomic number, and in vertical groups by chemical properties (usually with the exception of the lanthanides and actinides, which cannot be accommodated conveniently in this way and are conventionally listed separately).

pH—a measure of acidity or alkalinity, formally the common logarithm of the reciprocal of the hydrogen-ion activity. A value of 7 represents neutrality, while lower values indicate acid conditions and higher, alkaline.

Pin (fuel)—a tube packed with fuel pellets—used in suitably spaced groups of up to several hundred, variously termed clusters, elements, sub-assemblies etc.

Plume—the pattern of effluent dispersing from a discharge point.

Plutonium—the second transuranic element, atomic number 94.

Plutonium cycle—fission yielding neutrons that convert U-238 to plutonium which serves as the fissile component for the next round of fuel.

Poison (ion-exchange)—a substance that reduces capacity by irreversibly occupying absorption sites.

Poison (neutron)—a particularly powerful absorber of neutrons; may be introduced deliberately into equipment to prevent an accidental criticality, or into fuel to reduce the initial reactivity and so the range of control required during residence (in which case it is chosen to be *burnable*, i.e. converted by the absorption into a low-absorbing nuclide); alternatively, it may be an undesirable by-product of other processes.

Polysaccharide—a compound formed from linked sugar molecules, e.g. starch or cellulose.

Pond—a water-filled storage facility for irradiated nuclear fuel.

Positron—the anti-particle to the electron, with equal mass but opposite charge.

Prompt (criticality)—a fission chain sustained by prompt neutrons alone, and therefore uncontrollable.

Prompt (neutrons)—emitted immediately from a fission reaction (a small proportion is slightly delayed).

Proton—an ordinary hydrogen nucleus, with a mass of 1.0072765 on the atomic scale and unit positive charge. With neutrons, a component of all other atomic nuclei.

Pulp—finely-ground ore prepared for extraction of valuable constituents.

Pulsed column—equipment for solvent extraction in which the phases flow in opposite directions through a vertical column. A periodic pressure pulse forces them through transverse perforated plates and so maintains dispersion.

Purex—the currently dominant solvent extraction process for separating discharged fuel into uranium, plutonium and wastes.

Purging—replacing contaminated fluid (e.g. water in a cooling pond) in whole or more commonly in part by a clean supply.

PWR—Pressurised Water Reactor—cooled by water kept under enough pressure to prevent boiling, with a secondary steam circuit to which it transfers the heat drawn from the reactor.
Pyrohydrolysis—hydrolysis at high temperature.
Pyrolysis—decomposition by heat, usually intense.
Radioactivity—the spontaneous disintegration of an atomic nucleus with the emission of ionising radiation—see alpha, beta, gamma radiation. May be natural or artificially induced.
Radiography—use of high-energy radiation to form an image of internal discontinuities in an opaque body.
Radiology—the study of the effects of radiation.
Radionuclide—a radioactive isotope.
Radon—chemically the last of the noble gases; the principal isotope Rn-222 (alpha-emitting with a half-life 3.8235 days), a daughter of radium-226, seeps naturally from the ground and is responsible for a large proportion of radioactivity in the indoor environment.
Raffinate—the residue after removal of valuable components in a solvent extraction process.
Rare earth—see lanthanide.
Rating—see thermal rating.
RBMK reactor—a Russian design, graphite-moderated and water-cooled (as at Chernobyl). The possibility of losing coolant while retaining moderation confers an inherent tendency to instability, incompatible with usual safety standards.
Reactor (nuclear)—the complex of equipment to maintain and control a fission chain reaction. Two main classifications are:
    *thermal*—has a neutron flux more or less equilibrated with the core in general;
    *fast*—has a very much more energetic neutron spectrum;
        *breeder*—generates at least as much fissile material as it consumes by transmuting fertile isotopes, e.g. U-238.
Various types (not a comprehensive coverage) are listed individually.

Redox potential—a measure of oxidising or reducing conditions.
Refabrication—manufacture into new fuel of material recovered after discharge from a reactor.
Reprocessing—the separation of fuel discharged from a reactor into potentially useful products and waste.
Resonance—an enhanced probability of interaction between certain nuclei and neutrons of a particular energy.
Rheology—the study of flow properties.
Safeguards—administrative measures to ensure that fissile material is not diverted from civil to military purposes.

Salt-free (process)—leaving no involatile or indestructible residues in solution (desirable to allow high degrees of volume reduction in wastes that have to be stored).

Scrubbing—in solvent extraction, contacting a newly loaded extract with a clean portion of the source phase to remove unwanted contaminants; in gas cleaning, removal of impurities by intimate contact with an aqueous stream.

Shielding—material used to absorb radiation before it can cause damage or injury.

Sievert (Sv)—the S.I. unit of radiation dose. The dose from natural sources in the UK is on average in the region of 1/500 Sv (2mSv), but varies widely from place to place.

Sintering—heating to a point where particles begin to coalesce without actually melting.

Solubility product—in a saturated solution of an ionic substance, the product of concentrations (in molar units) of the constituent ions.

Solvent extraction—a separatory process in which one or more components of a mixture are transferred from one solvent to another essentially immiscible with it (usually from aqueous solution to an organic oil or the reverse).

Stoicheiometry—the numerical relationship between participants in a chemical reaction or between atoms of different elements in a compound. If in the latter case it is continuously variable over some range with only a gradual change in properties, the compound is said to be non-stoicheiometric.

Stope—a cavity formed by extraction of ore in a mine.

Stratigraphy—the layering of geological sediments.

Stripping—back-extraction of material after solvent extraction.

Sublimation—direct vaporisation from the solid state.

Super-criticality—an accelerating fission chain reaction in which more than one neutron from each fission goes on to cause another.

Supernova—an exploding star with intense temperatures and neutron fluxes sufficient to build up the heaviest elements.

Swarf—fragments of (Magnox) cladding peeled from fuel elements.

Tailings (mining or milling)—the residue after separation of valuable ore components.

Tails (enrichment)—the by-product depleted in the less valuable isotope.

TBP—tri-*n*-butyl phosphate—solvent used in reprocessing plants, usually diluted with an inert hydrocarbon.

Thermal (neutron)—at or near energetic equilibrium with the surroundings.

Thermal (reactor)—operating on an essentially thermal neutron spectrum.

Thermal rating (of fuel or reactor)—the total heat output as distinct from the electrical generating capacity, which is commonly about a third of it (depending on the steam temperature) owing chiefly to inherent thermodynamic limitations on efficiency.

Thermite—reaction between a metallic oxide and a highly electropositive metal (typically aluminium), producing intense heat.

Third phase—in solvent extraction, sometimes formed when the concentration of extracted species exceeds solubility in the diluent; may render the process ineffective or inoperable.

Thixotropic—gel-like until fluidised by shearing forces.

Thorium—element 90. Although an actinide, it behaves chemically more like a Group 4 element.

Thorium cycle—converting Th-232 to U-233 which serves as the fissile component for the next round of fuel.

THORP—the British THermal Oxide Reprocessing Plant at Sellafield, Cumbria.

Tonne—metric ton, 1000 kg or 2200 lb, c.f. imperial ton of 2240 lb.

Transmutation—the conversion of one element into another, for instance by absorbing a neutron and emitting a beta-particle to increase by one atomic number unit.

Transuranic elements—those beyond uranium (element 92) in the periodic table, essentially artificial although traces of plutonium (element 94) are found in nature.

Tritium—the hydrogen isotope of mass number 3 (two neutrons in the nucleus). Beta-emitting with a half-life of 12.3 years.

TRU—(waste) containing transuranic elements.

Unconformity (geological)—a break in the sequence of deposits.

Uranium—the heaviest element occurring naturally in significant amounts; atomic number 92.

Uranyl ion—the group $UO_2^{2-}$ which passes unchanged through many chemical processes; the usual form of hexavalent uranium—U(VI)—in solution.

Valency—the combining power of a chemical element, e.g. the number of hydrogen or chlorine atoms with which each atom can combine to form a neutral molecule. In many elements it is a variable number depending on the state of oxidation or reduction.

Vitrification—the conversion of wastes to a glassy form for permanent disposal.

VVER—Russian type of pressurised water reactor.

Yellowcake (uranium ore concentrate, UOC)—uranium oxide etc. largely purified at the mine site.

Zeolite—a mineral, natural or artificial, based on mixed oxides of aluminium and silicon, with a structure characterised by open channels of molecular dimensions; various types are used as catalysts in gas reactions and ion-exchangers in aqueous solution.

Zircaloy—an alloy chiefly of zirconium, used as fuel cladding on account of its resistance to corrosion and low neutron absorption.

# Appendix 1: elements of the periodic table

Nuclear data in this and the following appendices are taken from 'Nuclides and Isotopes,' 14th edition, General Electric Company, San Jose, California (1989).

| Atomic No. | Symbol | Name | Atomic Weight |
|---|---|---|---|
| 1 | H | Hydrogen | 1.0079 |
| 2 | He | Helium | 4.002602 |
| 3 | Li | Lithium | 6.941 |
| 4 | Be | Beryllium | 9.012182 |
| 5 | B | Boron | 10.811 |
| 6 | C | Carbon | 12.011 |
| 7 | N | Nitrogen | 14.0067 |
| 8 | O | Oxygen | 15.9994 |
| 9 | F | Fluorine | 18.998403 |
| 10 | Ne | Neon | 20.18 |
| 11 | Na | Sodium | 22.98977 |
| 12 | Mg | Magnesium | 24.305 |
| 13 | Al | Aluminium | 26.98154 |
| 14 | Si | Silicon | 28.0855 |
| 15 | P | Phosphorus | 30.973762 |
| 16 | S | Sulphur | 32.07 |
| 17 | Cl | Chlorine | 35.453 |
| 18 | Ar | Argon | 39.948 |
| 19 | K | Potassium | 39.0983 |
| 20 | Ca | Calcium | 40.078 |
| 21 | Sc | Scandium | 44.95591 |
| 22 | Ti | Titanium | 47.88 |
| 23 | V | Vanadium | 50.9415 |
| 24 | Cr | Chromium | 51.996 |
| 25 | Mn | Manganese | 54.93805 |
| 26 | Fe | Iron | 55.847 |
| 27 | Co | Cobalt | 58.93320 |
| 28 | Ni | Nickel | 58.69 |
| 29 | Cu | Copper | 63.546 |
| 30 | Zn | Zinc | 65.39 |
| 31 | Ga | Gallium | 69.723 |
| 32 | Ge | Germanium | 72.61 |
| 33 | As | Arsenic | 74.92159 |
| 34 | Se | Selenium | 78.96 |
| 35 | Br | Bromine | 79.904 |
| 36 | Kr | Krypton | 83.80 |

| Atomic No. | Symbol | Name | Atomic Weight |
| --- | --- | --- | --- |
| 37 | Rb | Rubidium | 85.4678 |
| 38 | Sr | Strontium | 87.62 |
| 39 | Y | Yttrium | 88.90585 |
| 40 | Zr | Zirconium | 91.224 |
| 41 | Nb | Niobium | 92.90638 |
| 42 | Mo | Molybdenum | 95.94 |
| 43 | Tc | Technetium | (99) |
| 44 | Ru | Ruthenium | 101.07 |
| 45 | Rh | Rhodium | 102.90550 |
| 46 | Pd | Palladium | 106.42 |
| 47 | Ag | Silver | 107.8682 |
| 48 | Cd | Cadmium | 112.41 |
| 49 | In | Indium | 114.82 |
| 50 | Sn | Tin | 118.71 |
| 51 | Sb | Antimony | 121.75 |
| 52 | Te | Tellurium | 127.60 |
| 53 | I | Iodine | 126.90447 |
| 54 | Xe | Xenon | 131.29 |
| 55 | Cs | Caesium | 132.90543 |
| 56 | Ba | Barium | 137.327 |
| 57 | La | Lanthanum | 138.9055 |
| 58 | Ce | Cerium | 140.115 |
| 59 | Pr | Prasaeodymium | 140.90765 |
| 60 | Nd | Neodymium | 144.24 |
| 61 | Pm | Promethium | (147) |
| 62 | Sm | Samarium | 150.36 |
| 63 | Eu | Europium | 151.96 |
| 64 | Gd | Gadolinium | 157.25 |
| 65 | Tb | Terbium | 158.92534 |
| 66 | Dy | Dysprosium | 162.50 |
| 67 | Ho | Holmium | 164.93032 |
| 68 | Er | Erbium | 167.26 |
| 69 | Tm | Thulium | 168.93421 |
| 70 | Yb | Ytterbium | 173.04 |
| 71 | Lu | Lutecium | 174.967 |
| 72 | Hf | Hafnium | 178.49 |
| 73 | Ta | Tantalum | 180.9479 |
| 74 | W | Tungsten | 183.85 |
| 75 | Re | Rhenium | 186.207 |
| 76 | Os | Osmium | 190.2 |
| 77 | Ir | Iridium | 192.22 |
| 78 | Pt | Platinum | 195.08 |
| 79 | Au | Gold | 196.96654 |
| 80 | Hg | Mercury | 200.59 |
| 81 | Tl | Thallium | 204.3833 |
| 82 | Pb | Lead | 207.2 |
| 83 | Bi | Bismuth | 208.98037 |
| 84 | Po | Polonium | (210) |
| 85 | At | Astatine | (210) |
| 86 | Rn | Radon | (222) |
| 87 | Fr | Francium | (223) |

# Appendix 1: elements of the periodic table

| Atomic No. | Symbol | Name | Atomic Weight |
|---|---|---|---|
| 88 | Ra | Radium | (226) |
| 89 | Ac | Actinium | (227) |
| 90 | Th | Thorium | 232.0381 |
| 91 | Pa | Protoactinium | (231) |
| 92 | U | Uranium | 238.0289 |
| 93 | Np | Neptunium | (237) |
| 94 | Pu | Plutonium | (239) |
| 95 | Am | Americium | (241) |
| 96 | Cm | Curium | (244) |
| 97 | Bk | Berkelium | (249) |
| 98 | Cf | Californium | (249) |
| 99 | Es | Einsteinium | |
| 100 | Fm | Fermium | |
| 101 | Md | Mendelevium | |
| 102 | No | Nobelium | |
| 103 | Lr | Lawrencium | |
| 104 | Rf | Rutherfordium | |

# Appendix 2: decay chains of heavy radionuclides

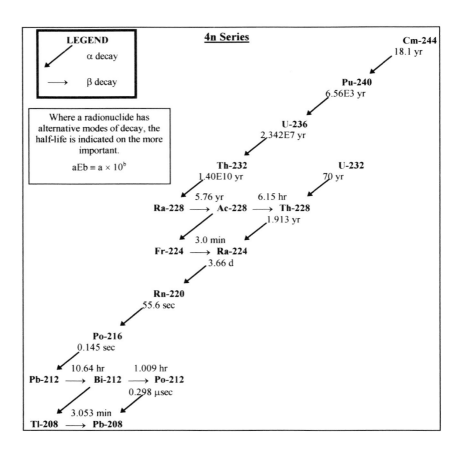

*Appendix 2: decay chains of heavy radionuclides* 311

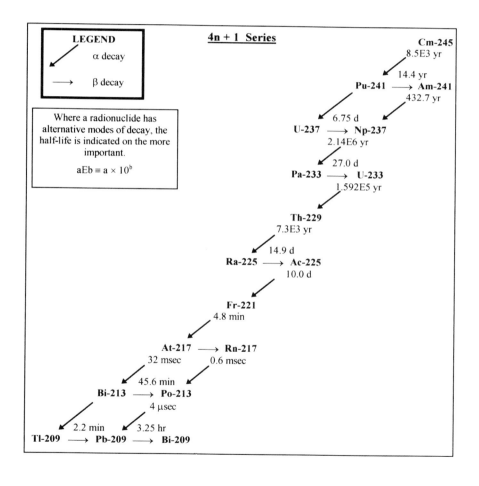

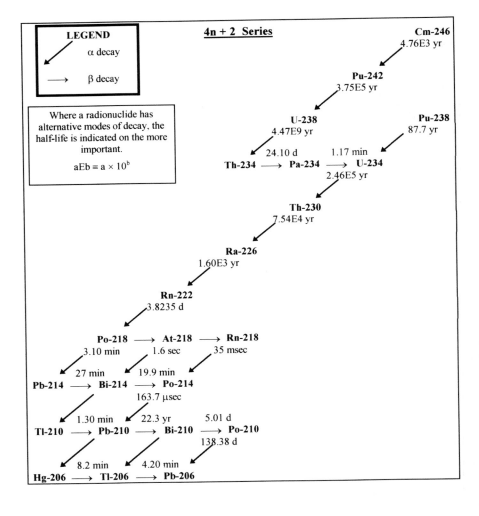

# Appendix 2: decay chains of heavy radionuclides

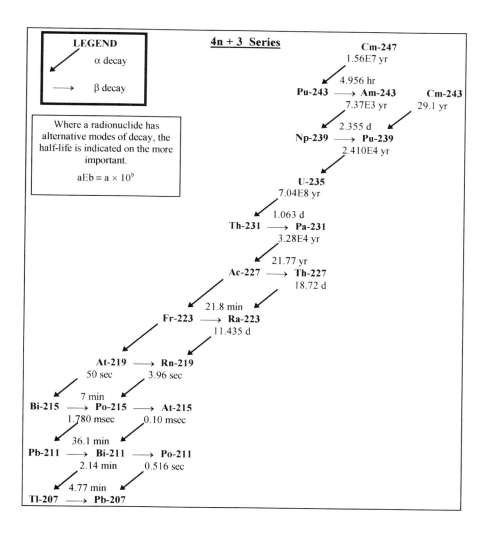

# Appendix 3: products from thermal fission of U-235

See Fig. 1.6 for a graphical illustration. Yields are from S. Katcoff, Fission-product yields from neutron-induced fission, *Nucleonics*, **18(11)** pp. 201–8 (November 1960).

N.B. Fission of other nuclei or by fast neutrons yields different proportions, although in a qualitatively similar pattern.

| Element | Mass no. | Yield (%) | Half-life | Daughter |
|---|---|---|---|---|
| Zn | 72 | 1.60E-05 | 46.5 hours | Ga-72 |
| Ga | 73 | 1.10E-04 | 4.87 hours | Ge-73 |
| Ga | 74 | 3.50E-04 | 8.1 minutes | Ge-74 |
| Ge | 77 | 0.0031 | 11.3 hours | As-77 |
| As | 77 | 0.0083 | 38.8 hours | Se-77 |
| Ge | 78 | 0.02 | 1.45 hours | As-78 |
| As | 78 | 0.02 | 1.512 hours | Se-78 |
| As | 79 | 0.056 | 9.0 minutes | Se-79 |
| Br | 80 | 1.00E-05 | 4.42 hours, 17.66 minutes | Kr-80 |
| Se | 81 | 0.148 | 57.3 minutes, 18.5 minutes | Br-81 |
| Br | 82 | 4.00E-05 | 1.471 days | Kr-82 |
| Se | 83 | 0.22 | 22.3 minutes | Br-83 |
| Br | 83 | 0.51 | 2.40 hours | Kr-83 |
| Kr | 83 | 0.544 | Stable | – |
| Br | 84 | 0.939 | 6.0 minutes | Kr-84 |
| Kr | 84 | 1 | Stable | – |
| Se | 85 | 1.1 | 32 seconds | Br-85 |
| Kr | 85 | 0.293 | 10.73 years | Rb-85 |
| Rb | 85 | 1.3 | Stable | – |
| Kr | 86 | 2.02 | Stable | – |
| Rb | 86 | 2.90E-05 | 18.65 days | Sr-86 |
| Se | 87 | 2 | 5.6 seconds | Br-87 |
| Rb | 87 | 2.49 | 4.8E10 years | Sr-87 |
| Sr | 88 | 3.57 | Stable | – |
| Sr | 89 | 4.79 | 50.52 days | Y-89 |
| Sr | 90 | 5.77 | 29.1 years | Y-90 |
| Sr | 91 | 5.81 | 9.5 hours | Y-91 |
| Y | 91 | 5.4 | 58.5 days | Zr-91 |
| Zr | 91 | 5.84 | Stable | – |
| Sr | 92 | 5.3 | 2.71 hours | Y-92 |

# Appendix 3: products from thermal fission of U-235

| Element | Mass no. | Yield (%) | Half-life | Daughter |
|---|---|---|---|---|
| Zr | 92 | 6.03 | Stable | – |
| Y | 93 | 6.1 | 10.2 hours | Zr-93 |
| Zr | 93 | 6.45 | 1.5E6 years | Nb-93 |
| Zr | 94 | 6.4 | Stable | – |
| Zr | 95 | 6.2 | 64.02 days | Nb-95 |
| Mo | 95 | 6.27 | Stable | – |
| Zr | 96 | 6.33 | Stable | – |
| Nb | 96 | 6.10E-04 | 23.4 hours | Mo-96 |
| Zr | 97 | 5.9 | 16.8 hours | Nb-97 |
| Mo | 97 | 6.09 | Stable | – |
| Nb | 98 | 0.064 | 51 minutes | Mo-98 |
| Mo | 98 | 5.78 | Stable | – |
| Mo | 99 | 6.06 | 2.75 days | Tc-99m |
| Mo | 100 | 6.3 | Stable | – |
| Ru | 101 | 5 | Stable | – |
| Ru | 102 | 4.1 | Stable | – |
| Ru | 103 | 3 | 39.27 days | Rh-103 |
| Ru | 104 | 1.8 | Stable | – |
| Ru | 105 | 0.9 | 4.44 hours | Rh-105 |
| Ru | 106 | 0.38 | 1.02 years | Rh-106 |
| Rh | 107 | 0.19 | 21.7 minutes | Pd-107 |
| Pd | 109 | 0.03 | 13.5 hours | Ag-109 |
| Ag | 111 | 0.019 | 7.47 days | Cd-111 |
| Pd | 112 | 0.01 | 21.04 hours | Ag-112 |
| Cd | 115 | 0.0104 | 44.6 days, 2.23 days | In-115 |
| Cd | 117 | 0.011 | 3.4 hours | In-117 |
| Sn | 121 | 0.015 | 1.13 days | Sb-121 |
| Sn | 123 | 0.0013 | 129 days | Sb-123 |
| Sn | 125 | 0.013 | 9.63 days | Sb-125 |
| Sb | 125 | 0.021 | 2.758 years | Te-125 |
| Sb | 127 | 0.13 | 3.84 days | Te-127 |
| Te | 127 | 0.035 | 109 days | I-127 |
| Sn | 128 | 0.37 | 59.1 minutes | Sb-128 |
| I | 128 | 3.00E-05 | 25.0 minutes | Xe-128 |
| Te | 129 | 0.35 | 33.6 days | I-129 |
| I | 129 | 0.8 | 1.57E7 years | Xe-129 |
| Sn | 130 | 2 | 3.7 minutes | Sb-130 |
| I | 130 | 5.00E-04 | 12.36 hours | Xe-130 |
| Te | 131 | 0.44 | 1.35 days | I-131 |
| I | 131 | 3.1 | 8.04 days | Xe-131 |
| Xe | 131 | 2.93 | Stable | – |
| Te | 132 | 4.7 | 3.26 days | I-132 |
| Xe | 132 | 4.38 | Stable | – |
| I | 133 | 6.9 | 20.8 hours | Xe-133 |
| Xe | 133 | 6.62 | 5.243 days | Cs-133 |
| Cs | 133 | 6.59 | Stable | – |
| I | 134 | 7.8 | 52.6 minutes | Xe-134 |
| Xe | 134 | 8.06 | Stable | – |
| I | 135 | 6.1 | 6.57 hours | Xe-135 |
| Xe | 135 | 6.3 | 9.10 hours | Cs-135 |
| Cs | 135 | 6.41 | 2.3E6 years | Ba-135 |

| Element | Mass no. | Yield (%) | Half-life | Daughter |
|---|---|---|---|---|
| I  | 136 | 3.1     | 1.39 minutes  | Xe-136 |
| Xe | 136 | 6.46    | Stable        | –      |
| Cs | 136 | 0.0068  | 13.16 days    | Ba-136 |
| Cs | 137 | 6.15    | 30.17 years   | Ba-137 |
| Ba | 138 | 5.74    | Stable        | –      |
| Ba | 139 | 6.55    | 1.396 hours   | La-139 |
| Ba | 140 | 6.35    | 12.75 days    | La-140 |
| Ce | 140 | 6.44    | Stable        | –      |
| La | 141 | 6.4     | 3.90 hours    | Ce-141 |
| Ce | 141 | 6       | 32.50 days    | Pr-141 |
| Ce | 142 | 6.01    | Stable        | –      |
| Ce | 143 | 5.7     | 1.38 days     | Pr-143 |
| Nd | 143 | 6.03    | Stable        | –      |
| Ce | 144 | 6       | 284.6 days    | Pr-144 |
| Nd | 144 | 5.62    | Stable        | –      |
| Nd | 145 | 3.98    | Stable        | –      |
| Nd | 146 | 3.07    | Stable        | –      |
| Nd | 147 | 2.7     | 10.98 days    | Pm-147 |
| Sm | 147 | 2.36    | 1.06E11 years | Nd-143 |
| Nd | 148 | 1.71    | Stable        | –      |
| Sm | 149 | 1.13    | Stable        | –      |
| Nd | 150 | 0.67    | Stable        | –      |
| Sm | 151 | 0.44    | 90 years      | Eu-151 |
| Sm | 152 | 0.281   | Stable        | –      |
| Sm | 153 | 0.15    | 1.929 days    | Eu-153 |
| Eu | 153 | 0.169   | Stable        | –      |
| Sm | 154 | 0.077   | Stable        | –      |
| Sm | 155 | 0.033   | 22.2 minutes  | Eu-155 |
| Eu | 155 | 0.033   | 4.71 years    | Gd-155 |
| Eu | 156 | 0.014   | 15.2 days     | Gd-156 |
| Eu | 157 | 0.0078  | 15.13 hours   | Gd-157 |
| Eu | 158 | 0.002   | 45.9 minutes  | Gd-158 |
| Gd | 159 | 0.00107 | 18.6 hours    | Tb-159 |
| Tb | 161 | 7.60E-05| 6.91 days     | Dy-161 |

## Notes

1. The symbolism $a\mathrm{E}b$ (where $a$ is a decimal number and $b$ an integer) represents $a \times 10^b$.

2. Where two half-lives are given, different nuclear isomers are involved; the first may decay to the second or directly to the daughter product.

# Appendix 4: decay properties of selected radionuclides

| Nuclide | Half-life | Main decay mode | Energy (MeV) |
|---|---|---|---|
| H-3 | 12.3 years | Beta | 0.0186 |
| C-14 | 5730 years | Beta | 0.157 |
| Co-60 | 5.271 years | Beta, gamma | 2.824 |
| Kr-85 | 10.73 years | Beta, gamma | 0.687 |
| Sr-90 | 29.1 years | Beta | 0.546 |
| Y-90 | 2.67 days | Beta | 2.282 |
| Zr-95 | 64.02 days | Beta, gamma | 1.125 |
| Nb-95 | 34.97 days | Beta, gamma | 0.926 |
| Tc-99 | 2.13E5 years | Beta | 0.293 |
| Ru-103 | 39.27 days | Beta, gamma | 0.764 |
| Ru-106 | 1.020 years | Beta | 0.039 |
| Rh-102 | 2.9 years | Electron capture, gamma | 1.10 or 2.28 |
| Rh-106 | 29.9 seconds | Beta, gamma | 3.54 |
| I-129 | 1.57E7 years | Beta, gamma | 0.191 |
| I-131 | 8.040 days | Beta, gamma | 0.971 |
| I-133 | 20.8 hours | Beta, gamma | 1.77 |
| Xe-133 | 5.243 days | Beta, gamma | 0.427 |
| Xe-135 | 9.10 hours | Beta, gamma | 1.16 |
| Ce-144 | 284.6 days | Beta, gamma | 0.319 |
| Tl-207 | 4.77 minutes | Beta | 1.43 |
| Tl-208 | 3.053 minutes | Beta, gamma | 4.998 |
| Pb-209 | 3.25 hours | Beta | 0.644 |
| Pb-210 | 22.3 years | Beta, gamma | 0.063 |
| Pb-211 | 36.1 minutes | Beta, gamma | 1.3 |
| Pb-212 | 10.64 hours | Beta, gamma | 0.571 |
| Pb-214 | 27 minutes | Beta, gamma | 1.03 |
| Bi-210 | 5.01 days | Beta | 1.162 |
| Bi-211 | 2.14 minutes | Alpha | 6.623 |
| Bi-212 | 1.009 hours | Beta, gamma | 2.252 |
| Bi-214 | 19.9 minutes | Beta, gamma | 3.28 |
| Po-210 | 138.38 days | Alpha | 5.3044 |
| Po-211 | 0.516 seconds | Alpha | 7.451 |
| Po-212 | 2.98E-7 seconds | Alpha | 8.7844 |
| Po-214 | 1.637E-4 seconds | Alpha | 7.6869 |
| Po-215 | 1.780E-3 seconds | Alpha | 7.386 |
| Po-216 | 0.145 seconds | Alpha | 6.7785 |

# Appendix 4: decay properties of selected radionuclides

| Nuclide | Half-life | Main decay mode | Energy (MeV) |
| --- | --- | --- | --- |
| Po-218 | 3.10 minutes | Alpha, gamma | 6.0024 |
| Rn-219 | 3.96 seconds | Alpha, gamma | 6.8193 |
| Rn-220 | 55.6 seconds | Alpha, gamma | 6.2882 |
| Rn-222 | 3.8235 days | Alpha | 5.4895 |
| Ra-223 | 11.44 days | Alpha, gamma | 5.7164 |
| Ra-224 | 3.66 days | Alpha, gamma | 5.6855 |
| Ra-226 | 1600 years | Alpha, gamma | 4.7844 |
| Ra-228 | 5.76 years | Beta | 0.046 |
| Ac-228 | 6.15 hours | Beta, gamma | 2.141 |
| Th-228 | 1.913 years | Alpha, gamma | 5.423 |
| Th-232 | 1.40E10 years | Alpha | 4.013 |
| Th-233 | 22.3 minutes | Beta, soft gamma | 1.244 |
| Th-244 | 24.10 days | Beta, soft gamma | 0.273 |
| Pa-233 | 27.0 days | Beta, gamma | 0.571 |
| U-233 | 1.592E5 years | Alpha | 4.824 |
| U-234 | 2.46E5 years | Alpha | 4.776 |
| U-235 | 7.04E8 years | Alpha, gamma | 4.400 |
| U-238 | 4.47E9 years | Alpha | 4.197 |
| Np-237 | 2.14E6 | Alpha | 4.788 |
| Np-239 | 2.355 days | Beta, gamma | 0.722 |
| Pu-238 | 87.7 years | Alpha | 5.4992 |
| Pu-239 | 2.410E4 years | Alpha | 5.156 |
| Pu-240 | 6.56E3 years | Alpha | 5.1683 |
| Pu-241 | 14.4 years | Beta | 0.021 |
| Pu-242 | 3.75E5 years | Alpha | 4.901 |
| Am-241 | 432.7 years | Alpha | 5.4857 |
| Am-242 | (1) 141 years | Isomeric transition | 0.0486 |
|  | (2) 16.02 hours | Beta | 0.663 |
| Am-243 | 7.37E3 years | Alpha | 5.276 |
| Cm-242 | 162.8 days | Alpha | 6.1127 |
| Cm-244 | 18.1 years | Alpha | 5.8048 |

# Index

*Italic numbers denote reference to illustrations*

accountancy, fissile material 123
actinides 7, 167–9, 215, 219–22
 minor 7, 101, 129, 279, 288–91
 extractability 129
activation 5, 84, 241
Airox process 288
ALARA principle 162, 191, 226
americium 7, 142, 156, 210, 215, 220–1
amine extractants 34
ammonium diuranate 55–6, 145–6, 181–2
ammonium uranyl carbonate 56–7
ammonium uranyl plutonyl carbonate 153


Brazil nuts 222, 270
breeding 96–7, 101


caesium 83, 104, 166, 207–15 *passim*,
 219, 279, 289–90
calcium–41: 241–2
cancer induction 226
carbon–14: 122, 179, 216, 241, 270
cask, fuel storage 114–15, 198
chain reaction 6–7, 79–80
Chernobyl 82, 96, 207, 219, 282
chlorine-36: 241–2
cladding, fuel 10, 12, 43, 44, 50, 62, 64, 65,
 117–21, 276, 287
clarification, liquor 122–3
clinoptilolite 166–7
CMPO 289
coal equivalence of uranium 15

cobalt 5, 104, 241
 dicarbollide 290
computer applications 282–3
conditioning 123, 132
contactor, centrifugal 135, 282–3
conversion to $UF_6$ 46–8
coolant 10, 81–2, 83, 87–98, 276, 284–5
critical group 215, 224, 259, 268–70
criticality 80, 124–5, 263–7, 275
crown ethers 290
curium 7–8; *see also* actinides, minor


decay, radioactive 3–5, 310–13, 317–18
 heat 9–10, 13, 83, 103, 139, 154, 164,
  181, 279
decladding, novel methods 287
decommissioning 15–16, 39, 229–51
 costs 233–4, 248
 material recycling 243–5
 planning and strategy 235–6, 249–50
 principles 229–35
 regulation 230–1
 techniques 236–42
 timing 232–5
 waste management 242–6
decontamination
 surface 237
 factor 116, 130, 167, 169
DEHPA 33–4
delay storage 164
denitration, thermal 144–5, 149–50
diluent 34, 126, 132, 286
discount rate 15–16, 194–5

dismantling 239
disposal, waste
  chemical environment 192
  costs 194–5
  decommissioning 242–6
  forms 196
  in France, Germany, Japan and Russia 182
  fuel, direct 11, 197–200
  high level 195–7
  intermediate level 190–5
  low level 185–90
  national practices 200–4
  options 185
  policy 195–6, 200–4
  principles 196–7
  repository design 194, 225
  safety 191–2
  site geology 186–7, 197
  solvent 169–73
  transuranic (TRU) 200–1
dissolution *see* leaching
distribution ratio 126–7, 209
Dounreay 117, 181–2
Drigg disposal site 186–90

EARP 167–9
economics 14–16, 273–7
efficiency, thermodynamic 93, 284
effluent 243, 269–71
  aerial 122, 178–81, 208–9
  liquid 37, 39, 163–9, 212–13
electron 1, 3–4
energy, nuclear 5–6, 15, 272, 283–4
enrichment, isotopic 67–77, 138, 265
  centrifuge 71–4
  diffusion 69–71
  electromagnetic 67
  laser 74–5, 159
  plants 68
  of recycled uranium 75–6, 147
  tails *70*, 99, 139
environment 189–90, 223–5
  marine 212–16
extraction factor 127

fault
  identification 256–7, 264
  tree 259–61
feedback 81–2, 96, 282
filtration 30, 36, 168–9, 180
fission 6–7, 78–82, 97, 263
  products 8–10, 11, 13, 83, 129, 161–3, 314–16
flask, transport 105–10
flocculation 167–9
fluidics 278
fluoride volatility process 45, 288
fluorides, melting and boiling points 48
food chain 211–12, 214–15, 224
fuel 11–12, 41–66, 87, 138–60
  breeding 96–7, 100–1
  carbide or nitride 149, 284–5
  cycle *Frontispiece*, 13–14
  dissolution, see leaching
  elements *42*, 49–50, 61–5
  fast breeder 99, 143, 154–5, 284–5
  fluid 276, 285
  irradiated 116
  management 11
  metal 43–4, 48–50
  MOX 99–100, 133, 139, 149–58
  oxide 44, 50–65
  pellets 57–61
  storage 102–5, 110–15, 198
  supply 12, 283–4
  transport 105–110
  unconventional 284–5
  various reactor types 61–5
fusion 6

generator, thermoelectric 99, 220
geology 186–7, 192–3, 211
ground water 187, 189, 192–3, 197, 201, 203, 210–11

half–life 5, 310–18
HAZOP 257, 264
hex (UF$_6$) 47–8, 51–7, 146–7
hulls 120–1

humic substances 210

IFR 275
  process 288, 290
incineration, nuclear 278–9, 288–91
iodine 83, 122–3, 179, 210, 218, 270, 278–9, 289
ion exchange 31–3, 166–7, 209, 289
ionisation 4
Irish Sea 212–14
iron-55: 242
isotope 3
  radio- 4, 5, 209–12, 216–23, 278–80, 288–91

key devices 262–3
krypton-85: 122, 180–1, 216–17, 270, 278

La Hague (UP2–800, UP3) 117, 120–1, 130, 134, 212
laverweed 215
leaching
  fuel 118–21, 277–8
  ore 26–30
leukaemia cluster 225–6
LOCA 282

Magnox 43, 48–50, 120, 133
  reactors 87, 91–3
mass–energy equivalence 1
Masterblend (U and Pu) 156, 159
milling 12, 21–37
minerals, uranium 18–19, 28
mining 19–21
mixer–settler 35, 124–5
mobility, in soil 209–11
model, environmental dispersion 224–5, 258
moderator 7, 10, 80–2, 86–7, 94, 96, 264–5
modifier, solvent 34, 289
molluscs, accumulation of radionuclides by 214–15
molybdenum 123

MOX, *see* fuel

neptunium 7, 129, 130–2, 142, 148, 219–20, 288, 290
neutron 1–3, 6–9, 81–4, 108
  amplifier 101, 291
  delayed 82
nitric acid, recovery of 122, 163
nitrogen oxides 122, 123, 132, 179
noble gas 122, 180–1, 216–17, 278
nucleon 3
nucleus 1–7

off–gas 122, 178–81
Oklo (natural reactor) 79, 85, 225
OMEGA initiative 280, 288–9

packing fraction 5–6
palladium 123, 290
partition and transmutation (P&T) 134, 279–80, 288–91
pelleting 57–61
periodic table 2, 307–9
phosphorus 215, 289
pins, fuel 12, 41, *42*, 61–2
platinum metals 122–3, 280
plutonium 210, 215, 220, 281
  burning 281
  formation 7–8, 83
  fuel 11, 12, 99–100, 138–41, 143, 149–59, 274–5, 283–4
  isotopes 7, 86, 97, 138–9, 148, 220
  military 86, 140–1, 159
  processing 123, 129, 130–2, 149–54, 159–60
POCO 237
poison
  ion exchange 32
  neutron 82, 89, 91, 116, 121, 125, 138, 265, 290
polonium 212
ponds 102–4, 111–13, *119*
positron 3, 4
potassium 4, 223

precipitator, electrostatic 180
process intensification 135
process simulation 282–3
proliferation, weapon 8, 140, 273
protoactinium 8
proton 1–3, 291
pulsed column 125, *126*
Purex process 117, 123–34, 285–7

radiation 3
  design criteria 267
  exposure 21, 200, 224–7, 252–3, 267–71, 277–8
  recommended limits 227
radioactivity 3–5, 20, 241–2
radionuclides 4, 5, 209–12, 216–23, 278–80, 288–91, 310–13, 317–18
radium 37, 222
radon 4, 21, 222–3
reactor 10–11, 80–4, 292; *see also* fuel
  AGR 91–3
  BWR 89–91
  CANDU 94–6
  control 10, 82–3
  data 87
  fast (neutron) 8, 96–9, 281, 283–5, 292
  gas–cooled 91–4, 285
  high temperature 93–4, 276
  Integral Fast 275
  Magnox 91–3
  maritime 99
  molten salt 276, 285
  natural 79, 84–5, 225
  passive safety 100
  pebble bed 94
  PWR 87–9
  RBMK 96
  space 99
  special–purpose 85–6
  thermal 86–96
  thorium–fuelled 8, 100–1, 284
  VVER 87
recycling 138–60, 243–5
reductant 131
regulation 230–1, 255, *268*

reliability data 259–61
reprocessing 11, 116–37, 158, 274–6, 292
  head end 117–23, 134–5, 287
  novel methods 287–8
  plants 117
  solvent extraction 123–35, 285–7
resonance 7
rhodium 123, 290
risk 255–63, 281–3
  criteria 261–2
rivers 213
ruthenium 122–3, 129, 132, 215, 216, 217–8

safety 193–4, 252–71, 273, 281–3
  culture 252–5
  key devices 262
  operating rules 262–3
sea 212–16
security 140, 280–1
sediments 212–15
Sellafield 117–33 *passim*, 161–80 *passim*, 192–4, 210–12, 224, 230, 247, 267–71
separation
  solid–liquid 30, 36, 122–3, 168–9, 180
  uranium and plutonium 130–2, 267
sequence construction 15
shellfish 214–15
shielding 102–15 *passim*
silver zeolite 179
sintering 60, 156
site, island 281
site remediation 39, 242
size reduction (dismantling) 239–40
soil 209–11
solvent extraction 33–5, 45–6, 117, 123–32, 135, 285–7, 289–90
solvent treatment 133, 134
solvent destruction 169–73
specifications, revision 274–5
SPIN programme 280
storage, fuel 102–5, 110–15, 198–9
  cask 114–15
  dry 104–5
  pond 103–4, 111–13

vault 113
well 113–14
strontium 166, 209–11, 214–15, 217, 279, 289–90
structure, atomic 1–3

tailings, mine 37–9
tails, enrichment 70, 139, 140, 155
TALSPEAK process 289
technetium 123, 129, 131–2, 217, 278–9, 289
thallium-208: 101, 138
third phase 127–8
thorium 4, 8, 41, 100–1, 222, 284, 291
THORP 117, 118, 119, 122, 128, 130, 131,133, 179, 267–71
Three Mile Island 83, 237–8, 283
toxic potential 142
tracers 224
transmutation 7–8, 290–1
tri-$n$-butyl phosphate (TBP) 34, 45–6, 125–8, 285–7
  degradation 128
tritium 86, 122, 216, 241, 269
TRUEX process 289

ultrafiltration 168–9
UOC 44–6
uptake, biological 211–12, 214–15
uranium 4, 6–8, 18–37, 41–61, 130–2, 221–2, 267, 283–4
  conversion to $UF_6$: 46–8
  hexafluoride (hex) 47–8, 51–7, 146–7
  isotopes 7, 8, 10–11, 67–76, 79, 138–9
  leaching 26–30, 120–1
  occurrence 18–19, 28
  ore beneficiation 23–6
  oxide 44, 46–7, 50–60, 120–1, 144–6
    ADU process 55–6, 145–6

AUC process 36, 56–7, 146
  dry routes 52–5
precipitation 35–6, 145–6
purification 31–5, 44–6
solvent extraction 33–5, 129
thermal denitration 144–5

ventilation 178, 238
vibro–compaction 152

waste 141–2, 161–205;
  aqueous 163–9
  classification 162–3
  compaction 188
  encapsulation 176–8, 182, 189
  management 12–13, 184–205, 242–6, 279–80, 291, 292
  solid 173–8
  solvent 169–73
  vitrification 174–6
  see also disposal
water
  heavy 7, 81, 94
  as storage medium 103–4, 112–13
  on waste disposal site 189
Windscale 84, 86, 241–2, 267

xenon poisoning of reactor 82
X–rays 4–5

yellowcake 35, 44
Yucca Mountain 201

Zircaloy 44, 63, 64, 88–9, 291
zirconium 44, 123, 129, 282